The Illusion of Reality

Howard L. Resnikoff

The Illusion of Reality

With 189 Illustrations

Springer-Verlag
New York Berlin Heidelberg
London Paris Tokyo

Howard L. Resnikoff
Aware, Inc.
One Cambridge Center
Cambridge, MA 02142
USA

Library of Congress Cataloging-in-Publication Data
Resnikoff, Howard L.
  The illusion of reality.
  Bibliography: p.
  Includes index.
  1. Information theory.   2. Information science.
I. Title.
Q360.R395   1987   001.53′9   87-9681

Typeset by Asco Trade Typesetting, Ltd., Hong Kong.
Printed and bound by Arcata Graphics/Halliday, West Hanover, Massachusetts.
Printed in the United States of America.

9 8 7 6 5 4 3 2 1

ISBN 0-387-96398-7 Springer-Verlag New York Berlin Heidelberg
ISBN 3-540-96398-7 Springer-Verlag Berlin Heidelberg New York

To Joan

Yes, I have tricks in my pocket, I have things up my sleeve. But I am the opposite of a stage magician. He gives you the illusion that has the appearance of truth. I give you truth in the pleasant guise of illusion.

—Tennessee WILLIAMS, *The Glass Menagerie*

Strangely enough, there was always a period when all hands had to get used to the fresh air. When it first came in, it smelled awful. For hours we had been accustomed to air impregnated with cooking odors and other smells, and the fresh air was almost disgusting for a few minutes.

—Captain George GRIDER, USN

Preface

The Illusion of Reality was conceived during my tenure as director of the newly established Division of Information Science and Technology at the National Science Foundation in 1979–1981 as a partial response to the need for a textbook for students, both in and out of government, that would provide a comprehensive view of information science as a fundamental constituent of other more established disciplines with a unity and coherence distinct from computer science, cognitive science, and library science although it is related to all of them. Driven by the advances of information technology, the perception of information science has progressed rapidly: today it seems well understood that information processing biological organisms and information processing electronic machines have something basic in common that may subsume the theory of computation, as well as fundamental parts of physics.

This book is primarily intended as a text for an advanced undergraduate or a graduate introduction to information science. The multidisciplinary nature of the subject has naturally led to the inclusion of a considerable amount of background material in various fields. The reader is likely to find the treatment relatively oversimplified in fields with which he is familiar and, perhaps, somewhat heavier sailing in less familiar waters. The theme of common principles among seemingly unrelated applications provides the connective tissue for the diverse topics covered in the text and, I hope, justifies the variable level of presentation.

Some of the material appears here for the first time. The axiomatic definition of the measure of information and its applications presented in Chapter 2, the analysis of subjective contours and the treatment of the duality between noise and aliasing in Chapter 5, and the theory of linear superposition of textons in Chapter 6 are the principal examples.

Early drafts of this book were used in my courses in the Computer Science program in the Division of Applied Science and in the Department of Psychology and Social Relations at Harvard University in 1983–1985. I am grateful to my students for their enthusiastic interest. I owe a particular debt of gratitude to my friend and colleague R. Duncan LUCE, who first encouraged me to introduce a course on information science at Harvard and to Joe B. WYATT, who, in other no less important ways, also made it possible. I am pleased to express my appreciation to John DAUGMAN for his continuing enthusiasm and encouragement, as well as for many valuable discussions.

The manuscript of the book was completed at the Center for Biological Information Processing in the Department of Brain and Cognitive Sciences at MIT, whose hospitality, and that of its director, Professor Tomaso POGGIO, it is my pleasure to acknowledge.

In a book that has had so long a gestation period, and touches upon, although it cannot adequately cover, so many different subjects, there are many to thank. My colleagues Richard C. ATKINSON, Bruce BENNETT, Richard HERNNSTEIN, Max KOECHER, David MUMFORD, Emma DUCHANE SHAH, Jayant SHAH, Charles S. SMITH, Charles STROMEYER, Edward C. WEISS, and R. O. WELLS, JR., have made important contributions to my thinking about the subject. They and many other friends and colleagues, too numerous to mention, have played parts greater than they may have realized in bringing this work to fruition.

Mary SHEPHERD typed the first draft of the manuscript, prepared many of the figures, and arranged the whole so that it could be used as a textbook for my course. Her expert and conscientious assistance was invaluable. Later drafts were typed in part by Mary BROCKENBROUGH, Chris MARUK, and Debbie WIDENER at Thinking Machines Corporation. Bruce NEMNICH wrote macros for printing the texton equations in Chapter 6 and Karl SIMS prepared some of the figures for Chapter 5. I appreciate their assistance and the encouragement and support of the company during that fruitful period.

I am pleased to record my obligation and gratitude to the staff of Springer-Verlag in general, and particularly to the members of the Computer Science Editorial and Book Production departments, who exercised exceptional patience and understanding when my other responsibilities too often conflicted with and overcame production deadlines.

My wife, Joan, to whom this book is dedicated, has become exceptionally skilled in distinguishing illusion from reality in the many years of our life together. Over those years and in more ways than I can say, I have relied on her, on that expertise, and on her knowing insights into the visual arts, and hope to continue to do so for some time more.

Cambridge, MA   Howard L. Resnikoff

Contents

Preface   vii

CHAPTER 1
Introduction: Purpose of This Book and History of the Subject   1

1.1	The Subject Matter of Information Science	1
1.2	History of Information Science.	3
1.3	Information and Thermodynamics.	4
1.4	Early Development of Sensory Physiology and Psychophysics.	6
1.5	Selective Omission of Information	9
1.6	The Quantum Mechanical Theory of Measurement.	10
1.7	Early Development of Electrical Communication Engineering and Its Formalization as Information Theory	12
1.8	The Development of Computers.	14
1.9	The Information Processing Viewpoint in Psychology.	16
1.10	Some Principles of Information Science.	19

CHAPTER 2
Mathematics of Information Measurement   22

2.1	Purpose of This Chapter	22
2.2	The Information Content of a Measurement	25
2.3	Formal Derivation of the Information Content of a Measurement	27
2.4	Application of the Information in a Measurement to Psychophysics. . . .	32
2.5	The Probability Interpretation of the Information Measure.	36
2.6	The Uncertainty Principle for Measurement.	39
2.7	Conservation of Information	42
2.8	Information Content of a Calculation	44
2.9	Generalization of the Notion of an Information Measure	54
2.10	Excursus on the Theory of Groups and Their Invariants	57

CHAPTER 3
Physical Measurements and Information   65

3.1   Purpose of This Chapter   65
3.2   The Uncertainty Principle of Heisenberg.   67
3.3   Theory of Measurement and the Conservation of Information   69
3.4   Entropy and Information   70
3.5   Information, Energy, and Computation   71
3.6   The Physics of Information Processing   73
3.7   What Is Life?.   74
3.8   The Theory of Measurement and Information   75
3.9   Time's Arrow   81
3.10 Information Gain and Special Relativity.   85

CHAPTER 4
Principles of Information-Processing Systems and Signal Detection . .   87

4.1   Purpose of This Chapter   87
4.2   Hierarchical Organization of Information-Processing Systems.   98
4.3   Extremal Systems and the Cost of Information   117
4.4   Signals, Modulation, and Fourier Analysis.   123
4.5   Shannon's Sampling Theorem and the Uncertainty Relation   132

CHAPTER 5
Biological Signal Detection and Information Processing   140

5.1   Purpose of This Chapter   140
5.2   Interconvertibility of Information Representations   144
5.3   Human Vision.   149
5.4   Continuity of the Visual Manifold   153
5.5   Stabilized Vision.   157
5.6   Information Content of Contours   166
5.7   Subjective Contours   178
5.8   Models of Human Color Perception   184
5.9   The Gaze as a Flying-Spot Scanner   208
5.10 Biological Echolocation Systems   212
5.11 A Catalog of Visual Illusions.   218
5.12 Hierarchical Sampling Systems and the Duality between Noise and
      Aliasing   245

CHAPTER 6
Pattern Structure and Learning.   264

6.1   Purpose of This Chapter   264
6.2   Texture and Textons   265
6.3   Edge Detection, Uncertainty, and Hierarchy   290
6.4   Pattern Structure and Learning   299

Biographical Sketches   305

References.   314

Index   331

CHAPTER 1

Introduction: Purpose of This Book and History of the Subject

Eadem mutata resurgo

1.1 The Subject Matter of Information Science

Information first began to assume an independent and quantifiable identity in the research of physicists and psychologists in the second half of the nineteenth century. It grew in importance with the development of electrical-based communication systems and, by the middle of the twentieth century, emerged as a unifying and coherent scientific concept along with the invention and development of computers.

The subject matter of a science of information is inherently abstract, for it concerns not the substance and forces of the physical world, but the arrangement of symbolic tokens or, as we may say, the structure of "patterns." Since symbolic tokens are necessarily represented by physical phenomena, as thoughts are represented by the electrical and biochemical states of the neural network, there is an intermixing of the physical sciences with the proper subject matter of information science. This makes it difficult to separate the pure properties of the latter from those of the former, which merely describe the physical embodiment of the informational patterns.

It is clear that particular information can be presented in a variety of forms which differ inessentially from one another. For example, a propositional statement can be expressed in any one of the hundreds of natural languages in current use, or via symbols recorded on paper, or by acoustic speech waveforms. We conclude that some information is independent of the form in which it is presented. The proper generalization of this remark is that *information is what remains after one abstracts from the material aspects of physical reality*. If one abstracts from the material aspects of the printed text, then all that is left is the organization of a sequence of symbols selected from a fixed but arbitrary finite inventory. In general, the concept of information is co-extensive with the notion of order or organization of material objects without regard to their physical constitution. Thus, the study of physically embodied information can be identified with the study of some type of relation of "equivalence" among classes of material configurations. Were it known how to axiomatize the properties of these equivalence classes, a formal and abstract "disembodied" theory of the structure of information would be the result.

In addition to studying the structure of patterns, a science of information must be concerned with the properties of information transfer. A relatively well worked out aspect of this problem falls under the heading of signal processing. Communication of information by real systems necessarily involves transfers of energy so, once again, there is an irreducible interplay between the physical substratum which carries the information and the patterned structures which constitute it.

The scientific study of information cannot be divorced from consideration of pairs of *inventories*, or systems which are capable of bearing information. This is responsible for certain important "relativistic" features of the subject. In vulgar circumstances, one may think of a pair of people conversing or of an information database and a user of information as exemplifying the constituents of a pair. In the sciences, one thinks of the information elicited from an experimental situation which involves the pair consisting of an experimental apparatus and an experimental observer. Regarding the latter, the role of the observer is well known to be critical in the realm of quantum phenomena. It is no less so in the cognitive sciences.

Until recently, the only known information processing systems were biological. Today they have been joined by electronic information processors based on computers and telecommunication networks which are, to be sure, still quite primitive in comparison with their biological cousins. Here we will be interested in studying that higher level of abstraction which encompasses both the "bioware" and the "hardware" implementations—with what men and machines have in common—rather than with the details of either electronic or biological implementations of information processing functions which exemplify their differences. Thus we will seek those general measures and principles which we may expect will apply universally to information processing systems, and perhaps to the fundamental processes of nature themselves.

The patterned structures about which the most is known are those associated with the processing of sensory signals by biological organisms, and especially by man. Neuroscience has uncovered much of value about bioware implementations which limit and thereby partially define the processes that govern the representation and processing of information. Psychology and linguistics, however, reveal our subject in a more nearly pure form. Thus it is appropriate to consider information science from the standpoint of the cognitive sciences. We should, however, also recognize the limitations of psychology and linguistics as experimental sciences. Although the initial conditions and externalities of the experimental arrangement can be modified by the experimenter, the structural properties of the system under investigation generally cannot be manipulated, nor can its internal state be prepared with any degree of certitude. These limitations do not necessarily deal a mortal blow to the role of experiment in the cognitive sciences: The paradigm of astronomy, most ancient and exact of the sciences, is limited to observation alone without control of even the initial conditions of the observed system. Nevertheless, the types of observations that can be made in the cognitive sciences are limited, and this constrains the types of theories that can be readily tested by the experimenter. It is here that the role of the computer in information science will be of crucial significance, for the computer scientist can prepare the internal state of the observed system as well as its initial conditions. For this reason, the computer is the most important experimental tool in the history of information science. It offers the promise of greatly accelerating progress in understanding the nature of information and the laws that govern it.

The role of the computer as an experimental machine and the abstract concepts which gave rise to it form the basis for an intellectual alliance between the cognitive and the computer sciences which, though still in its formative stages, has already begun to bear fruit for both disciplines.

1.2 History of Information Science

To say that information is "pattern" is to beg the question of definition, for no one can say what pattern is. However, our principal interest is not to define information, but to find measures of it; to delineate its properties; and to identify key experimental results which may help to uncover the laws which govern those properties.

The simplest way to present our approach to the subject and the content of this book is to give a brief outline of the history of information science.

The history of information science consists of four interwoven strands: the study of thermodynamics and the theory of measurement in physics; the study of sensory information processing and knowledge representation in biological systems; the rise of electrical communication engineering; and the study of computability and the development of computing machines. In recent years, the common features of these superficially disparate themes have begun to coalesce to form an integrated intellectual discipline whose principal problems blend elements of all four constituents fields.

1.3 Information and Thermodynamics

The recognition of information as a distinct entity in physical processes does not have a definite birthdate, but it appeared in an essential and ultimately highly developed form in the nineteenth century study of thermodynamics. Thermodynamics has always occupied a curious position in physics. It is a phenomenological subject and, despite revolutionary changes in the theories which underlie the composition and behavior of the fundamental constituents of matter and the material systems to which thermodynamics has been applied, the accumulation of experimental evidence gathered over a long period of time has provided overwhelming support for thermodynamical laws. It is as if thermodynamical relationships have, ultimately, very little to do with the underlying structural theories of matter and their interactions. Thermodynamics is valid for Newtonian physical systems; it remains valid and unchanged for systems governed by quantum mechanics; and it applies equally well and unchanged in the realm of relativistic phenomena. That this one part of physical theory could remain valid and essentially unaffected while the fundamental theories of physical composition and interaction underwent revolutionary gyrations suggests that perhaps thermodynamical descriptions are not, after all, descriptions of physical properties and relationships themselves but of some other still more fundamental property coeval with all regimes of physical inquiry, and abstractable from them. Were this true, thermodynamical relations might remain valid for all physical theories which describe an aspect of nature with an adequate degree of verisimilitude. This seems to be the case in the sense that thermodynamical quantities link physical entities to their organization, to their informational properties, and that the informational structures are in some sense abstractions independent of the physical mechanisms which embody them. This is the viewpoint we will adopt.

The celebrated Scottish physicist James Clerk MAXWELL was the first to recognize the connection between thermodynamical quantities associated with a gas, such as temperature and entropy, and the statistical properties of its constituent molecules. There is no point here in trying to explain the physical meaning of the terms *entropy* and *temperature* because the original physical interpretations are rather complicated and not directly germane to our purposes, but also because the most illuminating interpretation turns out to be the informational, rather than the physical, one. Nevertheless, it was in this way that probability entered into thermodynamical considerations and therewith opened the road to further abstractions and generalizations of the entropy concept which were more broadly applicable to systems quite divorced in conception and properties from the gases and other fluids of physics. The measure of the entropy of a state (and hence, as was later realized, of its information content) in terms of probability was supplied by Ludwig BOLTZMANN's famous formula, viz:

$$S = k \log W, \qquad (1.1)$$

where S denotes entropy, W the number of accessible microstates, and k BOLTZMANN's constant, which mediates between the physical units of entropy on the left side of the relation and the purely numerical quantity log W on the right; k is measured in units of energy per degree of temperature and has the value

$$k = 1.381 \times 10^{-16} \text{ erg/Kelvin.} \tag{1.2}$$

BOLTZMANN himself recognized that entropy is a measure of disorder of the configuration of states of the atoms or other particles which make up a thermodynamical system, and hence it can also be considered as a measure of the order or degree of organization of the system. This viewpoint was clarified and sharpened by SZILARD, who, in 1929, identified entropy with information, and the measure of information with the negative of the measure of entropy—an identification which is the foundation of the quantitative part of the theory of information.

SZILARD's contribution came about in the following way. An early hint that the statistical interpretation of thermodynamical properties must have something to do with information was already present in MAXWELL's discussion, in the midnineteenth century, of an apparent thermodynamical paradox. MAXWELL conceived two chambers separated by a common partition which could be removed in order to permit the objects in one of them to move freely into the other. If the two chambers are initially separated by the partition and one of them contains a gas (say, air under normal conditions of temperature and pressure), then, upon removal of the partition, the gas will rapidly diffuse to fill the entire chamber. But everyday experience tells us that the reverse sequence of events does not occur; the molecules of a gas distributed throughout a chamber will never congregate in one part so that, by slipping in a partition, the chamber can be divided into two, one of which contains all the gas and the other none of it. MAXWELL proposed a "demon" who, he suggested, would be capable of performing this counterintuitive feat. Here is MAXWELL's idea. Suppose that the combined chamber is filled with gas and that the partition is reinserted. Further suppose that the partition contains a small door just large enough to allow a molecule of gas to pass from one chamber to the other. This door, which can be opened and shut without frictional resistance, is attended by MAXWELL's demon, who observes the gas molecules in one of the chambers and, when in its random motion one of the particles heads toward the door and is about to rebound from it, opens the door briefly, permitting the molecule to pass into the other chamber. Very soon there will be an excess of molecules in one of the chambers and a deficiency in the other which will, as time passes, become increasingly extreme, tending toward a final state in which the entire gas occupies but one of the chambers and the other is empty. Since the demon requires an infinitesimal amount of energy to operate the door, this process appears to decrease entropy, in contradiction to the second law of thermodynamics.

This paradox was explained by SZILARD in his 1929 paper, whose importance only became apparent much later. It was obvious, he pointed out, that in order to perform its task, MAXWELL's demon had to be very well informed about the position and velocity of the molecules that approached the door so that it could judge when, and for how long, the door should be opened to enable a molecule to pass through and into the chamber which would ultimately have the excess of molecules without allowing any molecules to pass in the opposite direction. SZILARD was able to calculate the entropy gained by the demon during the process of letting a molecule pass through the door. Since the demon can play its role if and only if it has the necessary information, SZILARD was led to identify the negative of the demon's entropy increment as the measure of the quantity of information it used. This identification of information increments with entropy decrements extended BOLTZMANN's interpretation of entropy as a measure of the orderliness (i.e., the degree of organization) of a physical system. It changed the focus of attention from the physical to the abstract aspects of the concept.

1.4 Early Development of Sensory Physiology and Psychophysics

It is interesting to note that many of the scientists who were engaged in unraveling the mysteries of thermodynamical entropy and the role of information in the physical measurement process were also active contributors to the development of psychophysics, electrical communication engineering, and the invention of computers, which are the other three strands of our tale. Consider MAXWELL, the foremost mathematical physicist of the nineteenth century, creator of the theory of electromagnetism as well as founder of statistical mechanics, who also investigated some of the basic properties of human vision with instruments of his own invention. Such combinations of interests and talents were not at all rare in the scientists who investigated the properties of information.

While the physicists were busy interpreting the phenomenological notions of thermodynamics, first in terms of statistical mechanics and later in terms of more purely probabilistic and information-theoretic concepts, those scientists who studied biological information processing were not idle.

Ernst WEBER, one of the founders of psychophysics, and his two scientist brothers had concentrated their research efforts on applying the exact methods of mathematical physics to the study of the various physiological systems of the higher animals and man. In 1826 WEBER began a series of systematic investigations of the sensory functions which resulted in his formulation of the first general quantitative assertion of psychophysics. It can be stated as follows. Let the magnitude of some physical (objective) stimulus be denoted by S and the magnitude of the corresponding psychological (subjective) response be denoted by R. WEBER found that if S be increased by an amount ΔS sufficient to create a just noticeable different (*jnd*) in response to the different stimuli levels, then the initial stimulus magnitude S and the stimulus increment ΔS are related by the equation $\Delta S/S = c$, where c is a constant which depends on the sensory modality and on the individual; for typical compressive modalities, c is approximately equal to 1/30. It is known today that the "WEBER fraction" $\Delta S/S$ is approximately constant for only a limited range of sensory stimuli, but WEBER's observation was a key and fruitful result which itself was the stimulus for much later work.

In 1850 Theodore FECHNER, basing his analysis on these earlier investigations of WEBER, conceived the psychophysical function, which expresses the subjective measure of magnitude of the psychological response in terms of the objective measure of magnitude of the physical stimulus, and proposed that it could be represented by a logarithmic dependence of response magnitude upon stimulus magnitude. In terms of the stimulus magnitude S and the response magnitude R, this means that the psychophysical function $R = f(S)$ is given, according to FECHNER, by the formula $R = a \log S$, where a is a constant which characterizes the sensory modality and the individual; $1/a$ can be identified with the WEBER fraction. A function of this type compresses a great range of variation of the stimulus magnitude into a relatively small range of variation of the response magnitude; thus the latter is more readily assimilate by the biological information processing gear which is, after all, made from unreliable components that are slow and bulky and require special temperature and other environmental conditions in order to function.

FECHNER's result has been frequently challenged; perhaps most effectively by S. S. STEVENS. Indeed, it is obvious that the logarithm alone cannot account for the qualitative behavior of sensory response to either very large stimuli or, without suitable modifications for a "threshold" level of activity, to very small stimuli. In the former case, $\log S$ increases without bound as S does, but real sensory processing channels have a limited capacity. Nevertheless, subsequent studies have shown that the neural firing frequencies which encode sensory stimuli do appear to be proportional to the logarithm of the stimulus magnitude throughout much of the normal range of stimuli experienced by an organism. This provides a heretofore absent level of detailed confirmation of the essential validity of the pioneering work of both WEBER and FECHNER.

Hermann von HELMHOLTZ was the greatest figure in nineteenth century physiology and psychophysics, and one of the greatest scientists in general. He was interested in everything and made fundamental contributions to many aspects of physics, mathematics, and biology. It was HELMHOLTZ who formulated the law of conservation of energy what is essentially its modern form.

In 1850, the same fruitful year when FECHNER conceived the psychophysical function, HELMHOLTZ made the first measurement of the velocity of propagation of neural impulses. Earlier physiologists and physicists had held that the velocity of neural signals was so great as to never be measurably by experiment within the compass of an animal body. In fact, electrical impulses propagate along neurons at speeds which vary between about 10 meters per second and 100 meters per second, faster propagation corresponding to nerve cells whose axons are myelinated and have the greater diameter; the faster neurons have a channel capacity on the order of several thousand bits per second. Just as the discovery that light was propagated with a large but finite speed influenced theoreticians two centuries earlier, the measurement of the speed of neural impulses profoundly conditioned and limited the kinds of theories of sensory and mental activity that would henceforth be entertained.

HELMHOLTZ's results made it evident that not all information incident upon the sensory system could be processed by the neural net in *real time*; that is, rapidly enough for the organism to respond to sensed changes in its environment. If the speed of propagation were infinite, then an infinite amount of computation could be performed by a neural network in any brief interval, so there would be no computational barrier to the full processing of all sensory input signals. If the speed of propagation were finite but very great, say comparable with the speed of light, there would be theoretical limitations on computational capabilities and the time required to process complex signals, but they would not impose significant practical constraints. Neither of these two possibilities conditions the computational models of biological information processing systems, and neither has much explanatory power in helping the scientist to interpret experimental data or discriminate among alternative theories.

The truth of the matter is that neuron-based biological information processing is very slow. The limited speed of propagation of neural signals implies that it must take between 0.01 second and 0.1 second for a signal sent by the brain to reach the hand or foot, or as long as 3 seconds for one sensed by the tail of the dinosaur *Diplodocus* to reach its head. It is the low speed of neural signal propagation which makes the illusion of continuous motion possible in cinema pictures that successively display fewer than 60 still frames per second. This means, in particular, that the human vision system cannot distinguish continuous physical motion from a discontinuous succession of images if the rate of image presentation is greater than this small number per second. If neural signals propagated at infinite speed, we would have no difficulty in detecting discontinuous motion at any rate of presentation whatever, and the illusions so admirably created by television and motion picture photography would be impossible. HELMHOLTZ observed that

> Happily, the distances our sense-perceptions have to traverse before they reach the brain are short, otherwise our consciousness would always lag far behind the present, and even behind our perceptions of sound.

Taking the small finite value of the rate of transmission of neural impulses into account, an elementary estimate of the quantity of information that is, for example, incident upon the retina of the human eye shows that it greatly exceeds the transmission capacity of the neurons that lead from it to the higher cognitive centers. A similar assertion can be made about the capacity of the ear compared with that of the channels that conduct aural information to the brain. Most of the incident information must be omitted by the higher processing centers.

Several years after these discoveries of HELMHOLTZ and FECHNER, in 1853, the mathematician GRASSMANN showed that the space of physical light signals of all visible intensities and frequencies, which is an infinite-dimensional *objective* space, is compressed by the human vision system into a region in an abstract 3-dimensional *subjective* space of color perception, the so-called color cone.

Taken together, these results of HELMHOLTZ, FECHNER, and GRASSMANN suggest that sensory data is corrupted and modified by the sensing organism in ways which greatly reduce its effective quantity and substantially modify its original form.

1.5 Selective Omission of Information

These corruptions and compressions of sensory data have the effect of reducing an unmanageable glut of input information to an amount that can be processed by the mental computing equipment with sufficient rapidity to be useful for responding to changing environmental circumstances. There are many different ways in which the incident information could be reduced in quantity. The receptive system could, for instance, filter the input to limit the amount of data received through each input channel uniformly. This would preserve some fixed fraction of the information in the original signal without giving preference to any particular kind, but, because the total amount of information sent forward to the higher processing centers would be reduced, the reduction would be noticed more in those informational constituents which are the more "essential" for the organism's purposes. Were it possible to selectively omit information, retaining the "essential," or at least the "more essential," portions while rejecting the "less essential" ones, then the computational load on the mental computer could be substantially reduced and its ability to respond to sensory inputs in a timely way would be correspondingly increased. This is just what all biological information processing systems do. But how the discrimination between "more essential" and "less essential" is made is not yet fully understood although, as we shall see, certain basic features of the process are reasonably clear. This discrimination, which occurs to a considerable degree in the peripheral elements of the neural net adjacent to and connected to the sensory organs themselves, embodies a most important pattern classification mechanism which operates prior to the activities of the higher processing centers and limits the information which is available to the latter. It would be a particularly valuable contribution to understand the principles that govern this preliminary pattern classification which creates a fundamental discrimination between that part of the signal carrying information of general use to the organism and the remainder which is of equal physical significance but can be treated as "noise" from its informational standpoint.

The need to selectively omit information occurs in internal mental processing as well as the processing of sensory stimuli. Consider, for example, the mental analysis of a chess position. The game tree, which determines all possible consequences of alternative plays, has a combinatorial complexity which far surpasses the ability of the human brain or any computing machine to explore it completely; it is estimated that there are as many as 10^{120} different possible games. Humans reduce the complexity of this search problem by selectively omitting most of the pathways. The omitted ones are expected to be unproductive, but the process by which a person makes the decision to omit particular paths remains unknown. The computer scientist usually calls selective omission in the context of a decision problem which involves searching through a complex tree the application of *heuristic* procedures. But this is just a term to name neatly an essential process which is still largely a mystery.

It is important for us to realize that selective omission occurs not only at the level of omitted data, but also—perhaps more importantly—at the level of omitted logical procedures, procedures which could be and should be applied to data but, as the result of a systematic process of exclusion, are not applied.

1.6 The Quantum Mechanical Theory of Measurement

The heading of this section refers to the physical basis for measurement, which underlies not only measurements of microscopic physical phenomena, but all measurements, including those which are based on information gathered by the sensory apparati of organisms. Apart from the general argument that macroscopic information processing and measurement constraints are ultimately consequences of the limitations which exist at the most fundamental level of physical phenomena, many of the direct macroscopic measurements, made either by means of laboratory and other instruments or by the physiological detectors of sensation, operate with the same materials and means that support the foundations of physical measurement. For example, the eye is sensitive to photons of light which are governed by the wave-particle duality of quantum physics at the level of their interaction with the photosensitive chemical receptors in the rods and cones of the retina. Therefore it will be appropriate for us to summarize the development of certain aspects of the theory of measurement from the viewpoint of quantum physics in order that we may be able to form an impression of the nature of the physical limitations which appear to govern information processing at the most basic physical level.

The breakdown of classical physics, which became increasingly apparent at the end of the nineteenth century, was due to the improper extension of naive concepts of measurement and measurability from the realm of everyday life to circumstances increasingly distant from their origins. For our purposes, one of the central new developments which emerged in the first 30 years of the twentieth century was the recognition of the interdependence of measurements of variables of different kinds. In sharp contrast to ideas that had been accepted as the basis for science for more than 200 years, it was discovered, for example, that the accuracy of measurement of the *position* of an object interfered with the accuracy of measurement of its *velocity* of motion (more precisely, of its *momentum*), and that intervals of time and space have only a relative meaning, depending upon the motion of the observer. Thus, there is an inherent ambiguity, and uncertainty, in the possible knowledge that we can have of our physical surroundings. This restriction has important practical as well as philosophical consequences and affects the stability and structure of chemical combinations in a direct way. Erwin SCHRÖDINGER, one of the creators of quantum mechanics, saw in this property (coupled to certain other aspects of the new physics) the secret to the existence of life itself: that is, why life is possible amid the incessant battering of the genetic material in a sea of thermal agitation.

That energy was bundled in discrete units, the *quanta*, was discovered by Max PLANCK in 1900 in the course of his investigations into the inadequacies of the classical description of the thermodynamical properties of radiation confined in an enclosure. This key discovery led, through a complex and indirect pathway, to the creation of a coherent new fundamental theory of physical phenomena by Werner HEISENBERG in 1925. It was HEISENBERG who discovered that the accuracy of measurement of "conjugate" physical variables was mutually constrained by the inequalities of uncertainty which bear his name, and that all physical measurements are, ultimately, statistical in the sense that repeated measurements of identical physical situations will not lead to identical results, but only to measurements which are collectively related to one another by a statistical law.

These discoveries seem to place the basic structure of physical nature on the same slippery footing as human affairs; variable subjective responses are the consequences of fixed objective stimuli. Accurate statistical descriptions of, for example, the distribution of longevity for a collection of people are easily come by, but it is not possible to predict the life span of an individual.

The deep questions about the nature and limitations of measurement were taken up by the polymath John VON NEUMANN who, in 1932, attempted to formalize the new physical theory. His path-breaking book included an extensive analysis of the problem of measurement and showed how it was related to the thermodynamical and informational notions that had independently grown from the considerations of nineteenth century physcists such as BOLTZMANN and SZILARD. In effect, VON NEUMANN showed that the observer must be taken into account in the process of measurement, but that the dividing line between observer and observed is fluid. If the combined system consisting of the observed portion and the observing portion is treated according to quantum mechanics, then the total amount of information in the course of a measurement is preserved. But at some ultimate stage, the psychological "external" consciousness which assimilates the observation leads to an irreversible increase in entropy. This is an unusually subtle matter. It means that information about some variable or event which is gained must be compensated by a precisely equal amount of information about some other variable or process which is lost and, possibly, by some additional irreversible information loss as well. Although at the most fundamental level of measurement theory, such a "law" of conservation of information may hold, at the macroscopic level, it is not possible for us to trace all the information flows which result from a measurement. The consequence is that some information "gets lost" whenever a measurement is performed, and that the degree of organization of the universe continually decreases.

In recent years, with the penetration of computer technology into the microworld where quantum phenomena are the governing commonplace rather than theoretical curiosities, questions related to quantum restrictions on the process of measurement and hence on the limits of information processing have been studied with increasing interest, vigor, and relevance.

1.7 Early Development of Electrical Communication Engineering and Its Formalization as Information Theory

The third of the historical threads which combine to form the science of information is the development of electrical communication engineering. *Communication engineering* refers to communication by means of electromagnetic signals, including the transmission of electrical currents through wires, as in telegraphy and telephony, and the transmission of electromagnetic waves through space, as in broadcast radio or television. The subject grew from the practical needs of engineers to understand how to design complex systems for efficient performance. Because variations in electric currents and electromagnetic fields cannot be directly sensed by people, except for the small band of radiant electromagnetic energy called "light," the methods employed in this practical field are more theoretical and more dependent on measuring instruments than are those of other engineering fields. Indeed, the boundary between "science" and "engineering" is often scarely definable. As a result, the interplay between theory and practice has been unusually intense and has stimulated the exceptionally rapid development of the subject.

As it turns out, the general principles which govern the transmisson of information by electrical means extend to other media as well; moreover, the discovery that the neural network transmits information by encoding it as chemically generated electrical impulses suggests that the tools built by the communications engineer should be of great value in the general study of neurologically mediated information communication. This was first clearly anticipated by Norbert WIENER and has, indeed, turned out to be the case.

The most decisive contribution of communication engineering was the discovery of the proper measure of information in the context of the transmission of information. Although arrived at from an entirely different standpoint, the result coincided with what the statistical thermodynamicists had found: the measure of information is formally identical to the negative of the entropy measure, assuming, of course, an appropriate interpretation for the corresponding concepts. This will be described in greater detail in Chapter 2, but here we may simply recall that a communication system functions by successively transmitting signals which are drawn from a certain inventory, such as the collection of letters of the alphabet. If the kth signal in the inventory is transmitted with a probability p_k, then the measure of information provided by the transmission of a single signal selected from the inventory is

$$I = -\sum p_k \log_2 p_k, \tag{1.3}$$

where the summation runs over all signals. R.V.L. HARTLEY of the Bell Telephone Laboratories offered a logarithmic definition of the quantity of information in 1928, but it did not take into account the case of unequal probabilities. He was unable to erect a predictive theory which made significant use of the definition and it languished until, 20 years later, its deficiencies were remedied by SHANNON. But HARTLEY understood the engineering issue at stake quite precisely. In the introduction to his paper he described his goal by stating:

> What I hope to accomplish ... is to set up a quantitative measure whereby the capacity of various systems to transmit information may be compared.

Other researchers were thinking about similar problems. In 1922 the statistican Ronald FISHER had attempted to define the information content of a statistical distribution and suggested the reciprocal of the variance as a candidate. World War II provided the impetus for bringing scientists and engineers together to focus their intellectual energies on the problems of communication in the context of the development of radar, which may have provided the main stimulus to the creation of an essentially complete theory of communication by Claude SHANNON. SHANNON was a student of Norbert WIENER, who appears to have independently suggested the measure of information which today is usually associated with SHANNON's name. It was the latter, however, who used this measure to create a theory which could be applied to a wealth of problems of the most general variety to determine the efficiency of a communication channel and to aid in the design of communication systems. SHANNON's pioneering work was published in a series of three papers in 1948. Although his work has been the subject of innumerable investigations since then, within its framework, no fundamental new idea or result has been added to the elegant edifice he constructed.

1.8 The Development of Computers

> The whole of the developments and operations of analysis are capable of being executed by machinery.
>
> —Charles BABBAGE

The fourth and final strand in the history of information science is the story of the invention and development of the stored program digital computer. The idea of a special purpose machine which could compute and perform certain restricted information processing tasks is very old. In antiquity, *analog* devices were used to aid astronomical calculations. An ancient calendar computer from *circa* 80 B.C. has been recovered from the wreck of a Greek trading vessel off Antikythera. The abacus, another ancient device, is an example of a *digital* aid to computation. The concept of a machine that could perform an unrestricted range of mathematical calculations dates at least to the time of the Thirty Years War. In 1623 Wilhelm SCHICKARD, a professor of astronomy, mathematics, and Hebrew at Tübingen, designed and built a mechanical device which performed addition and subtraction completely automatically and partially automated multiplication and division. This machine appears to have had no discernible effect on his contemporaries. A generation later, in 1642–44, Blaise PASCAL, then 20 years of age, invented and built a simple machine which performed addition and subtraction. PASCAL's calculating engine seems to have been inferior to SCHICKARD's, but it nevertheless received considerable attention and was later described in detail in the famous *Encyclopédie* of DIDEROT.

LEIBNIZ appears to have been the first to conceive of a universal logical machine which would operate by a "general method in which all truths of reason would be reduced to a kind of calculation." The earliest attempt to put this idea into practice was made by Charles BABBAGE, who, not later than 1822, was already toying with the notion of a steam-driven machine that could be applied to the onerous general calculations which arise in astronomy. The British government subsidized BABBAGE's work for 20 years, but his machines were never completed. Although the ideas behind them were correct, complex computing machines cannot be built from mechanical parts because of the inherent inaccuracy of their relative positions and the cumulative effects of wear, limitations which are both more fundamental than they may at first appear.

The principal theoretical impetus to the development of the general purpose digital computer can be traced back to the last decades of the nineteenth century. While scientists were struggling with the newly uncovered inadequacies of classical physics, logicians and mathematicians were becoming correspondingly uneasy about paradoxes and contradictions which were emerging from the logical processes of reason and the foundations of mathematics. As a result of attempts to arithmetize mathematical analysis, a general theory of sets was created by Georg CANTOR between 1874 and 1897, but it was soon cast into doubt by the discovery of numerous paradoxes, of which one discovered by Bertrand RUSSELL in 1902 is perhaps the most widely known. The RUSSELL paradox concerns the set of all sets which are not members of themselves. Is this set a member of itself? Straightforward application of the rules for reasoning show that it is; but then, according to its definition, it cannot be. Paradoxes of this kind raise the question of the definability of mathematical objects. At about the same time, mathematicians became concerned about the *consistency* of mathematical systems such as classical number theory and elementary geometry—whether the application of logical processes of proof to a collection of axioms might ever result in contradictory statements—and about their *completeness*—whether all true statements about, say, elementary geometry, could in principle be logically derived from a collection of consistent axioms. These and related questions exercised mathematicians such as David HILBERT and John VON NEUMANN as well as logicians and philosophers who, in order to cope with the increasing technical nature of the problems, found themselves increasingly adopting mathematical formulations and tools.

In 1925 HEISENBERG discovered the physical limitations to our ability to observe natural phenomena that are expressed by the uncertainty principle. A different kind of limitation and uncertainty principle for the processes of reason was revealed by the mathematician Kurt GÖDEL in 1931. GÖDEL showed that any mathematical system whose structure is rich enough to express the ordinary arithmetic of integers is necessarily incomplete; there will be true statements which cannot be proved within the system. This lack cannot be made good by incorporating such an unprovable true statement as an axiom in the system, for then other unprovable truths emerge. This is another way of saying that there are inherently undecidable statements, whose truth cannot be established by merely drawing out the logical consequences of collection of axioms.

GÖDEL's method, as well as his result, was important because it established a link between arithmetic and logic by coding the propositions and logical operations in arithmetic formulae and expressions, thereby effectively reducing questions about logic to questions about computation. The complete identification of logical reasoning with computation was proposed by Alonzo CHURCH in 1936 and implicitly adopted by Alan TURING in 1936–7. TURING's conception of a universal calculating machine (the *TURING machine*) marks the birth of the modern theory of the digital computer. This abstract machine is able, as TURING proved, to compute anything that any special machine can; in this sense it is "universal" and, consequently, a valuable model for the theoretical analysis of computability.

The course from a fledgling theory of computability to the deployment of practical and powerful computing machines has been run with remarkable rapidity. Spurred on by the needs and means provided by World War II, researchers made great strides in the design of computing machines and in the underlying electronic technology needed to implement those designs. By 1945 the key concept of the stored program computer had been created by the same John VON NEUMANN who a decade earlier had been analyzing the nature of observation and measurement in quantum mechanics and who would, a decade later, concentrate his mental powers on investigations of the workings of the brain and of self-reproducing automata—that is, of life.

It did not take very long for the concept of the TURING machine and its partial realization by practical and increasingly powerful general purpose logical (computing) machines to lead to the idea that the brain is perhaps nothing more than a spectacularly ingenious computer built from error-prone components wired together in a highly variable way, and that the continuity of existence of living species from one generation to the next is maintained by a kind of "program" coded in the macromolecular proteins of the genetic material. These are special instances of the more general idea that all of the features peculiar to living organisms and to life itself can be understood in terms of information processing principles inherent in the computer concept. Freeman DYSON expressed this view when he wrote:

> as we understand more about biology, we shall find the distinction between electronic and biological technology becoming increasingly blurred.

1.9 The Information Processing Viewpoint in Psychology

The computational interpretation of the process of thought and the signal processing viewpoint of the communications engineer both spread rapidly, even before computers having significant capabilities had been constructed. Within 5 years of the publication of SHANNON's basic papers and the construction of the first stored program computer, many investigators had begun to explore the implications of both developments for psychology.

As early as 1947, Walter PITTS and Warren McCULLOCH published a seminal paper with the provocative title "How We Know Universals: The Perception of Auditory and Visual Forms" in which they proposed a mechanism, based upon ideas from mathematical group theory, by which the neural net extracts universals from sensed data. Very much in the spirit of signal processing and the "brain as a computing machine" viewpoint, their work established the path that many others would later follow, and which continues to influence research today.

The next year, 1948, saw the publication of the first edition of Norbert WIENER's influential and remarkably prescient *Cybernetics: or Control and*

Communication in the Animal and the Machine, which identified the important role of feedback in governing the activities of machines and animals and adopted, for the first time, a completely comprehensive view of animal performance and capabilities from the computing and communication engineering standpoints. A discussion of how information can be quantified already appears here from the standpoint most suited for general application to the measurement process, the standpoint we adopt in Chapter 2.

By 1950 the question of whether the brain is merely a bioware implementation of a computing machine had become specialized into a spectrum of particular issues with "whether" giving way to a more experimentally oriented "how":

● *How could a machine be programmed to solve general problems?*

later addressed by NEWELL, SHAW, and SIMON, and

● *How could the behavior of a suitably powerful and properly programmed machine be distinguished from the behavior of a person?*

The latter question directed attention to the problem of deciding whether the abilities of computing machines were for all practical purposes indistinguishable from those of humans, which is different from the theoretical question of whether the brain is in fact merely a computing machine. TURING took up the matter and in 1950 proposed his famous "test." In 1951 Marvin MINSKY produced an early analog simulation of a self-reproducing system. In 1953 RIGGS and his colleagues, in their investigations of human vision, made the remarkable discovery that images stabilized on the retina vanish; this counter-intuitive result has both signal processing and higher categorical implications for the way the brain processes sensory information to produce the universals for which PITTS and McCULLOCH had been searching. In 1945 ATTNEAVE showed that the information content of 2-dimensional images was nonuniformly distributed and identified some of the preferred features in which it resides. His famous "cat" (Figure 1.9.1) demonstrated that the major part of the information contained in a contour is concentrated near points of extreme curvature. We will examine this result and its consequences in Chapter 5. Two years later George MILLER published a fundamental paper which described

Figure 1.9.1   Attneave's "cat."

the short term memory information processing capacity of the human brain. It is surprisingly limited. As the catchy title of his paper expressed it, only "seven plus or minus two" chunks of information can be held in short term memory. This is not the place to make the meaning of the term "chunk" precise; suffice it to say that if the input items are a sequence of bits, then about seven of them can be retained in short term memory, but that the same statement could be made with equal validity for other larger units, such as ordinary decimal digits (MILLER's result tells us why local telephone numbers consist of seven digits, and it also tells the post office why there is considerable resistance to eleven digit zip codes) or words drawn from the vocabulary of a natural language.

The year 1958 saw the appearance of VON NEUMANN's book titled *The Computer and the Brain*, and in 1959, Leon BRILLOUIN's ambitious attempt to incorporate all of scientific endeavour within the formal precincts of SHANNON's information theory was published. Although BRILLOUIN was not successful in providing a completely coherent theory capable of predicting new results as well as explaining old, his viewpoint acted as a stimulant and a signpost, hinting at the kind of theory that might ultimately be created. In the same year, LETTVIN, MATURANA, McCULLOCH, and PITTS analyzed the signal processing capabilities of the frog's eye and were able to determine what it tells the frog's brain. Not much, as it turns out, although the frog gets along very well, thank you, and therein lies the great significance of this important paper for the development of a realistic theory of knowledge representation.

In recent years, evidence about how knowledge is represented in animal brains and on the pattern classification capabilities of organisms has steadily accumulated. Concerning human abilities, attention has been focused on the use of visual illusions as an experimental tool for investigating the aggregated nature of image processing (a topic which will be taken up in Chapter 5). In addition to the traditional illusions known for many years, new ones based on the study of stabilized retinal patterns and on subjective contours provide interesting insights into the information processing mechanisms which the brain exploits to reduce the quantities of data it must process in order to sustain the illusion of reality. Recent work of KOSSLYN and of PINKER has identified some of the key features of the mental storage and manipulation of 3-dimensional image information and provided evidence about the computational mechanisms they involve. CRICK, MARR, and POGGIO have coupled sophisticated results from communication theory to neurophysiological investigations and produced a striking information processing explanation of visual acuity and hyperacuity in human beings. Using a computational theory of early vision based upon analysis of zero crossings of filtered retinal images, their work suggests that the fine grid of small cells in layer $IVC\beta$ of the striate cortex performs an approximate reconstruction of the image by means of a sampling theorem due to LOGAN. This work may stand as a guidepost to how the information processing approach can be applied to biological systems. Other remarkable, although less noted and quite different, studies of the pattern classification capabilities of pigeons and other lower animals by HERRNSTEIN and his colleagues suggest that there exists a universal pattern classification algorithm of sufficient simplicity to be implemented within the relatively limited brains of these animals. This seminal result should encourage researchers to investigate the details of the putative algorithm and to be guided by them in understanding the general properties of pattern classification by biological systems.

1.10 Some Principles of Information Science

Several fundamental problems and general principles of information science have emerged from the century and a half of research described in the previous sections. The central problem concerns the nature of patterned structures. Within the realm of human information processing, this is realized in two main questions:

● *How is knowledge represented in the brain and how can it be represented in the machine?*
● *What is the nature of learning?*

The first question raises two important collateral problems:

● *How is information that is captured by the sensory systems converted into the abstract forms that the brain actually uses?*
● *To what extent do artificially constructed information-bearing forms of communication, such as language, mathematics, and music, reflect the internal organization and structure of knowledge and the (mental or machine) means for employing them?*

The second main question raises the following collateral problem:

● *What is the relationship between the properties of the information to be learned and the internal state of the learner?*

A satisfactory solution to this problem would include a characterization of the learner's knowledge base requirements and processing rules as a function of the knowledge being acquired.

In the chapters that follow, we will concentrate on applying some general principles which have emerged from the wide-ranging investigations summarized in the preceding sections to these two main problems of human information processing.

As has already been observed, there appears to be a general

Principle of selective omission of information

at work in all biological information processing systems. The sensory organs simplify and organize their inputs, supplying the higher processing centers with aggregated forms of information which, to a considerable extent, predetermine the patterned structures that the higher centers can detect. The higher centers in their turn reduce the quantity of information which will be processed at later stages by further organization of the partly processed information into more abstract and universal forms so that the representatives of inputs to different sensory organs can be mixed with each other and with internally generated and symbolic information-bearing entities.

There are at least three particular principles which appear to play important roles in this process of successive selective omission and aggregation into more abstract forms of coded information. Each of these will be examined in the following chapters. The first is the

Principle of adjustment of information-processing systems to extremize the quantity of information relative to some "processing cost" constraints.

In Chapter 2 we will define a numerical measure of the quantity of information which is gained from an observation, provided by a mathematical calculation, or transmitted through a communication system. Designers of computers and other information processing systems naturally attempt to maximize certain measures of information processed relative to the cost of processing it. But biological information-processing systems and nature in general appear to be arranged so that information is extremized (not necessarily maximized) subject to constraints which depend upon the problem under consideration and are analogous to "costs."

The second particular principle is the

Principle of invariance of information processing structures and measures under appropriate group actions.

This statement of the principle is, unfortunately, phrased in terms of technical terminology whose clarification will require a considerable amount of explanation. We will make a summary start here. Information-processing situations often have natural symmetries associated with them. For example, as will be shown in Chapter 2, a measure of the quantity of information gained by an observation should not depend on the unit of measurement marked on a measuring rod or instrument gauge, nor upon where the zero-point of the measurement scale is placed upon the gauge. Arbitrary relocations of the zero-point and selection of the unit of measurement can be thought of as "symmetries" of the measuring instrument. They form an instance of the mathematical structure known as a *group*. A measure of the quantity of information which is independent of these "symmetry" changes is called a *group invariant*, and it is the identification of the invariants of the groups which arise in information processing situations which is called for by the second principle.

The use of group invariants is a powerful way of selectively omitting information. This was clearly recognized in the 1947 paper of PITTS and McCULLOCH which sought to discover universals in visual and auditory perception. In Chapter 2 we will show how it can be used to define the information measure. In the same chapter we will apply the idea of group invariance to study the psychophysical function (that is, the function which relates a measure of physical stimulus intensity to a measure of psychological response intensity) and will show that the functions proposed by FECHNER and STEVENS are the only ones which are compatible with certain natural group invariance requirements. Later, in Chapter 5, we will find that group invariance plays an important role in human color vision.

The third particular principle is the

Principle of hierarchical organization of information-processing structures.

It has been long recognized that the introduction of a hierarchical organization generally increases efficiency. The assembly line method of manufacture is one important socioeconomic example. The organization of governments, military forces, and large commercial firms tends to be hierarchical because this is an efficient structure for the information-processing components of decision making. Herbert SIMON has cogently stressed this point of view. Hierarchical structures are even more widespread. The architecture of contemporary computers, and especially of their memory systems, is largely hierarchical. Biological information processing often can be viewed as hierarchically organized. The relationship of short term and long term human memory offers one example. In humans, the relationship of foveal to peripheral vision (and in echolocating bats, the relationship of the auditory fovea to the rest of the auditory receiver) provides striking examples of common goals achieved by abstractly similar hierarchical mechanisms functioning in grossly different physical circumstances.

CHAPTER 2
Mathematics of Information Measurement

What is a number, that a man may know it, and a man, that he may know a number?

—Warren S. McCulloch

2.1 Purpose of This Chapter

Everyone recognizes the vital role that numbers and numerical relationships play in civilized life. In addition to their crucial position within science, numerical measures also are important in everyday activities although their formal properties are usually kept in the background. Nevertheless, the estimation and comparison of magnitudes are among the most common of occurrences, and the use of a numerical measure—money—to form equivalence classes of goods and services characterized by their common value, and its arithmetic properties to establish a means for comparing the value of inequivalent commodities, surely ranks as one of the most abstract ideas that has received general public acceptance.

Numbers are used in two fundamentally different ways. One way is as a means for describing measurements of observed external phenomena, such as a distance or elapsed interval of time, or the magnitude of a sound or the strength of a force. The other way numbers are used is in mathematical calculations which reflect internal mental processes rather than observations of external phenomena. We shall see, however, that there are types of mathematical processes which have features in common with observational measurements. The main difference between these two uses of numbers is that, in a sense which will be made more precise below, it is in principle impossible to know a number which represents an observed measurement with complete precision, whereas it is in principal possible to know some numbers which arise as part of a mental mathematical process.

The purpose of attaching a number to an observed magnitude is to supply some of the information gained from the observation in a convenient form. It is equally true that the purpose of a mathematical calculation is to provide certain information even if the calculation does not arise from an observation or measurement. Thus, we may conclude that numbers and calculations carry information. The principal purpose of this chapter is to identify and measure the information content of the numbers that arise from measurement and as a consequence of certain mathematical processes. The measure of information will not depend on any physical process or model of communication and consequently it will be broadly applicable to all physical processes, measurements of sensed and otherwise observed magnitudes, and internally generated mental processes which can be represented as a "calculation" of a general kind.

Let us begin our investigation by saying what we mean by "numbers." What is required is primarily a restatement of well known facts about the use of decimal (and binary) expansions to represent numbers in positional notation systems, and about their relationship to other methods for denoting numbers, such as fractions and special notations such as $\sqrt{2}$ and π.

By *numbers* we shall always mean the *real numbers* of mathematical analysis. A definition which is convenient for our purposes and familiar to everyone is provided by the representation of numbers as *decimal expansions*. One says that a denotes a real number if a can be represented in the form

$$a = a_{-N} \ldots a_{-1} a_0 . a_1 a_2 \ldots a_n \ldots, \tag{2.1}$$

where each of the symbols a_i denotes an integer between 0 and 9, and the leading digit a_{-N} is not zero if $N > 0$. This decimal expansion is a shorthand notation for the infinite series

$$a = a_{-N} 10^N + \cdots + a_{-1} 10 + a_0 + a_1 10^{-1} + \cdots a_n 10^{-n} + \cdots. \tag{2.2}$$

Each real number corresponds to a unique decimal expansion of the type (2.1) if it is agreed to exclude the expansions, such as $1.25999\ldots$, in which nines ultimately repeat. If 0 ultimately repeats indefinitely, as in the number $5.4000\ldots$, we agree to omit the infinitely many repeating zeros from the notation, and simply write 5.4.

Every number can be expressed as such a decimal expansion, but there are other common ways of describing certain classes of numbers. Consider fractions: these numbers are the ratios of integers (for this reason they are also called *rational* numbers) such as 2/5, 1/3, 13/70, etc. In general, a rational number can be expressed as p/q where p and q are integers and $q > 0$ and, by successively dividing by the denominator q, one can work out the decimal expansion that corresponds to p/q:

$$p/q = a_{-N} \ldots a_{-1} a_0 . a_1 a_2 \ldots a_n \ldots. \tag{2.3}$$

For instance,

$$2/5 = 0.4,$$

$$1/3 = 0.333\ldots3\ldots,$$

$$13/70 = 0.1857142857142857142\ldots.$$

A decimal expansion corresponds to a fraction if and only if some sequence of digits in the decimal expansion ultimately repeats indefinitely. In the expansion 0.4, the repeating digit is the (suppressed) 0 which follows the digit 4; in 0.333...3... the repeating digit is 3; and in the decimal expansion 13/70 the ultimately repeating sequence of digits is 857142. Moreover, the number of digits in the repeating sequence must be less than the denominator q of the fraction. This implies that for any fraction p/q we can in principle work out the value of each digit in its decimal expansion with only a finite amount of effort, for after a certain (known) stage, some sequence of digits will repeat. Thus we know the values of all the infinitely many digits in the expansion in this case.

There are other types of numbers for which we can know the value of every digit in their decimal expansion. For instance, the decimal expansion 0.1010010001..., where successive instances of the digit 1 are separated by a sequence of zeros and the number of zeros in a sequence is greater by one than the number in the previous sequence of zeros, is a number. It certainly is not rational because no sequence of digits ultimately repeats, but we know everything about its decimal expansion in a quite explicit form. (Those who are amused by such things can readily calculate the condition that determines whether a_n is the digit 1).

For numbers such as $\sqrt{2}$ or π the situation is more complicated. We "know" $\sqrt{2}$ in the sense that it is uniquely specified as that positive number whose square is equal to 2; but there is no ready way to state the value of each of the digits in its decimal expansion

$$\sqrt{2} = 1.41421356\ldots;$$

the digits must be worked out one by one. Similarly, the number denoted by π is uniquely specified by its definition as the ratio of the circumference of a circle to its diameter, but this does not directly help us to know the digits in the expansion

$$\pi = 3.14159265\ldots;$$

they too must be calculated one by one. In this case, however, all known methods of calculation require much greater effort than calculation of the digits of $\sqrt{2}$.

These differences in the categories of numbers correspond to differences in the amount of information they convey, and also in the effort or cost required to obtain that information. The next section will make these ideas more precise and introduce the measure of information which plays a fundamental role in all that follows.

2.2 The Information Content of a Measurement

> large, small, long, short, high, low, wide, narrow, light, dark, bright, gloomy, and everthing of the kind which philosophers term accidents, because they may or may not be present in things,—all these are such as to be known only by comparison.
>
> —Leon Battista ALBERTI

Knowledge of a decimal expansion means that we know each of its infinitely many digits a_i. We have seen that in certain circumstances it is possible to know all the digits of the expansion of a number explicitly. For instance, in principle, we can know the decimal expansion of any fraction exactly, and we make use of this fact in daily life. But this simplification never occurs in the process of *measurement* or *observation* of a quantity, for in those circumstances, we are not so fortunate as to know the result ahead of time. What actually transpires is that successively refined and increasingly precise variants of the measurement process yield the sequence of digits, ordered from most significant digit (i.e., leftmost) to least significant digit (i.e., the rightmost among the digits which the measuring apparatus can provide) of the decimal expansion of the measured quantity. Thus we conclude that in principle as well as in practice, the complete measurement of a quantity cannot be accomplished by finite measuring devices nor recorded in finite memory units. It is not possible for the human mind or any other conceivable measuring apparatus to distinguish all the real numbers from one another; equivalently, it is impossible for any physical measuring device to record a measurement with arbitrarily great precision. But certain numerical quantities which arise from mental considerations do lead to numbers, such as fractions, about which everything can be known with full precision. In particular, actual measurements of physical phenomena which may be read from a gauge or measuring rod or be printed by a computer are actually fractions no matter in what units the gauge or rod may be graduated.

A numerical observation or measurement of a physical quantity, say M, may be thought of as associating some real number with M. But in view of what has been said above, what actually occurs is that the measurement provides a finite sequence of digits which can be interpreted as the initial (i.e., most significant) digits of the decimal expansion of a certain real number, and actually is the decimal expansion of a fraction. Successive refinements of the measuring process will lead to better and better approximations consisting of longer and longer sequences of digits, but we have no way of being certain what values the as yet unmeasured digits will turn out to have. Thus we are

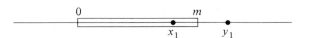

Figure 2.2.1   Measurement of a length.

Figure 2.2.2   Increasing the precision of a measurement.

naturally led to inquire how much additional information is provided by increasing the precision of a measurement.

Suppose, for instance, that a measurement of a quantity M is made and that its "true" numerical value is m. The decimal expansion of this value has infinitely many digits and hence cannot be determined precisely by the measurement. Instead, the measurement merely provides a range of values in which m will be contained. That is, the measurement provides a pair of numbers, x_1 and y_1, such that

$$x_1 < m < y_1. \tag{2.4}$$

If we think of the measurement of a length, then Figure 2.2.1 illustrates this situation. The rod in the figure has true length m, but the inherent inaccuracy of the measurement only specifies m within the limits given by (2.4). In terms of decimal expansions, this means that the measurement has determined a certain (finite) number of digits of the expansion of m. If, for instance, $m = \sqrt{2}$ and $x_1 = 1.41$, $y_1 = 1.42$, then $x_1 < m < y_1$ becomes $1.41 < m < 1.42$, which determines two digits of the expansion of $\sqrt{2}$ but nothing more.

Suppose greater precision is desired and a second more exact measurement is made which yields numbers x_2 and y_2 such that

$$x_1 < x_2 < m < y_2 < y_1.$$

This situation is illustrated in Figure 2.2.2, which shows how the uncertainty in the value of m has been reduced. The inequalities $x_2 < m < y_2$ determine some additional number of digits of the decimal expansion of m.

It is reasonable to suppose that the quantity of information gained by the second observation compared with the quantity of information known at the end of the first observation is measured by the number of digits of the decimal expansion of m known after the second observation, less the number of digits known after the first observation. That is, the increment of information is measured by the number of new digits provided by the more accurate second observation.

It can be shown that the gain in information, which we will denote by I_{10}, can be expressed quantitatively by the formula

$$I_{10} = \log_{10}\left(\frac{y_1 - x_1}{y_2 - x_2}\right) \quad \text{decimal digits;} \tag{2.5}$$

here $y_1 - x_1$ is the range of error of the first observation and $\log_{10}(y_1 - x_1)$ is (approximately) equal to the number of digits in its decimal expansion; $(y_2 - x_2)$ is the range of error of the more accurate second observation, and $\log_{10}(y_2 - x_2)$ is (approximately) equal to the number of digits in its decimal expansion. By a well known property of the logarithm, the formula

$$I_{10} = \log_{10}\left(\frac{y_1 - x_1}{y_2 - x_2}\right) = \log_{10}(y_1 - x_1) - \log_{10}(y_2 - x_2)$$

shows that the information gain I_{10} is the number of decimal digits supplied by the increased accuracy of the second observation.

The formula (2.5) measures information in units of *decimal digits*. Just as lengths can be measured in different units—feet or meters, etc.—so too there is a certain freedom of choice in selecting the unit of measure of information. Indeed, the argument presented above does not depend in any essential way on the choice of *decimal* expansions for expressing numbers; expansions relative to any base would serve as well. If binary expansions, that is expansions relative to the base 2, were used, the unit of information would be the *binary digit* or *bit*, and the formula for information measured in bits becomes

$$I_2 = \log_2\left(\frac{y_1 - x_1}{y_2 - x_2}\right) \text{ bits.} \tag{2.6}$$

In general, if base B expansions were used, the formula for information gain measured in base B digits would be

$$I_B = \log_B\left(\frac{y_1 - x_1}{y_2 - x_2}\right) \text{ base } B \text{ digits.}$$

The relationship amongst measurements in these different units is expressed by the formula

$$I_B = \log_B\left(\frac{y_1 - x_1}{y_2 - x_2}\right) = \log_2\left(\frac{y_1 - x_1}{y_2 - x_2}\right)/\log_2 B = I_2/\log_2 B. \tag{2.7}$$

Throughout the remainder of this book we will measure information in bits unless otherwise noted, and will use the symbol I in place of I_2, suppressing the subscript 2.

2.3 Formal Derivation of the Information Content of a Measurement

If nothing else, let us learn this, that the estimation which we commonly make of the size of things is variable, untrustworthy, and fatuous insofar as we believe that we can eliminate every comparison and can discern any great difference in size merely by the evidence of our own senses. Let us in short be aware that it is impossible to call anything "little" or "large" except by comparison.

—Constantijn HUYGENS, *circa* 1629–31

Of course we do not know the precise value *m* of the measured variable; the measurement merely assures us that it lies between *x* and *y*.

We are interested in the gain in information provided by a measurement. Let us denote this quantity by *I*. *I* will depend on the numbers *x* and *y* that express the numerical bounds on the accuracy of the measurement of *m*.

Let us begin by supposing that the information depends *only* on the endpoints *x* and *y* of the observed interval. We will find that such a measurement does not produce an increase in information. Then we will generalize the measurement process so that it does lead to a gain in information.

Howsoever the information content of a measurement depends on *x* and *y*, it surely cannot depend on the choice of the zero-point of the number line (measuring rod) used to measure *x* and *y*. This means that in its dependence on *x* and *y*, *I* must remain unchanged (remain *invariant*) when *x* and *y* are simultaneously replaced by $x + b$, respectively, where *b* is any real number whatsoever. This condition is expressed by the equation

$$I(x + b, y + b) = I(x, y). \tag{2.8}$$

Moreover, the information content of a measurement cannot depend on the scale, i.e., the unit of length, of the measuring rod used. Hence *I* must remain unchanged when the measurements *x* and *y* are simultaneously replaced by ax and ay, respectively, where *a* is any number different from 0. This is expressed by the relation

$$I(ax, ay) = I(x, y). \tag{2.9}$$

Let us try to determine the functions *I* that satisfy (2.8) and (2.9).

From (2.8), by setting $b = -x$, we obtain

$$I(x, y) = I(x + b, y + b) = I(0, y - x), \tag{2.10}$$

and from (2.9), by setting $a = 1/x$ (where we may suppose $x \neq 0$; otherwise add some nonzero *b* to *x* and to *y*), we obtain

$$I(x, y) = I(ax, ay) = I(1, y/x). \tag{2.11}$$

Application of (2.11) with $a = 1/(y - x)$ to (2.10) yields

$$I(x, y) = I(0, 1) = \text{constant}.$$

It is evident that this function does not remain unchanged when *x* is replaced by $x + b$ and *y* is replaced by $y + b$ unless it is constant, i.e., does not depend on *x* and *y* at all. Hence we conclude that the only functions of *x* and *y* which posses the desired properties are the constant functions.

This analysis does not provide a usable measure of information because

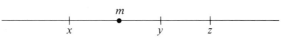

Figure 2.3.1   The measurement of a quantity.

the information gain does not depend on the results of the measurement. However, when the same argument is applied to functions of more than two variables, it yields the desired measure of information. Suppose that I were a function of 3 variables, x, y, z, which were the results of observation. We may think of the relation of these numbers to the measured quantity m as illustrated by Figure 2.3.1. Pursuing our previous line of attack, the conditions that information remain unchanged when the zero point and unit of measurement vary are expressed by

$$I(x, y, z) = I(x + b, y + b, z + b) \qquad (2.12)$$

and

$$I(x, y, z) = I(ax, ay, az), \qquad (2.13)$$

where b is any real number and a is any real number different from 0. Let us test whether such a function I can exist. Setting $b = -x$ in (2.12) we find

$$I(x, y, z) = I(0, y - x, z - x).$$

Setting $a = 1/y$ in (2.13) we obtain

$$I(x, y, z) = I(x/y, 1, z/y),$$

and combination of these results yields

$$I(x, y, z) = I(0, 1, (z - x)/(y - x)). \qquad (2.14)$$

We have shown that the information content of a measurement which yields the numbers x, y, z can be expressed as a function of the ratio of the difference $z - x$ to the difference $y - x$.

How shall we interpret the triple of numbers x, y, z? It describes *two* measurements: the first yields the bounds $x < m < z$ for m, and the second yields the better estimate $x < m < y$ for the value of m, where $x < y < z$; the left hand endpoints of the two observed intervals coincide in this case. $I(x, y, z)$ measures the *incremental information* provided by the estimate $x < m < y$ relative to the less accurate estimate $x < m < z$.

In order to discover more about the function I, some further properties must be postulated. Suppose that a sequence of three measurements of m is made, yielding the estimates

$$x < m < z, \qquad x < m < y, \qquad x < m < w,$$

where $x < w < y < z$; this situation is displayed in Figure 2.3.2. The information content of the second measurement relative to the first is $I(0, 1, (z - x)/$

Figure 2.3.2   Three increasingly precise measurements.

$(y - x)$), whereas the information content of the third measurement relative
to the second is expressed by $I(0, 1, (y - x)/(w - x))$. We may think of the
third measurement as the result of two steps; first, a refined measurement
produces an interval whose endpoints are x and y, and then a further measurement produces an interval whose endpoints are x and w. We will assume
that the information content of the third measurement (x, w) relative to
the first (x, z) must remain unchanged whether it is calculated directly as
$I(0, 1, (z - x)/(w - x))$, or expressed as the sum of the information yield of the
two composite steps, viz.:

$$I(0, 1, (z - x)/(y - x)) + I(0, 1, (y - x)/(w - x)).$$

At this point it will be convenient to write $I((z - x)/(w - x))$ in place
of $I(0, 1, (z - x)/(w - x))$ to simplify notation. With this abbreviation, our
assumption is expressed by the equation

$$I((z - x)/(w - x)) = I((z - x)/(y - x)) + I((y - x)/(w - x)).$$

Writing $z - x = t$, $y - x = u$, $w - x = v$, this relation becomes

$$I(t/v) = I(t/u) + I(u/v)$$

and if $s = 1/v$, $u = 1$, it further simplifies to

$$I(st) = I(t) + I(s). \tag{2.15}$$

It is well known that a function which satisfies this relation (and certain other
very weak technical mathematical conditions) must be a multiple of the
logarithm. Thus

$$I(s) = c \log s$$

where log denotes the logarithm with respect to a conveniently chosen base,
and c is a constant which will depend on the base and on the unit used for
measuring information content.

If we adopt the base 2 logarithm, $\log_2$, and set $c = 1$, we obtain

$$I(s) = \log_2 s;$$

in this case information is measured in *bits* (binary digits). It follows that the
result of the pair of estimates $x < m < z$, $x < m < y$, with $x < y < z$ yields a
gain of

$$I((z - x)/(y - x)) = \log_2((z - x)/(y - x))$$

bits of information.

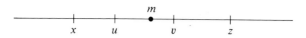

Figure 2.3.3   A pair of nested measurements.

It remains to discuss the case where an initial measurement yields an estimate $x < m < z$ for the measured variable m and a more precise subsequent measurement yields, say, $u < m < v$, where $x < u < v < z$; that is, the left hand endpoint of the second measurement does not coincide with x but improves the precision of the lower bound.

Additivity of the information function I provides the formula for this case. Consider

$$I((z - x)/(z - u)) + I((z - u)/(v - u)).$$

The first term is the information increment obtained by narrowing the estimate $x < m < z$ to $u < m < z$ where $x < u$, and the second term is the information increment gained by narrowing the estimate $u < m < z$ to $u < m < v$ where $v < z$. The sum of these terms can therefore be interpreted as the information increment gained by passing from the estimate $x < m < z$ to the more precise estimate $u < m < v$ with $x < u < v < z$. Since I is proportional to the logarithm function, we find for the sum the expression

$$I((z - x)/(z - u)) + I((z - u)/(v - u)) = I((z - x)/(v - u)).$$

Hence $I((z - x)/(v - u))$ is the information increment which results from refining the measurement estimate $x < m < z$ to $u < m < v$, where $x < u < m < v < z$.

The function $I((z - x)/(v - u))$ remains unchanged when each of the variables u, v, x, z is increased by an arbitrary number b or multiplied by an arbitrary non-zero number a, as was also true for $I((z - x)/(y - x))$. These two types of transformations of the variable x can be combined. The most general form will be

$$x \mapsto ax + b, \qquad a \neq 0, \qquad b \text{ arbitrary.} \tag{2.16}$$

We may say that x is *transformed*, or *mapped*, to $ax + b$. The collection of all such transformations forms a *group* which is known as the *affine group*, and the information function I which remains unchanged by the action of the group is called an *invariant* of the group or a *group invariant*. We have seen that a non-constant function I cannot be an invariant of the affine measurement group unless it involves at least three variables. This expresses the *relative* nature of information measures: *a single measurement does not provide information*.

We note in passing that the affine group is isomorphic to the group of 2×2 matrices of the form $\begin{pmatrix} a & b \\ 0 & 1 \end{pmatrix}$ where, as before, b is arbitrary and $a \neq 0$. The action of a group element $g = \begin{pmatrix} a & b \\ 0 & 1 \end{pmatrix}$ on a variable x is obtained by identifying

x with with the column vector $\binom{x}{1}$ and allowing g to act by matrix multiplication on the left. This is the first, and simplest, example of the role of group theory in information science.

2.4 Application of the Information in a Measurement to Psychophysics

The ideas of the preceding section can be applied to investigate the possible forms of the psychophysical function.

Suppose that two measurements have accuracies Δx_1 and Δx_2, respectively (that is, the lengths of the respective measurement intervals are Δx_1 and Δx_2), but that the measurement intervals are not necessarily related by containment of one within the other. In this situation the definition of information gain given in the previous section cannot be directly applied. We can, however, employ an artifice which suggests how these intervals should be compared with regard to their relative information yield.

Let Δx be an interval large enough to contain both Δx_1 and Δx_2. Then the information gain produced by the measurement of Δx_i relative to Δx is $\log_2(\Delta x/\Delta x_i)$, and the difference in the information gained is

$$\log_2(\Delta x/\Delta x_1) - \log_2(\Delta x/\Delta x_2) = \log_2(\Delta x_2/\Delta x_1); \qquad (2.17)$$

which does not depend on Δx. We will call this quantity the information gain of Δx_1 relative to Δx_2 even when Δx_2 does not contain Δx_1.

The *psychophysical function* relates the psychological response of an organism to a physical stimulus. Let S denote the magnitude of the physical stimulus and R the magnitude of the psychological response. The function f such that $R = f(S)$ is the psychophysical function. What can be said about f? From the relation $R = f(S)$ it follows that a small change ΔS in the stimulus magnitude is related to a small change ΔR is the response magnitude by the formula $\Delta R \simeq f'(S)\Delta S$, where $f'(S) = df/dS$ is the derivative of f with respect to S. If $\Delta R = 1$ jnd, (i.e., 1 *just noticeable difference*), then for given values S_1 and S_2 of S, there will be corresponding values ΔS_1 and ΔS_2 of the stimulus increments needed to yield a just noticeable difference at the given stimulus levels. Consequently the corresponding ratio $\Delta R_2/\Delta R_1$ satisfies

$$\frac{\Delta R_2}{\Delta R_1} = 1 = \frac{f'(S_2)\Delta S_2}{f'(S_1)\Delta S_1}$$

and, therefore, the information gain upon measuring ΔS_1 relative to ΔS_2 can be calculated from

$$\log_2\left(\frac{\Delta R_2}{\Delta R_1}\right) = \log_2 1 = 0 = \log_2\left(\frac{f'(S_2)\Delta S_2}{f'(S_1)\Delta S_1}\right):$$

thus

$$\log_2\left(\frac{\Delta S_2}{\Delta S_1}\right) = -\log_2\left(\frac{f'(S_2)}{f'(S_1)}\right). \tag{2.18}$$

The meaning of this condition is twofold: first, measurement of a 1 jnd difference yields the same relative information no matter where in the stimulus continuum it occurs, and second, this relation among measurements constrains the psychophysical function, as provided by equation (2.18).

Equation (2.18) is equivalent to

$$\Delta S_2/\Delta S_1 = f'(S_1)/f'(S_2).$$

We easily check that both FECHNER's psychophysical function and STEVENS's psychophysical function satisfy this relation. Indeed, if $f(s) = c \log s$ (FECHNER) then

$$1 = \Delta R_2/\Delta R_1 = (\Delta S_2/\Delta S_1)(S_1/S_2)$$

whence

$$\Delta S_2/\Delta S_1 = S_2/S_1 = (S_1/S_2)^{-1},$$

and if $f(s) = cs^k$ (STEVENS), then

$$1 = \Delta R_2/\Delta R_1 = (S_2/S_1)^{k-1}(\Delta S_2/\Delta S_1)$$

whence

$$\Delta S_2/\Delta S_1 = (S_1/S_2)^{k-1}.$$

Now suppose that the unit of scale of the stimulus measure is changed. This can be interpreted either as a change of measuring unit, as from meters to feet, or as a uniform modification of the stimulus, as when the iris of the human eye expands or contracts to admit a greater or lesser fraction of the incident light. In either event, it is evident that the relative information obtained from measurements at two positions in the stimulus continuuum must remain unchanged. Indeed, with a an arbitrary nonzero constant, we find from (2.18) the relation

$$\frac{\Delta S_2}{\Delta S_1} = \frac{f'(S_1)}{f'(S_2)} = \frac{f'(aS_1)}{f'(aS_2)};$$

that is, the function

$$g(S_1, S_2) = f'(S_1)/f'(S_2)$$

satisfies $g(S_1, S_2) = g(aS_1, aS_2)$. Select $a = 1/S_1$ to obtain

$$g(S_1, S_2) = g(1, S_2/S_1) = h(S_2/S_1),$$

where the last equality defines the function h. We have shown that

$$\frac{f'(S_1)}{f'(S_2)} = h\left(\frac{S_2}{S_1}\right);$$

the ratio of derivatives depends only on the ratio of stimulus intensities.

It follows that

$$f'(S_1) = h(S_2/S_1)f'(S_2). \tag{2.19}$$

Set $S_1 = 1$ to find $f'(1) = h(S_2)f'(S_2)$; that is, $h(S) = f'(1)/f'(S)$. After division by $f'(1)$, (2.19) becomes:

$$h(S_2) = h(S_2/S_1)h(S_1).$$

Writing $S_2 = uv$ and $S_1 = u$, this can be expressed as

$$h(uv) = h(u)h(v). \tag{2.20}$$

Thus, the function h preserves the multiplicative structure of the group of non zero numbers. Such functions have been studied by the mathematicians who have shown that (subject to certain technical conditions) $h(u)$ is necessarily proportional to a power of u:

$$h(u) = cu^k. \tag{2.21}$$

Expressing h in terms of f, this becomes

$$f'(u) = cu^k,$$

where we have absorbed the constant $f'(1)$ in c. The solution of this equation is

$$f(u) = \begin{cases} \dfrac{c}{k+1} u^{k+1} & \text{if } k+1 \neq 0 \\[2ex] c \log u & \text{if } k+1 = 0. \end{cases} \tag{2.22}$$

The former is STEVENS's psychophysical function, and the latter is FECHNER's psychophysical function. These are the only possibilities for psychophysical functions which:

1. yield constant relative information gain for 1 jnd responses, and
2. yield relative information that is invariant under changes of scale for the stimulus measure.

An application of the measure of information gained from an observation which combines psychological features of human information processing and formal properties of the information measure arises from considerations of the apparent correlation of variables displayed on a scatterplot. In an interesting 1982 paper, CLEVELAND, DIACONIS, and MCGILL showed that "variables on scatterplots look more highly correlated when the scales are increased." Figure 2.4.1 displays reductions of two scatterplots used in their experiments. The point-cloud data was generated by the same probability distribution, but it is displayed using different scales. The tendency is for judged association of

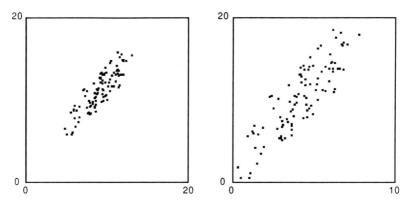

Figure 2.4.1   Data point-clouds drawn from the same probability distribution but displayed using different scales.

linearity to increase as the size of the point-cloud is desceased by an increase in scale size of the coordinate system. Thus it appears that direct subjective judgement of the displayed data may be a poor guide to the intrinsic relationships that exist among the data points. The authors of the referenced work do not propose a reason for the peculiar results they obtained, although they do remark that it is known that subjective judgements of area and linear extent tend to differ from the geometrically correct values.

These results assume a somewhat different aspect when considered from the standpoint of information processing and the information content of a display.

Let scatterplot data be displayed within a rectangular graph whose area (measured in fixed but arbitrary units) is G. The data will be concentrated in some subset of the rectangle of area S. According to the theory of the information content gained from an observation, the information (measured in bits) that corresponds to the point-cloud of data *considered relative to the context presented by the graph* will be

$$I_{G/S} = \log_2(G/S).$$

The scale used for the graph context need not have anything to do with the actual experimental conditions under which the data points were observed or generated. Let the actual ranges of the independent and dependent variables which determine the context of the actual observations correspond to a data context rectangle of area D. Then the gain in information provided by the observations relative to their actual context will be

$$I_{D/S} = \log_2(D/S).$$

Suppose that the actual data context rectangle, of area D, is a subset of the context of the graphical display, of area G. Then the information gained by observing the actual experimental context relative to the graphical display context is

$$I_{G/D} = \log_2(G/D).$$

The relation

$$I_{G/S} = I_{G/D} + I_{D/S} \tag{2.23}$$

holds; indeed, it is valid even if the experimental context is larger than the graphical display context.

This relation tells us that $I_{G/S}$ will increase if the scale used to present the point-cloud of data increases. In the absence of other information, it is not unreasonable for the viewer of a graph to assume that the range of the scales corresponds to the range of variation of the parameters in the experimental observation. If this be so, $I_{G/D} = 0$ and $I_{G/S}$ will correctly estimate the information gained from the observed data. However, if the graphical display uses another scale, equation (2.23) specifies the difference between $I_{G/S}$ and the information gain $I_{D/S}$; the latter is, of course, what the display is intended to convey.

CLEVELAND, DIACONIS, and MCGILL point out that the standard measure of linear association is the correlation coefficient, which, of course, is independent of the units used for the graphical display. The interpretation based on information gain does not suggest that the subjective perception of linear association would agree with the measure of association provided by the correlation coefficient, but it does identify a property of graphical displays which must be taken into account if displays are not to mislead the viewer.

2.5 The Probability Interpretation of the Information Measure

We have seen that the pair of measurements which yields the estimate $x < m < y$ followed by the improved estimate $u < m < v$, where $x < u < m < v < y$, yields $\log_2((y - x)/(v - u))$ bits of information. Since $y - x$ measures the length of the numerical interval with endpoints x and y and $v - u$ measures the length of the subinterval with endpoints u and v, the ratio $(v - u)/(y - x)$ can be interpreted as the probability that a randomly selected point in the interval between x and y will actually lie between u and v. Denoting this probability by p,

$$p = (v - u)/(y - x),$$

we can express the information increment provided by the pair of measurements as

$$I = \log_2((y - x)/(v - u)) = \log_2\left(\frac{1}{p}\right) = -\log_2 p. \tag{2.24}$$

Recall the BOLTZMANN formula mentioned in section 1.3 which expresses the entropy S of a system in terms of the number of (equally probable) miscrostates W by the formula

$$S = k \log W, \tag{2.25}$$

where k is a constant and log denotes the natural logarithm. Suppose that a measurement of the system is made which decreases the number of possible microstates from W to W'; the corresponding final entropy will be S'. Then $p = W'/W$ can be interpreted as the probability that a randomly selected microstate from amongst the W initial microstates will actually be one of the W' final microstates. Then

$$S' = k \log W' \quad \text{and} \quad S = k \log W$$

imply $S' - S = k(\log W' - \log W) = k \log(W'/W) = k \log p$; one obtains

$$(S - S')/k = -\log p = -(\log 2) \log_2 p$$

whence the information gained from the measurement is

$$I = -\log_2 p = -(S' - S)/(k \log 2) \text{ bits.} \tag{2.26}$$

Thus we have identified the negative of the change in entropy with the change in information, apart from a conversion factor from physical units of entropy to bits of information.

Now suppose we are observing a system which can be in one of a (finite) number of "states" $\underline{S}_1, \ldots, \underline{S}_n$ (whose order of labelling is entirely arbitrary), and that the probability of the system being in the kth state is p_k. Because p_k is a probability distribution, each p_k satisfies $0 \le p_k \le 1$. Moreover, since the system must be in some state, the probabilities sum to 1:

$$p_1 + \cdots + p_n = 1.$$

Suppose that we make an observation of the system and discover it is in state $\underline{S}_k$, say. How much information have we gained? This question can be answered in the spirit of our previous discussion. Let us suppose that our observing equipment associates with each state of the system an interval on a measuring rod according to the following prescription. State 1, which occurs with probability p_1, will be associated with the interval $0 < x < p_1$; state 2 with the interval $p_1 < x < p_1 + p_2$; and in general the kth state will be associated with the interval

$$p_1 + p_2 + \cdots + p_{k-1} < x < p_1 + p_2 + \cdots + p_k. \tag{2.27}$$

In particular, the Nth state, which is the last one in our list, will be associated with the interval

$$p_1 + \cdots + p_{N-1} < x < p_2 + \cdots + p_N = 1,$$

where the last equality just expresses the fact that the sum of the probabilities is equal to 1.

Thus each state corresponds to an interval in the range $0 \leq x \leq 1$, no two intervals overlap, and the sum of the lengths of the intervals is equal to 1: except for their endpoints (which can be ignored without loss of generality), the intervals associated with the states fill out the range $0 \leq x \leq 1$.

Before observing the system we know *a priori* (before "plugging the machine in") that the pointer on our measuring instrument will fall somewhere in the interval $0 \leq x \leq 1$; this just means that the system must be in *some* one of the preassigned states. In order to determine which state the system is in, we will make an observation and examine the measuring device to see in which interval the pointer falls.

Suppose that we "plug the machine in" to observe the system, and that the pointer on our measuring instrument falls in one of the intervals, say that which corresponds to the state $\underline{S}_k$. The length of this interval is p_k, so it follows that the gain in information compared with the state of affairs before we made the observation is, according to our previous analysis,

$$\log_2(1/p_k), \tag{2.28}$$

where $1/p_k$ is the ratio of the length 1 to the length p_k of the interval in which the pointer fell after the measurement.

Now suppose that the system does not remain in a single state but varies from one state to another with the passage of time (or with the variation of some other parameter with which the changes of state are correlated), and suppose that we measure a state, reset our measuring apparatus, and then measure the subsequent state when the system produces one. This sequence of measurements will yield a sequence of gains in information about the evolution of the system. This procedure can be represented analytically as follows.

Let the succession of states in which the system is found be denoted

$$S(1), S(2), \ldots, S(n), \ldots.$$

Each of the states $S(n)$ must be one of the fixed inventory of possible states $\underline{S}_1, \ldots, \underline{S}_N$, so we must carefully distinguish between *subscripts* in our notation, which index the *possible* states the system can occupy, and the parenthesized numbers in $S(1), S(2), \ldots$ which indicate the succession of states assumed by the system as it evolves.

Bearing these notational conventions in mind, the nth occurring state $S(n)$ in the evolution of the system will be one of the $\underline{S}_k$ and, since $\underline{S}_k$ occurs with probability p_k, p_k is exactly the probability that this state will occur as the system evolves, and the information it provides when it does occur will be $\log_2(1/p_k)$.

The *average information gained* when the system evolves from one state to the next is an important quantity. For simplicity, let us suppose that a given state of the system is independent of its prior state. Since the possible states $\underline{S}_1, \ldots, \underline{S}_n$ occur with probability $p_1, \ldots, p_n$ respectively, and since the information gained when state $\underline{S}_k$ occurs is $\log_2(1/p_k)$, it follows that the average information gained when the system passes from one state to another is the average value of the numbers $\log_2(1/p_k)$, each weighted by its probability of occurrence, that is:

$$I = \text{Average information}$$

$$= -p_1 \log_2 p_1 - \cdots - p_k \log_2 p_k - \cdots - p_N \log_2 p_N \qquad (2.29)$$

$$= \sum_1^N p_k \log_2(1/p_k),$$

using the summation notation of mathematics.

This expression is the famous formula for the measure of information introduced by Claude SHANNON in the context of telephonic (electrical) communication and displayed in section 1.7, and is the key idea in his development of information theory.

There are numerous natural examples of systems of the kind described above, including many which are familiar in human communication. For instance, let the "states" of the system be the symbols found on a standard typewriter keyboard or in the conventional ASCII alphanumeric symbol inventory used in computing, and let the probabilities p_k be the probabilities of occurrence of the various symbols in a natural written language such as English. Here a "measurement" corresponds to the observation of a particular character in a sequence of text symbols, and observation of the succession of states which exhibits the evolution of the system is just what we do when we read. The quantitative measure of information gained is described by eq(2.29), assuming of course, that successive text characters are independent of each other. This assumption is false, but the idea that the information content of text can be measured in some such way is correct, and it is not difficult to modify the discussion above to take context dependencies into account. These questions have been explored at length; cf., e.g. SHANNON (1951).

2.6 The Uncertainty Principle for Measurement

Let us suppose that a measuring rod is marked in equispaced intervals of which there are N to the unit of measurement. If the measurement variable is denoted t, then a small increment Δt of the variable t will be detectable by the measurement system if t and $t + \Delta t$ lie on opposite sides of a division on the measuring rule or, what amounts to the same thing, if t and $t + \Delta t$ do not fall within the same measurement interval. This can be expressed symbolically by the condition that $\Delta t \geq 1/N$, that is

$$N\Delta t \geq 1. \tag{2.30}$$

Let $\Delta N = \frac{1}{2}N$; ΔN is the halfwidth of the range of the number of divisions per unit which provides accuracy of measurement not greater than that provided by N. Thus, eq(2.30) can be written as eq(2.31) and interpreted as asserting that if measurement accuracy within Δt is desired, it will be necessary to have at least N uniform divisions per unit of measurement

$$\Delta N \Delta t \geq \frac{1}{2}. \tag{2.31}$$

As an example, suppose it is desired to measure with an accuracy of 1 part in 10^k. Then there must be at least 10^k divisions per unit; eq(2.31) asserts no more than this.

In certain cases it is possible to give eq(2.31) a further meaning. Suppose we think of the variable Δt as representing time. Then N denotes the number of equal subintervals into which the unit of time has been partitioned by a measuring apparatus. N consequently has the dimensional units of a pure number per unit time, that is, of frequency. We can think of the uniformly spaced N ticks on the measuring rod or gauge as produced by a periodic "clock" of frequency N. With this interpretation, in order to have measurements accurate to within Δt units of time, the frequency of the measuring clock must be great enough to satisfy eq(2.31). This is a simple but fundamental mathematical relationship in signal processing. In order to distinguish brief time intervals, high frequencies—a large "bandwidth"—must be used.

Equation (2.31) has another, and more powerful, interpretation. It asserts that there is a trade-off between the accuracy of a measurement and the range of variability of the measuring instrument. Thus, the uncertainty of the observed value and the uncertainty (i.e., range of values) of the frequency range required to perform the measurement are related; if one is small, then the other must be large.

This simple idea has been elaborated and refined into an important tool by mathematicians, physicists, and communications engineers. Although we will not need the details of the derivation for our work, an exact statement of the relationship between measurement accuracy and measuring instrument accuracy will be important to us later, so we will make a brief digression to describe it.

With the exception of mathematically pathological examples, any function of a real variable can be expressed as a superposition of simple periodic functions of varying frequencies. This is called the FOURIER *representation* of the function. Our ears "calculate" the FOURIER representation of an acoustic signal function, which describes the variation with time of air pressure at the eardrum. This analysis picks out the frequencies—"pitches"—which occur in the sound and determines the amplitude of the signal—quantity of sound— which corresponds to each of them.

The analytical tool for performing a frequency analysis of a signal is the

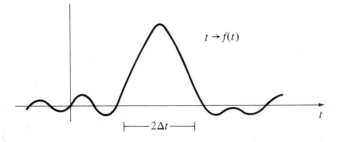

Figure 2.6.1   Graph of a signal.

FOURIER *transform*. We cannot develop its properties here (see sections 4.5 and 4.7 for a summary description); however, without reproducing the mathematical details, we can describe their consequences.

Suppose that a function, or signal, f, has a graph similar to that shown in Figure 2.6.1. A "width" can be assigned to this function which measures how much it spreads. This "width", which is denoted Δt, generalizes the Δt previously discussed to a wider class of signals. It corresponds to the root-mean-square error in the statistical analysis of experimental data.

The FOURIER transform of f is a function of frequency v, which describes the extent to which the various frequencies contribute to the makeup of f. The graph of the FOURIER transform of f may look similar to Figure 2.6.1; in any event, it too will have a "width" Δv which generalizes the frequency "width" mentioned.

It can be proved that, no matter with what function f we start, the uncertainty relation

$$\Delta t \Delta v \geq 1/4\pi \qquad (2.32)$$

is valid. If Δt is small, this means that f is concentrated on some small interval; the smaller Δt, the more concentrated f is and the smaller the interval. In this case, the uncertainty inequality eq(2.32) asserts that a large range of frequencies will be necessary in order to represent such a narrowly concentrated function. Similarly, if we try to build a function using a narrow spread of frequencies, so that Δv is small, then the uncertainty relation eq(2.32) asserts that Δt will be large; this means that the corresponding function will be broad, not concentrated on a small interval; it will endure for a long time.

We may express these relations in an equivalent but slightly different language. If only low frequencies are used to detect signals, then only those changes that occur over long periods of time can be detected. If brief periods of time must be distinguished, then high frequencies must be used to detect the brief intervals.

These relationships will be discussed in greater detail in Chapter 4.

2.7 Conservation of Information

> An addition to knowledge is won at the expense of an addition to ignorance.
>
> —Arthur Stanley EDDINGTON

Let us relate the considerations of the previous section to the information gain provided by measurement. We have already found that information results from the *comparison* of two measurements of a quantity. Let the first measurement limit the value m of the measured quantity to the interval $t_1 < m < t_2$ and put $\Delta t = t_2 - t_1$. Let the second measurement refine the first by further restricting the measured quantity to the interval $t_1' < m < t_2'$ where $t_1 < t_1' < m < t_2' < t_2$, and put $t' = t_2' - t_1'$ (Figure 2.7.1). Then the information gained is

$$I = \log_2((t_2 - t_1)/(t_2' - t_1')) = \log_2(\Delta t/\Delta t').$$

We have seen that a measurement which is accurate to within Δt implies the use of measurement frequencies N (subdivision ticks on the measurement rod) that satisfy the uncertainty relation

$$\Delta N \Delta t \geq \tfrac{1}{2};$$

similarly

$$\Delta N' \Delta t' \geq \tfrac{1}{2}$$

where $\Delta N'$ refers to the second measurement.

If the measurements are made in the most economical way, i.e., in that way which minimizes the joint uncertainty, then we will have $\Delta N \Delta t = \tfrac{1}{2}$ and $\Delta N' \Delta t' = \tfrac{1}{2}$; let us suppose these equalities are valid, although they cannot always be attained in theory or in practice. Then $\Delta t/\Delta t' = \Delta N'/\Delta N$, whence

$$I(\Delta t, \Delta t') = \log_2(\Delta t/\Delta t') = -\log_2(\Delta N/\Delta N'). \tag{2.33}$$

The observational arrangement described is completely symmetrical with respect to the "time" variable t and the "frequency" variable N. That is, we may use the ticks N of the frequency clock to measure time t, or we may use the elapsed intervals of t to measure N. In the latter case, the information resulting from the succession of measurements which narrows the value of N to within ΔN in the first measurement and to within $\Delta N'$ in the second measurement, yields an information increment which is measured by

$$I(\Delta N, \Delta N') = \log_2(\Delta N/\Delta N').$$

Figure 2.7.1   Refinement of a measurement.

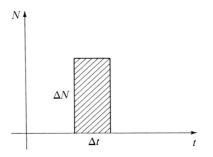

Figure 2.7.2   Uncertainty rectangle.

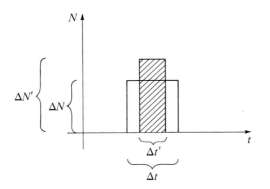

Figure 2.7.3   Effect on the uncertainty rectangle of refining a measurement.

Comparison of this equation with eq(2.33) yields:

$$I(\Delta t, \Delta t') + I(\Delta N, \Delta N') = 0;$$

that is, information about one of the variables t, N is *gained* at the expense of an equal *loss* of information about the other variable. This is the *principle of conservation of information.*

If t and N are represented along the axes of a Cartesian coordinate system, then the uncertainty relation eq(2.33) asserts that the area of the rectangle with sides Δt and ΔN is at least $\frac{1}{2}$ (Figure 2.7.2). In the case of optimally efficient measurement, the area is exactly equal to $\frac{1}{2}$. A second measurement will change the dimensions of the rectangle while preserving its area. In this case, if the t is measured more precisely, the rectangle will be narrower (corresponding to $\Delta t' < \Delta t$) but taller (corresponding to $\Delta N' > \Delta N$) (Fig. 2.7.3).

If the first measurement is optimal, so that $\Delta t \Delta N = \frac{1}{2}$, but the second one is not, so that $\Delta t' \Delta N' > \frac{1}{2}$, then

$$\Delta t \Delta N / \Delta t' \Delta N' < 1$$

whence

$$I(\Delta t, \Delta t') + I(\Delta N, \Delta N') < 0; \qquad\qquad (2.34)$$

the pair of measurements results in a net loss of information although information about the t variable or the N variable (but not about both) may be gained.

If the first measurement is not optimal (so that $\Delta t \Delta N > \frac{1}{2}$), but the second one is optimal (so that $\Delta t' \Delta N' = \frac{1}{2}$), or at least an improvement compared with the first, then

$$\Delta t \Delta N / \Delta t' \Delta N' > 1$$

so

$$I(\Delta t, \Delta t') + I(\Delta N, \Delta N') > 0; \tag{2.35}$$

there is a net gain of information which results from the improved measurement technique which makes it possible, in this instance, to refine the accuracy of both the t and the N variable measurements.

2.8 Information Content of a Calculation

We have considered the information gained (or lost) by a pair of measurements of some variable. While this notion applies in the first place to the measurement of some external physical variable represented by the position of a pointer on a meter or measuring rod, it also can be applied to mathematical calculations themselves. We will not enter deeply into this topic, but a few brief remarks are in order to indicate the breadth of application of the idea of the information content of a measurement, and also to show how at least one class of mental activity can be brought under its umbrella.

Consider, for example, knowledge of the number π, the ratio of the circumference of a circle to its diameter. Although the *definition* of π is known and many *formulae* for computing the value of π are also known, this knowledge is not exactly equivalent to knowledge of the digits of the decimal expansion of π—we may call the latter *detailed knowledge* of the number—because application of the formulae requires the investment of substantial computational effort, and determination of the infinitely many digits of the decimal expansion of π would entail an infinite amount of effort. This being clearly impossible, we may turn to ask how much information is gained about the expansion of π when one of the formulae for computing it is used to a given extent. For instance, the well known series of GREGORY,

$$1 - 1/3 + 1/5 - 1/7 + \cdots + (-)^n/(2n + 1) + \cdots,$$

converges to $\pi/4$. Let us write

$$S = \pi/4 = a_1 - a_2 + a_3 - a_4 + \cdots$$

where $a_n = 1/(2n + 1)$ and inquire how much information is gained when we pass from the approximation

$$S_n = a_1 - a_2 + \cdots + (-)^{n-1} a_n \tag{2.36}$$

to

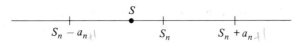

Figure 2.8.1 Estimating the sum of an alternating series.

$$S_{n+1} = a_1 - a_2 + \cdots + (-)^{n-1}a_n + (-)^n a_{n+1} = S_n + (-)^n a_{n+1}, \quad (2.37)$$

i.e., by taking into account the "next term" in the series.

Here we must rely upon some basic facts about the convergence of series. If a series consists of terms of alternating sign which decrease to 0, as GREGORY's series does, then it is well known that the error made by omitting all of the terms after a_n is limited according to the formula

$$S_n - a_{n+1} < S < S_n + a_{n+1}. \quad (2.38)$$

We may interpret this to mean that a "measurement" of S which takes the first n terms of the series into account provides the estimate eq(2.38); compare Figure 2.8.1. If a further "measurement" is made which takes the next term, a_{n+1}, into account, then we similarly obtain

$$S_{n+1} - a_{n+2} < S < S_{n+1} + a_{n+2}. \quad (2.39)$$

According to our general principles, the information gained will be

$$\log_2 \left(\frac{(S_n + a_{n+1}) - (S_n - a_{n+1})}{(S_{n+1} + a_{n+2}) - (S_{n+1} - a_{n+2})} \right) = \log_2 \left(\frac{a_{n+1}}{a_{n+2}} \right) \text{ bits.} \quad (2.40)$$

At the first stage we have $S_1 = 1$ as an initial estimate for $\pi/4$ and here $a_2 = 1/3$; the next approximation yields $S_2 = 1 - 1/3 = 2/3$ and here $a_3 = 1/5$, whence the information gained upon passing from the approximation $S_1 = 1$ to the approximation $S_2 = 2/3$ will be

$$\log_2(a_2/a_3) = \log_2((1/3)/(1/5)) = \log_2(5/3) = 0.74 \text{ bits.}$$

The total number of bits of information obtained by successively taking into account the terms $a_1, a_2, \ldots, a_n$ is equal to

$$\log_2 \left(\frac{a_2}{a_3} \right) + \log_2 \left(\frac{a_3}{a_4} \right) + \cdots + \log_2 \left(\frac{a_n}{a_{n+1}} \right)$$

$$= \log_2 \left(\frac{a_2}{a_3} \times \frac{a_3}{a_4} \times \cdots \times \frac{a_{n-1}}{a_n} \times \frac{a_n}{a_{n+1}} \right)$$

$$= \log_2 \left(\frac{a_2}{a_{n+1}} \right)$$

$$= \log_2 \left(\frac{1/3}{1/(2n+1)} \right)$$

$$= \log_2 \left(\frac{2n+1}{3} \right). \quad (2.41)$$

This quantity increases without bound as n increases: each term of the series adds new information. But the function increases very slowly, which tells us that GREGORY's series will be a very costly method for calculating the digits of the expansion of π. In fact, passing from S_n to S_{n+1} yields

$$\log_2(a_{n+1}/a_{n+2}) = \log_2((2n + 5)/(2n + 3))$$
$$= \log_2(1 + 2/(2n + 3))$$
$$\sim (2/(n + 3/2)) \log 2 \text{ bits}$$

of information. We find, for instance, $\log_2(a_3/a_4) = 0.49$ bit, $\log_2(a_4/a_5) = 0.36$ bit, $\log_2(a_5/a_6) = 0.29$ bit, $\log_2(a_{49}/a_{50}) = 0.03$ bit, and the term a_{100} yields only about 0.007 bits of information relative to a_{99}.

A more familiar and simpler example is provided by the alternating geometric series

$$S = 1 - r + r^2 - r^3 + \cdots + (-)^n r^n + \cdots$$

where $0 < r < 1$. A well known formula from high school algebra informs us that the sum of this series is exactly given by the relation

$$S = 1/(1 + r).$$

If we write

$$S_n = 1 - r + r^2 - \cdots + (-)^n r^n,$$

then

$$S_n - r^{n+1} < S < S_n + r^{n+1}$$

and, following the procedure described previously, we obtain

$$I = \log_2(r^{n+1}/r^{n+2}) = \log_2(1/r) \text{ bits}$$

for the information gained by passing from the approximation S_n to the approximation S_{n+1} of the sum S. Observe that in contrast to GREGORY's series for π, the amount of information gained by the addition of one new term to the series is a constant, independent of n. In particular, if $r = \frac{1}{2}$ so that

$$S = 1 - \frac{1}{2} + \left(\frac{1}{2}\right)^2 - \cdots + (-)^n \left(\frac{1}{2}\right)^n + \cdots,$$

then

$$\log_2(1/\tfrac{1}{2}) = \log_2 2 = 1 \text{ bit}$$

is added for every summand considered. We therefore receive a greater informational return per term calculated than for the terms of GREGORY's series.

These ideas can be adapted to calculate the information contributed by the nth digit in a decimal (or binary) expansion of a number. Let

$$S = a_0 + \frac{a_1}{b} + \frac{a_2}{b^2} + \cdots + \frac{a_n}{b^n} + \cdots$$

where b is the base of the number system and $0 \le a_i < b$. Set

$$S_n = a_0 + \frac{a_1}{b} + \cdots + \frac{a_n}{b^n}.$$

Since the series S is not alternating, we cannot assert that S lies between

$$S_n - \frac{a_{n+1}}{b^{n+1}} \quad \text{and} \quad S_n + \frac{a_{n+1}}{b^{n+1}}.$$

But obviously $S_n < S$ and it is also evident that

$$\frac{a_{n+1}}{b^{n+1}} + \cdots + \frac{a_{n+k}}{b^{n+k}} + \cdots < \frac{b}{b^{n+1}} + \cdots + \frac{b}{b^{n+k}} + \cdots$$

$$= \frac{1}{b^n}\left\{1 + \frac{1}{b} + \frac{1}{b^2} + \cdots\right\}$$

$$= \frac{1}{b^n}\left(\frac{1}{1 - 1/b}\right) = \frac{1}{b^{n-1}(b-1)}.$$

Hence $S_n < S < S_n + 1/(b^{n-1}(b-1))$ and $S_{n+1} < S < S_{n+1} + 1/(b^n(b-1))$. The information increment corresponding to the passage from the approximation S_n to S_{n+1} is therefore

$$I = \log_2\left(\frac{1/(b^{n-1}(b-1))}{1/(b^n(b-1))}\right) = \log_2 b;$$

for binary expressions, each binary digit contributes 1 bit of information.
We can also write

$$S = \frac{b^{1/2}}{b^{1/2} - 1}\left\{a_0 - \frac{a_0}{b^{1/2}} + \frac{a_1}{b^{2/2}} - \frac{a_1}{b^{3/2}} + \frac{a_2}{b^{4/2}} - \frac{a_2}{b^{5/2}} + \cdots\right\},$$

an alternating series. The information gained by calculating an additional term of this series is, as we have seen earlier,

$$I = \log_2(AB/CD)$$

$$= \log_2 b^{1/2} = \tfrac{1}{2}\log_2 b,$$

where

$$A = \frac{b^{1/2}}{b^{1/2} - 1}, \qquad B = \frac{a_k}{b^{2k/2}},$$

$$C = \frac{b^{1/2}}{b^{1/2} - 1}, \qquad D = \frac{a_k}{b^{(2k+1)/2}}$$

and we have selected the paired terms corresponding to the same coefficient a_k since both members of the pair arise from a *single* term of the original expansion S. Therefore, the information gained by passing from a_n/b^n to

a_{n+1}/b^{n+1} in S is equal to the information gained by passing from $a_n/b^{2n/2}$ to $a_{n+1}/b^{(2n+3)/2}$ in the alternating series, which corresponds to 2 pairs of alternating series terms, each of which yields $\frac{1}{2}\log_2 b$ bits of information for a total of $\log_2 b$ bits, in agreement with the previous calculation.

Consider another series example. Let $\zeta(s) = 1 + 1/2^s + 1/3^s + \cdots + 1/n^s + \cdots$ be the RIEMANN zeta function. Then we readily obtain

$$\zeta(s) = 1 - 1/2^s + 1/3^s - 1/4^s + 1/5^s + \cdots + 2\{1/2^s + 1/4^s + 1/6^s + \cdots\}$$

$$= 1 - 1/2^s + 1/3^s - 1/4^s + \cdots + 1/2^{(s-1)}\{1 + 1/2^s + 1/3^s + \cdots\}$$

so

$$1 - 1/2^s + 1/3^s - 1/4^s + \cdots = (1 - 1/2^{(s-1)})\zeta(s);$$

this expresses $\zeta(s)$ in terms of the alternating series

$$\tau(s) = 1 - 1/2^s + 1/3^s - 1/4^s + \cdots .$$

Using our standard techniques, we find that the information gain on passing from

$$\tau_n(s) = 1 - 1/2^s + 1/3^s - \cdots + (-)^n 1/n^s$$

to $\tau_{n+1}(s)$ is

$$I = \log_2\left(\frac{(n+1)^{-s}}{(n+2)^{-s}}\right) = s\log_2(1 + 1/(n+1)) \simeq \frac{s}{(n+1)\log_e 2}.$$

Thus, the behavior of the zeta function is similar to that of GREGORY's series (to which $\tau(s)$ reduces for $s = 1$). Note that if $s = 1$, the formula $\tau(s) = (1 - 1/2^{s-1})\zeta(s)$ fails because the left side is the convergent GREGORY's series whereas $S(1) = 1 + 1/2 + 1/3 + \cdots + 1/N + \cdots$ is the divergent harmonic series and the factor $(1 - 1/2^{s-1})$ is, in this case, 0.

The series expansion of the important exponential function can also be treated by these techniques. Recall that

$$\exp x = 1 + x + \frac{x^2}{2!} + \cdots + \frac{x^n}{n!} + \cdots$$

is valid for all x. In particular, for negative values of the argument variable, this series is alternating:

$$\exp(-x) = 1 - x + \frac{x^2}{2!} - \frac{x^3}{3!} + \cdots, \tag{2.42}$$

where x is positive, and the value of $\exp x$ can be expressed in terms of the value of $\exp(-x)$ by the formula $\exp(-x) = 1/\exp x$. Thus it suffices to calculate the information gained by using the series eq(2.42) to calculate values of the exponential function. If we think of this series as expressed in our by now standard form

$$\exp(-x) = a_0 - a_1 + a_2 - \cdots, \tag{2.43}$$

then the information increment provided by the term a_{n+1} compared with that provided by the term a_n is

$$I = \log_2(a_n/a_{n+1})$$

$$= \log_2\left(\frac{x^n/n!}{x^{n+1}/(n+1)!}\right)$$

$$= \log_2((n+1)/x) \text{ bits,}$$

whence the information gain is ultimately positive and increases without bound as n increases, regardless of the fixed value of x. This is a rapidly convergent series whose use is efficient because the information yield per term increases as the number of terms increases.

Let us consider an *algorithm* rather than a series and examine how the notion of information content of a calculation can be interpreted in this case. Perhaps the simplest and most ancient example is the specialization of NEWTON's formula to the calculation of square roots, known at least implicitly to the Babylonians almost 4,000 years ago. Suppose it be desired to calculate $a^{1/2}$ for some positive number a different from 1. The square root can be approximated by use of the following recursive algorithm, which is a special case of NEWTON's method for finding the real roots of a polynomial.

Set $a_1 = 1$. Given a_n, calculate

$$a_{n+1} = \left(a_n + \frac{a}{a_n}\right)\Big/2. \tag{2.44}$$

The a_n form a sequence of increasingly accurate approximations to $a^{1/2}$:

$$a^{1/2} = \lim_{n\to\infty} a_n.$$

In general, a_n and a_{n+1} do not lie on opposite sides of $a^{1/2}$ so the method for calculating information gain presented in the previous sections cannot be directly applied. But notice that a_n and a/a_n lie on opposite sides of $a^{1/2}$, for if not, then either $a_n \le a^{1/2}$ and $a/a_n \le a^{1/2}$, or the reverse inequalities hold. The former case implies $a_n^2 \le a$ and $a^2/a_n^2 \le a$, i.e., $a \le a_n^2$, which is a contradiction. The case of reversed inequalities is treated similarly.

Not only do a_n and a/a_n lie on opposite sides of $a^{1/2}$, but a/a_n and a_{n+1} do also if $n > 1$. This is the same as asserting that a_n and a_{n+1} lie on the same side of $a^{1/2}$. For suppose otherwise. If $a^{1/2} < a_n$, then $a^{1/2} > a_{n+1}$. Hence $a < (a_{n+1})^2$, $4a > (a_n + a/a_n)^2 = a_n^2 + 2a + (a/a_n)^2$, so $0 > a_n^2 - 2a + (a/a_n)^2$, a contradiction.

With these preliminaries in hand we can construct an algorithm for approximating $a^{1/2}$ that produces a sequence of terms which bracket $a^{1/2}$ and to which the measure of information gain can be applied. Consider the sequence

$$1 = a_1, a/a_1, a_2, a/a_2, \ldots a_n, a/a_n, a_{n+1}, \ldots. \tag{2.45}$$

Since the numbers a_n approach $a^{1/2}$ as n approaches infinity, the numbers a/a_n also approach $a^{1/2}$. Moreover, successive terms in this approximating sequence bracket $a^{1/2}$, and they therefore yield information according to the formula

$$I = \log_2 \left(\frac{a_n - a/a_n}{a_{n+1} - a/a_n} \right)$$

for the succession of approximations a_n, a/a_n, a_{n+1}. This expression can be simplified by using eq(2.44). From $a_{n+1} = (a_n + a/a_n)/2$ obtain $a_{n+1} - a/a_n = (a_n - a/a_n)/2$ so

$$I = \log_2 2 = 1 \text{ bit per iteration.}$$

For the succession of approximations a/a_n, a_{n+1}, a/a_{n+1} the information gain is

$$I = \log_2 \left(\frac{(a/a_n) - (a_{n+1})}{(a/a_{n+1}) - (a_{n+1})} \right).$$

This expression can also be simplified. From eq(2.44) we obtain

$$a_{n+1} - a/a_n = (a_n - a/a_n)/2$$

so

$$I = \log_2 \left(\frac{1}{2} \frac{((a/a_n) - (a_n))}{((a/a_{n+1}) - (a_{n+1}))} \right).$$

Set $a_n = a^{1/2}(1 + \varepsilon_n)$. Then $a/a_n \sim a^{1/2}(1 - \varepsilon_n)$ if a_n is close to $a^{1/2}$, so

$$a/a_n - a_n \sim a^{1/2}(1 - \varepsilon_n) - a^{1/2}(1 + \varepsilon_n) = 2a^{1/2}\varepsilon_n.$$

Hence

$$I \sim \log_2 \left(\frac{\varepsilon_n}{2\varepsilon_{n+1}} \right).$$

The ratio $\varepsilon_n/\varepsilon_{n+1}$ can be estimated. We have

$$a_{n+1} = a^{1/2}(1 + \varepsilon_{n+1}) = \frac{1}{2}(a_n + a/a_n)$$

$$= \frac{a^{1/2}}{2}((1 + \varepsilon_n) + (1 - \varepsilon_n + \varepsilon_n^2))$$

$$= a^{1/2}\left(1 + \frac{\varepsilon_n^2}{2}\right)$$

$$= \varepsilon_n/\varepsilon_{n+1} = 2/\varepsilon_n$$

$$= \varepsilon_{n+1} = \varepsilon_n^2/2.$$

Hence

$$I \sim \log_2\left(\frac{\varepsilon_n}{2\varepsilon_{n+1}}\right) = \log_2\left(\frac{1}{\varepsilon_n}\right).$$

Finally, observe that $\varepsilon_{n+1} = \varepsilon_n^2/2$ implies

$$\log_2\left(\frac{1}{\varepsilon_{n+1}}\right) = \log_2\left(\frac{2}{\varepsilon_n^2}\right) = 1 + 2\log_2\left(\frac{1}{\varepsilon_n}\right); \qquad (2.46)$$

this provides a recursion formula for estimating the growth in the gain of information as n increases. If we write $I(n) = \log_2(1/\varepsilon_n)$, then eq(2.46) shows that

$$I(n+1) \simeq 2I(n) + 1$$

from which we can conclude that the number of bits of information (or accuracy) gained grows exponentially with n. Indeed, if m is any positive integer, then

$$I(m+n) \simeq 2^n I(m).$$

An example will help to make clear how rapidly information is gained by using this algorithm. Consider calculation of $2^{1/2}$. Here $a = 2$. Table 2.8.1 illustrates the corresponding gain in information.

Now let us consider an example that shows how to calculate the information gained by computing the roots of a given function by means of an iterative algorithm. Suppose that a function $x \mapsto f(x)$ is given and that it is desired to calculate the value of an x^* such that $f(x^*) = 0$. We will base our analysis on the well known algorithm of NEWTON, but the method is applicable to other numerical approximations as well.

Suppose that two numbers u_0 and v_0 such that $u_0 < x^* < v_0$ are given, and suppose further that the graph of $y = f(x)$ is convex downward through-

Table 2.8.1 Information Gained in Calculating $\sqrt{2}$

$a_1 = 1$		
	$+1$	
$a/a_1 = 2$		$I = 1$ bit
	-0.5	
$a_2 = 1.5$		$I = 1.58$ bit
	-0.16	
$a/a_2 = 1.3$		$I = 1$ bit
	$+0.083$	
$a_3 = 1.416\ldots$		$I = 4.09$ bits
	-0.0049	
$a/a_3 = 1.41176470\ldots$		$I = 1$ bit
	$+0.0024$	
$a_4 = 1.41421568\ldots$		$I = 9.15$ bits
	-0.0000042	
$a/a_4 = 1.41421143\ldots$		

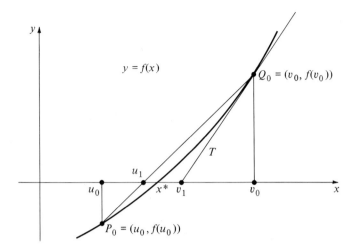

Figure 2.8.2   Zero of a function in a convex downward neighborhood.

out the interval $u_0 < x < v_0$. This condition will be assured if $f''(x) > 0$ for $u_0 < x < v_0$, where f'' denotes the second derivative. In this case, the graph of $y = f(x)$ lies below the straight line that joins the points with coordinates $((u_0, f(u_0))$ and $(v_0, f(v_0))$, as is displayed in Figure 2.8.2. This line segment $P_0 Q_0$ intersects the x-axis at a point denoted u_1 in the figure; since the line lies above the curve,

$$u_0 < u_1 < x^*. \tag{2.47}$$

NEWTON's method for approximating the root x^* of $f(x)$ is based on the relationship of the slope of the tangent line to the graph of $y = f(x)$ to the value of the derivative of f. In the figure, the slope of the line T tangent to the graph of $y = f(x)$ at Q_0 can be calculated in two ways: it equals $f'(v_0)$ but can also be expressed as

$$\frac{f(v_0)}{v_0 - v_1}.$$

Equating these expressions and solving for v_1 yields

$$v_1 = v_0 - \frac{f(v_0)}{f'(v_0)}.$$

The number v_1 satisfies the inequalities $x^* < v_1 < v_0$, which can be combined with eq(2.47) to produce the chain

$$u_0 < u_1 < x^* < v_1 < v_0.$$

Thus we have constructed a pair of intervals that both contain x^*. The gain in information upon passing from (u_0, v_0) to (u_1, v_1) is

$$\log_2 \left(\frac{v_0 - u_0}{v_1 - u_1} \right) \text{bits.}$$

The process of passing from the interval (u_0, v_0) to (u_1, v_1) can be iterated. The formulae that express the coordinates of the new interval in terms of the old are:

$$u_{n+1} = u_n - \frac{f(u_n)}{f'(u_n)} \tag{2.48}$$

and

$$v_{n+1} = \frac{u_n f(v_n) - u_n f(u_n)}{f(v_n) - f(u_n)}. \tag{2.49}$$

The information gained by passing from the nth stage $((u_n, v_n)$ known) to the $(n + 1)^{st}$ stage $((u_{n+1}, v_{n+1})$ known) is

$$I_{n:n+1} = \log_2\left(\frac{v_n - u_n}{v_{n+1} - u_{n+1}}\right). \tag{2.50}$$

To illustrate the method, let us approximate the positive root of $f(x) = x^2 - 2$, that is, $x^* = \sqrt{2}$. Select $u_0 = 0$ and $v_0 = 2$. It is easily verified that $0 < \sqrt{2} < 2$. Moreover, $f''(x) = 2$, so the convexity condition is satisfied. Recursive application of equations (2.48) and (2.49) yields the values in Table 2.8.2.

The corresponding gains in information (measured in *decimal digits*) are:

$$\log_{10}\left(\frac{2}{0.5}\right) = 0.60205\ldots,$$

$$\log_{10}\left(\frac{0.5}{0.01666\ldots}\right) = 1.47712\ldots,$$

$$\log_{10}\left(\frac{0.01666\ldots}{0.0000145\ldots}\right) = 3.06042\ldots.$$

The total information gained through the $n = 3$ iteration of the algorithm relative to the initial "guess" interval $0 < x^* < 2$ is the sum of the three numbers, viz.:

$$I_{0:3} = 0.60205\ldots + 1.47712\ldots + 3.06042\ldots$$

$$= 5.13959\ldots\text{decimal digits.}$$

Table 2.8.2 Newton's Algorithm for $\sqrt{2}$

	u_n	v_n	$v_n - u_n$
$n = 0$	0	0	2
1	1	1.5	0.5
2	1.4	1.41666...	0.01666...
3	1.41420118	1.41421568	0.00001450
⋮			

This means that the estimated value of $\sqrt{2}$ is correct to 5 decimal digits at this stage. Indeed, $\sqrt{2} = 1.41421356\ldots$ and the $n = 3$ iteration provides the bounds

$$1.41420118 < 1.41421356\ldots < 1.41421568\ldots.$$

The rate at which this algorithm produces information should be compared with the special method for calculation of square roots that was previously described.

2.9 Generalization of the Notion of an Information Measure

This section presents a mathematical generalization of the notion of an information measure intended for readers who are acquainted with the theory of topological groups and measure spaces.

Suppose that Ω is a set which has a measure m defined on it (in the sense of that branch of mathematical analysis called measure theory). Then there is a distinguished family of subsets of Ω called measurable sets such that for every measurable set $\omega \subset \Omega$, $m(\omega)$ is defined and is a non negative real number. If $\{\omega_k\}$ is a countable collection of pairwise disjoint measurable subsets of Ω, then

$$m(\omega_1 \cup \omega_2 \cup \cdots \cup \omega_k \cup \cdots) = m(\omega_1) + m(\omega_2) + \cdots + m(\omega_k) + \cdots.$$

The empty set $\varnothing$ and the set Ω itself are measurable, and $m(\varnothing) = 0$. There may be nonempty, measurable subsets of Ω whose measure is 0.

If m is a measure on Ω and if c is a positive number, then cm is another measure for which the measurable subsets of Ω are the same as the measurable subsets of Ω with respect to m.

Let us think of Ω as a parameter space of possible outcomes of an ideal observation, and of the measurable subsets of Ω relative to the measure m as the possible outcomes of a real observation. The measure $m(\omega)$ of a subset will be thought of as a measure of the size of the set of possible outcomes of an observation of some element of the set Ω. We will construct a numerical measure of the information gained by an observation. This measure of information, which will be denoted I, should not be confused with the measure m defined on Ω; the latter sense of the term conforms to usage in the branch of mathematical analysis known as measure theory but has nothing to do with measures of information *per se*.

Suppose that an observation is made, resulting in the measurable set $\omega \subset \Omega$. We will assume that the information associated with an observation of ω depends only on the value $m(\omega)$; that is, all measurable subsets which have a given value of the measure m yield the same quantity of information. Then we can write $I(m(\omega))$ for this quantity of information.

Following the line of reasoning set forth in section 2.3, the quantity of information gained by the observation should be independent of the units in which $m(\omega)$ is evaluated. This implies that for every $c > 0$ and measurable subset $\omega \subset \Omega$ such that $m(\omega) \neq 0$,

$$I(cm(\omega)) = I(m(\omega)).$$

By selecting $c = 1/m(\omega)$, it follows that $I(m(\omega)) = I(1)$, whence I is constant. As in section 2.3, we conclude that such a measure of information gain is hardly useful, and turn to consider the possibilities for evaluating the *relative* gain in information when one observation is followed by another. Let ω_1 and ω_2 be measurable subsets of Ω corresponding to two observations. It is helpful to initially confine ourselves to the case where $\omega_2 \subset \omega_1$, for this has the simple interpretation of a second observation which refines the first by restricting the possibilities for the unknown point in Ω that the observations are intended to determine. In this case the gain in information can be denoted by the function of two variables $I(m(\omega_1), m(\omega_2))$, and its independence of the units in which the measure m is evaluated is expressed by the equation:

$$I(cm(\omega_1), cm(\omega_2)) = I(m(\omega_1), m(\omega_2))$$

for all $c > 0$ and measurable ω_1, ω_2 such that $\omega_2 \subset \omega_1$. Suppose that $m(\omega_2) \neq 0$. Select $c = 1/m(\omega_2)$ to obtain

$$I\left(\frac{m(\omega_1)}{m(\omega_2)}, 1\right) = I(m(\omega_1), m(\omega_2));$$

hence I depends only on the ratio $m(\omega_1)/m(\omega_2)$, so we can simplify the notation by writing

$$I\left(\frac{m(\omega_1)}{m(\omega_2)}\right)$$

in place of $I(m(\omega_1), m(\omega_2))$.

Now suppose that a sequence of three observations is made, resulting in a nested sequence of measurable subsets $\omega_3 \subset \omega_2 \subset \omega_1$. Following the line of reasoning in section 2.3, we will assume that the information gained from ω_3 relative to ω_1 is the same as the sum of the information gained from ω_2 relative to ω_1, and the information gained from ω_3 relative to ω_2. In symbols,

$$I(m(\omega_1/m(\omega_3)) = I(m(\omega_1)/m(\omega_2)) + I(m(\omega_2)/m(\omega_3)). \qquad (2.51)$$

If we write $t = m(\omega_1)$, $u = m(\omega_2)$, $v = m(\omega_3)$, (2.51) is the same as the unnumbered equation preceding (2.15). Furthermore, if $m(\omega)$ assumes all positive natural numbers as values as ω runs through the measurable subsets of Ω (a condition which depends on the choice of Ω and m, and which must be verified for each choice), then we are in precisely the circumstances of equation (2.15), and the discussion given there applies. Hence, in this case,

$$I(m(\omega)) = c \log m(\omega), \qquad (2.52)$$

where c is a constant.

EXAMPLE 2.9.1. The case treated in section 2.3 concerned observations on a line which yielded intervals whose endpoints, say x and y, satisfy $x < y$. In this case ω is the interval determined by these endpoints and $m(\omega) = y - x$ is just the length of the interval.

In section 2.3 the measure $y - x$ of the interval ω determined by the endpoints x and y was fixed by its property of invariance under changes of the origin of coordinates of the measuring rod. This is a special case of the general situation when the measure m is invariant with respect to the action of a group of "motions" of the space Ω. In the case treated in section 2.3, Ω is the set of real numbers and the group is the group of "translations" $x \mapsto x + a$ of the numbers x in Ω by a fixed but otherwise arbitrary real number a.

More generally, let Ω be a homogeneous space of a group G, and let m be a G-invariant measure on Ω. Thus, for every $\omega \in \Omega$ and $\omega \in G$,

$$m(g\omega) = m(\omega).$$

This condition means that subsets of Ω can be "moved about" by the "motion" g without changing their "volume." The phrases in quotation marks are meant to bring to mind the particular case of 3-dimensional Euclidean space with ordinary volume as the measure m and the usual distance-preserving motions, composed of rotations, translations, and reflections, comprising the group G.

If Ω itself is a (topological) group, then there is an invariant measure on Ω called HAAR measure, and the preceding argument shows that the information gained by one observation compared with another is proportional to the logarithm of the ratio of the HAAR measures of the sets produced by the two observations. This can be applied to the problem, which will be treated from first principles in section 5.6, of quantifying the gain in information which results from the observation of an *angle*. We will show how this question is easily resolved by the general method introduced previously. The circle of radius 1 can be identified with the group G of complex numbers of modulus 1, where the group operation is multiplication. If z is such a complex number, then the HAAR measure on G is proportional to $-i(dz/z)$, where dz denotes an infinitesimal change in z (the *differential* of z) and i denotes $\sqrt{-1}$. This measure is certainly invariant under the group action. For if w is another fixed complex number of modulus 1, then

$$\frac{d(wz)}{wz} = \frac{w\,dz}{wz} = \frac{dz}{z}.$$

The measure dz/z can be expressed in a simpler form by recalling that $z = e^{i\theta}$ for some angle θ. Then $dz = ie^{i\theta}\,d\theta$ whence

$$-i\frac{dz}{z} = -i\frac{(ie^{i\theta}\,d\theta)}{e^{i\theta}} = d\theta.$$

The finite form of this measure is simply $\Delta\theta$, the usual measure of an angle. Hence equation (2.52) becomes

$$I = c \log(\Delta\theta_1/\Delta\theta_2)$$

for the information gained by measuring the angle $\Delta\theta_2$ after having first measured $\Delta\theta_1$. If we take $\Delta\theta_1 = 2\pi$ (this does not require an observation nor does it limit generality since no observed angle will exceed 2π) and measure information in bits, then the *a priori* information obtained by observing the angle $\Delta\theta$ is

$$I = \log_2(2\pi/\Delta\theta), \tag{2.53}$$

which is the result arrived at in equation (5.3).

2.10 Excursus on the Theory of Groups and Their Invariants

The theory of groups is a far-ranging generalization of the properties that are common to both ordinary addition and multiplication. We will not have need for the deeper aspects of group theory, but an acquaintance with the general concept and certain particular examples will be valuable in the chapters that follow.

Group theory plays a central role in physics and in chemistry, as well as in mathematics, but it has thus far entered information science, communication engineering, and cognitive psychology only peripherally although some researchers believe that it will ultimately play a fundamental role in these fields as well.

Consider ordinary addition of numbers. If x and y are two integers, then $x + y$ is another integer and the operation of addition possesses the following properties:

$$(x + y) + z = x + (y + z); \tag{2.54}$$

there is an integer 0 such that

$$x + 0 = x \tag{2.55}$$

for any integer x; if x is any integer, then there exists another integer y such that

$$x + y = 0. \tag{2.56}$$

These properties do not *characterize* addition but they do form the basis for most computational procedures that involve its use. It is these properties that the theory of groups generalizes.

A *group* G consists of two objects: a set (which is usually also denoted by G without engendering confusion) and a *binary operation* or *relation* which generalizes the operation of addition. The group operation, which will be denoted by juxtaposition of the group elements on which it acts, associates an element of the set G with each ordered pair of elements of G. If x and y are elements of G, then xy denotes the element of G with which they are associated by the relation. It is important to notice that yx may differ from xy. In the case of addition, the set G is the set of integers, the relation is the operation "$+$" of addition, and we usually write $x + y$ in place of xy.

Corresponding to the properties of addition expressed by eqs(2.54–2.56) are the following generalizations, which can be taken as axioms characterizing the group G:

G1. For every x, y, and z belonging to G,

$$(xy)z = x(yz).$$

This is the *associative law* for the group.

G2. There exists an element e in G such that for every x in G,

$$xe = x \quad \text{and} \quad ex = x.$$

e is called the *unit element* of the group; it generalizes the role played by 0 for ordinary addition.

G3. For any element x in G there is a unique element denoted x^{-1} such that

$$xx^{-1} = e \quad \text{and} \quad x^{-1}x = e.$$

x^{-1} is called the *inverse* of x; it generalizes $-x$ in the case of ordinary addition.

Addition possesses an additional property: $x + y = y + x$ for any integers x and y. For an arbitrary group, the analogue of this equation is

$$xy = yx,$$

and the group is said to be *commutative* if this relation is valid for arbitrary elements x and y of G. Although the simplest groups are commutative, in general groups are not commutative, and most of the complexity and interesting applications of group theory are consequences of their noncommutativity. We will consider examples of groups that are not commutative later.

The operation of ordinary multiplication leads to another elementary example of a group. In this case, the set G consists of the positive real numbers and the juxtaposition operation stands for multiplication. If x any y are positive real numbers, then xy is also positive, so xy belongs to G. The associative axiom **G1** asserts that $(xy)z = x(yz)$. The unit element e is the number 1 since $(x)(1) = x$ and $(1)(x) = x$; and finally, x^{-1} is the positive number $1/x$, since $x(1/x) = 1$ and $(1/x)x = 1$.

Now let us consider some less familiar examples which illustrate the properties of groups and show how they arise in contexts other than arithmetic or other purely mathematical concerns.

EXAMPLE   2.10.1   (THE GROUP OF ROTATIONS IN THE PLANE). Consider the set G of rotations in the Euclidean plane about a given point O.

G can be turned into a group by defining the group operation to be composition of rotations. That is, if g and h are rotations about O, then gh will denote the rotation about O which results from first applying the rotation h and then following it by the rotation g (it is conventional to interpret the action of a succession of rotations or other group elements as beginning with the rightmost symbol). A rotation in the plane is completely determined by the angle through which points are rotated. If the rotation g corresponds to an angle θ, then we will express this by writing $g(\theta)$ for g.

If $g(\theta_1)$ and $g(\theta_2)$ are rotations, then the rotation $g(\theta_2)g(\theta_1)$ is a rotation which corresponds to the angle $\theta_1 + \theta_2$; that is,

$$g(\theta_2)g(\theta_1) = g(\theta_1 + \theta_2). \tag{2.57}$$

Since $g(0)$ is a rotation by the angle 0, it leaves each point fixed in place whence, according to eq(2.57), $g(\theta)g(0) = g(\theta)$ for any rotation $g(\theta)$. Thus $g(0)$ is the unit element e of the rotation group G: $e = g(0)$.

Formula (2.57) also implies that axiom **G1** holds for rotations; indeed,

$$(g(\theta_3)g(\theta_2))g(\theta_1) = g(\theta_3)(g(\theta_2)g(\theta_1)) \tag{2.58}$$

because both sides of the equation are equal to $g(\theta_3 + \theta_2 + \theta_1)$.

The inverse of the rotation $g(\theta)$ is the rotation which returns points to their original position, i.e., a rotation by an angle $-\theta$. Indeed, $g(\theta)g(-\theta) = g(0) = e$ and $g(-\theta)g(\theta) = g(0) = e$ by eq(2.57). If we prefer to confine ourselves to nonnegative angles, then, using radian measure, we can write $g(2\pi - \theta)$ in place of $g(-\theta)$.

Finally, observe that the group of rotations in the plane is commutative since $g(\theta_2)g(\theta_1) = g(\theta_1)g(\theta_2)$ for any two rotations $g(\theta_1)$ and $g(\theta_2)$.

EXAMPLE 2.10.2 (THE GROUP OF ROTATIONS IN 3-DIMENSIONAL SPACE). Now consider rotations in 3-dimensional Euclidean space about a given point O. In this case a rotation is prescribed by the axis about which points rotate and the angle of rotation about that axis. It will be evident that a complete quantitative description of rotations in 3-dimensional space will be substantially more complicated than the description for rotations in the plane because information about the axis of rotation must be carried along. Nevertheless, it is still true that when rotations h and g are performed (in that order), the result, denoted gh, is a rotation about *some* axis in space. Moreover, it is not too difficult to convince one's self that the axioms **G1**, **G2**, **G3** are satisfied: here too the unit element corresponds to the rotation by an angle 0 (about any axis; these rotations are all the same!) and, if $g_A(\theta)$ denotes the rotation by an angle θ about the axis A, then the inverse of this rotation is the rotation by an angle $-\theta$ *about the same axis*, i.e., $g_A^{-1}(\theta) = g_A(-\theta)$.

The group of rotations in 3-dimensional space does have something essentially new to offer: it is *not* commutative. To demonstrate this, consider 3-dimensional space equipped with the usual $x-$, $y-$, and $z-$ axes of a Cartesian coordinate system, and consider a rigid configuration consisting of

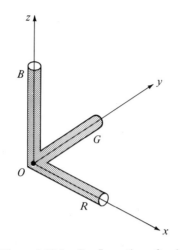

Figure 2.10.1   Configuration of rods.

a red rod R, a green rod G, and a blue rod B joined together at a common point. Suppose that the common point is at the origin $O = (0, 0, 0)$ of coordinates and that the red (*respectively* green, blue) rod points along the positive x (*respectively*, y, z) axis; this situation is illustrated in Figure 2.10.1.

Let g_x denote a rotation by an angle $\pi/2$ about the x-axis in the direction which carries G to B; similarly let g_y denote a rotation by an angle $\pi/2$ about the y-axis which carries B to R, and g_z denote a rotation about the z-axis by the same angle that carries R to the original position of G.

These motions can be described by tables which show the axis on which a colored rod lies before and after the rotation. If a rod lies along the negative direction of an axis, we will precede its symbol by a minus sign: thus $-R$ will mean that the rod colored R lies on the negative side of the axis. With this convention, the effects of g_x and g_z are described by the following tables.

Condition	x	y	z
Initial	R	G	B
After g_x	R	$-$B	G

Condition	x	y	z
Initial	R	G	B
After g_z	$-$G	R	B

The tables which describe the composite rotations $g_z g_x$ and $g_x g_z$ can easily be obtained from these. We find

Condition	x	y	z
Initial	R	G	B
After g_x	R	$-$B	G
After $g_z g_x$	B	R	G

Condition	x	y	z
Initial	R	G	B
After g_z	$-$G	R	B
After $g_x g_z$	$-$G	$-$B	R

The last lines of these tables are different, indicating that $g_z g_x$ and $g_x g_z$ are different, thus demonstrating that the group of rotations in 3-dimensional space is *not* commutative.

EXAMPLE 2.10.3 (GROUPS OF LINEAR TRANSFORMATIONS). Rotations in 2- or 3-dimensional spaces can be represented by linear equations which relate the coordinates of a point prior to the application of a rotation transformation (the "old" coordinates) to their "new" coordinates after the transformation has been applied. In 2-dimensional space, a (counter clockwise) rotation by an angle θ about the origin of a cartesian coordinate system can be expressed by linear equations as follows:

$$x' = x \cos \theta - y \sin \theta, \qquad y' = x \sin \theta + y \cos \theta, \qquad (2.59)$$

where the point $P = (x, y)$ is carried to the point $P' = (x', y')$ by the rotation. This rotation is completely specified by the collection of coefficients

$$\begin{pmatrix} \cos \theta & -\sin \theta \\ \sin \theta & \cos \theta \end{pmatrix}. \qquad (2.60)$$

This arrangement of the coefficients is called a *matrix*.

When we combine two rotations, how are the matrices to which they correspond combined? The operation of combining matrices defines a new kind of "product" which turns the collection of matrices into a group that generalizes the ordinary multiplication of numbers. For 2-dimensional rotations, the multiplication of matrices corresponding to the combination of rotations can be expressed as follows.

Let the rotation g correspond to the matrix

$$M = \begin{pmatrix} a & b \\ c & d \end{pmatrix}$$

and the rotation g' correspond to

$$M' = \begin{pmatrix} a' & b' \\ c' & d' \end{pmatrix}.$$

Then $g'g$ will correspond to a matrix M'' which is the *matrix product* defined by

$$M'' = \begin{pmatrix} a'a + b'c & a'b + b'd \\ c'a + d'c & c'b + d'd \end{pmatrix}.$$

This correspondence can be more compactly expressed as $M'' = M'M$, where the right side stands for the product of matrices.

It is easy to check that matrix multiplication is associative: for any three matrices,

$$(M_1 M_2)M_3 = M_1(M_2 M_3). \tag{2.61}$$

Moreover, the matrix

$$E = \begin{pmatrix} 1 & 0 \\ 0 & 1 \end{pmatrix}$$

is the unit element for matrix multiplication.

Do matrices have "inverses"? That is, can we find a matrix M^{-1} such that $MM^{-1} = E$ and $M^{-1}M = E$? It is easy to check that the matrix

$$M = \begin{pmatrix} a & b \\ c & d \end{pmatrix}$$

has as inverse

$$M^{-1} = \begin{pmatrix} d/(ad - bc) & -b/(ad - bc) \\ -c/(ad - bc) & a/(ad - bc) \end{pmatrix}$$

but the coefficient elements of this matrix only make sense if the common denominator $ad - bc$ is different from 0. (Notice that $ad - bc$ is the determinant of the matrix M.)

Putting these results together, we can conclude that the set of all matrices

$$M = \begin{pmatrix} a & b \\ c & d \end{pmatrix}$$

for which $ad - bc$ is different from 0 forms a group with respect to the operation of matrix multiplication, and that the group of rotations in 2-dimensional space can be represented as a group of matrices.

There is a completely analogous correspondence between groups of rotations in 3-dimensional space and groups of 3-rowed matrices. Indeed, any group can be represented as a group of matrices in an appropriate way so, in principle, when we are concerned with groups, we can limit our attention to groups of matrices.

In order to see how these ideas can be applied in the context of the measure of information, let us reconsider the discussion presented in section 2.3, and especially eq(2.16), which described the effect of simultaneous transformation of the zero-point of the measurement scale and of the unit of measurement:

$$x \mapsto ax + b, \qquad a = 0, \qquad b \text{ arbitrary.} \tag{2.62}$$

It was noted in section 2.3 that transformations of this type can be combined by composition and form a group called the *affine group* G_{aff}. We will show how this group can be expressed as a group of matrices. Consider eq(2.16) as a transformation of the point in the plane with coordinates $(x, 1)$ to the point with coordinates $(x', 1)$. Then (2.16) can be written

$$x' = ax + b$$

$$1 = 0x + 1.$$

The matrix that corresponds to this system of equations is

$$M = \begin{pmatrix} a & b \\ 0 & 1 \end{pmatrix}$$

where $a \neq 0$, b arbitrary. If

$$M' = \begin{pmatrix} a' & b' \\ 0 & 1 \end{pmatrix}$$

is a matrix corresponding to another such change of scale and unit, then

$$M'M = \begin{pmatrix} a' & b' \\ 0 & 1 \end{pmatrix}\begin{pmatrix} a & b \\ 0 & 1 \end{pmatrix} = \begin{pmatrix} a'a & a'b + b' \\ 0 & 1 \end{pmatrix},$$

which is the matrix that corresponds to application of the transformation $x \mapsto x' = ax + b$ followed by $x' \mapsto x'' = a'x' + b'$; indeed, $x'' = a'x' + b' = a'(ax + b) + b' = a'ax + (a'b + b')$ corresponds to the displayed coefficient matrix as claimed.

The measure of information gained by an observation is unchanged by the action of the measurement group G_{aff}. In general, a function which is independent of the action of the elements of a group is said to be an *invariant* of that group.

The matrices of a linear group act on a column of numbers (which will be called a *vector*) in the way prescribed by the corresponding linear transformation of the coordinates of a point (in a space of a suitable number of dimensions). If the coordinates of the point x in a 2-dimensional space are (x_1, x_2), and if M is the matrix of the linear transformation

$$x_1' = ax_1 + bx_2$$

$$x_2' = cx_1 + dx_2,$$

then this relationship can be concisely expressed as

$$x' = Mx.$$

A highly developed part of the mathematical theory of groups is concerned with the characterization of group invariant functions. Rather than approaching the problem of finding invariant functions directly for each group as it is encountered in our analysis, we can, in what follows, appeal to the general theory for the results we need.

When the matrices M belonging to a group are applied to a variable point x, x is transformed into Mx. If $f(x, y,...)$ is a function of the coordinates of the points x, y, ..., then f is said to be *invariant* if

$$f(x, y,...) = f(Mx, My,...)$$

for every matrix $M \in G$.

Applications of these ideas to the measure of information gained from an observation, where the information measure must be an invariant of the affine group, proceeds in the following way. From the general mathematical theory, it is known that if the group is represented by matrices acting on points with coordinates $(x, 1)$, $(y, 1)$, ..., then the simplest invariant will be a function of the ratio

where

$$\begin{vmatrix} a & b \\ c & d \end{vmatrix} = ad - bc$$

denotes the determinant of the matrix. The ratio of determinants is equal to

$$(x - y)/(x - z);$$

hence the measure of information corresponding to observation of the interval (x, y) followed by observation of (x, z) is a function of $(x - y)/(x - z)$, as was shown directly in section 2.3.

CHAPTER 3
Physical Measurements and Information

it is inherently entirely correct that the measurement or the related process of the subjective perception is a new entity relative to the physical environment and is not reducible to the latter. Indeed, subjective perception leads us into the intellectual inner life of individual, which is extra-observational by its very nature (since it must be taken for granted by any conceivable observation or experiment). Nevertheless, there is a fundamental requirement of the scientific viewpoint—the so-called principle of the psychophysical parallelism—that it must be possible so to describe the extra-physical process of the subjective perception as if it were in reality in the physical world—i.e., to assign to its parts equivalent physical processes in the objective environment, in ordinary space.

—John VON NEUMANN

3.1 Purpose of This Chapter

Toward the end of the nineteenth century the accumulation of inconsistencies of theory with experiment had set the stage for a far-reaching reconsideration of the foundations of classical physics. As a result, by 1925 two great new theories had grown up to supplant and extend NEWTONian dynamics and MAXWELLian electromagnetism. One of these, the general theory of relativity, had profound observable consequences in the realm of the very large and provided a revolutionary new approach to the origin, development, and structure of the universe on the cosmic scale. The equally profound consequences of the other new theory, quantum mechanics, were primarily observable in the realm of the very small, of atomic and subatomic spatial dimensions and the brief time intervals characteristic of the changes in the configuration of these microscopic systems. Although these theories appear to deal with disjoint extremes along the continuum of physical size, both of them were the result of novel and profound investigations of the nature of measurement and a re-assessment of the extent to which information can be acquired as the result of experimental observations.

The special theory of relativity was based on an examination of the notion of simultaneity of events in a space-time where the maximum speed of propagation of information-bearing signals is finite. The hypothesis that there is an upper limit to the rate at which signals can be propagated was suggested by theoretical considerations and powerfully supported by the MICHELSON-MORLEY experiments on the constancy of the speed of light (A similar reconsideration, although in the more limited context of living systems, should have resulted from HELMHOLTZ's measurement of the speed of neural propagation, but the physiologists did not make as much of this as they could have at the time. Another "missed opportunity" in the history of science!). The general theory of relativity added this premise to the revolutionary conception that the geometry of the universe depends upon its material contents.

Quantum mechanics, built on MAX PLANCK's discovery, in 1900, that certain physical quantities, such as energy, can change only by discrete jumps, was based on counterintuitive but incontrovertible evidence that physical measurements have an intrinsically probabilistic character: repetitions of an observation utilizing the same experimental apparatus, subject to the same initial conditions, will in general yield different measurements of the observed variable, although the totality of a large number of observations is governed by a known statistical law. At the heart of this strange but fundamental and unavoidable behavior lay another unexpected property of nature. Physically observable variables come in "conjugate" pairs—position and momentum, energy and time—both of which cannot be simultaneously measured with arbitrarily high accuracy. This celebrated uncertainty principle of Werner HEISENBERG can be formulated more sharply: if p and q, say, are conjugate pairs of variables and if Δp and Δq denote the (root-mean square) error of observation of p and q, respectively, then the product $\Delta p \Delta q$ of these uncertainties cannot be smaller than a certain positive universal constant. Thus, a very precise observation of the value p can be made only at the expense of generating a corresponding imprecision in the knowledge of the conjugate variable q: if the position of a particle is accurately measured, its corresponding momentum cannot be very accurately known; if the instant when an atomic system emits a light quantum is accurately known, the energy of the emitted light cannot be known with precision. Moreover, if a system is prepared so that p is known precisely, then q cannot be known at all, and if a second observation is made to determine q, at its conclusion the value of p will be uncertain. Thus we see that information *gained* about one variable as the result of a series of observations will in general correspond to information *lost* about the conjugate variable. This leads to a kind of weak law of "conservation of information" for quantum mechanical observations, and by inference, for macroscopic observations as well, which asserts that the algebraic sum of the information that is gained by a measurement of some physical variable and the information that is lost about the conjugate variable in the course of the measurement is negative. As EDDINGTON put it in his characteristically succint way:

> An addition to knowledge is won at the expense of an addition to ignorance.

Since macroscopic physical systems are constructed from atoms, the quantum laws of measurement have indirect macroscopic consequences. Moreover, modern fabricated microelectronic structures are so small that certain quantum phenomena have directly measurable effects. In this regard we should not forget that the photosensitive receptors in the retina of the human eye are comparable in size to the wavelength of the photons of visible light which stimulate them, and that the vision system is sensitive to the capture of a single photon (this is not to say that the capture of a single photon is necessarily registered as a distinct visual event); thus biological vision systems stand at the crossroads between the macroscopic and the microscopic worlds.

The purpose of this chapter is to discuss the relationship of the uncertainty principle to the mathematical principles of measurement presented in Chapter 2, and to explore some of its consequences for physical and biological information processing systems.

3.2 The Uncertainty Principle of Heisenberg

In section 2.6 we found that, when considered from a purely mathematical standpoint divorced from any particular physical or psychological embodiment, the process of measurement is constrained by a mathematical uncertainty relation of the form

$$\Delta t \Delta v \geq 1/4\pi \tag{3.1}$$

where t denotes the variable which is the object of the measurement (and which, in aid of intuition, it is convenient to think of as *time*) and v denotes the mathematically conjugate variable which is formally defined by means of the mathematical FOURIER transform (but which, in aid of intuition, it is convenient to think of as *frequency*). The symbols Δt and Δv denote the half-width of the range of variation of the variables t and v in the circumstances under consideration (which is an intuitive substitute for the root-mean-square deviation of these variables for the functions which prescribe the variation of t and v in the given situation); see section 4.5 for a precise definition.

The inequality (3.1), which arises from purely mathematical considerations of the concept of measurement, contains a substantial portion of the physical uncertainty principle of HEISENBERG which governs the accuracy of measurements of conjugate physical variables. We will show how the latter can be derived by combining (3.1) with two basic experimental physical facts, one arising from quantum mechanics and the other from relativity.

It has already been remarked that in 1900 PLANCK discovered that energy is bundled in discrete packets called *quanta*, and that every energy quantum is associated with an oscillatory phenomenon, or vibration, having a certain frequency. When a physical system passes from one energy state to another, it either emits or absorbs a quantum of energy whose magnitude ΔE is equal to the energy difference of the two states. PLANCK's formula expresses the energy difference in terms of the difference Δv in frequencies of the two states. The relation is

$$\Delta E = h\Delta v, \tag{3.2}$$

where PLANCK's constant $h = 6.62 \times 10^{-27}$ erg-second is a universal constant of nature. We can solve (3.2) for Δv and substitute the result in the mathematical uncertainty relation (3.1) to find

$$\Delta t \Delta E \geq h/4\pi; \tag{3.3}$$

this is the celebrated HEISENBERG uncertainty principle for time and energy. It informs us that we cannot have unlimited simultaneous precision in the measurement of both time and energy corresponding to a single physical event.

According to the principle of special relativity, space and time have equal standing in the equations of physics. When given its proper formal interpretation, this means that there is a definite way to "transform the coordinates" in the 4-dimensional space-time so as to automatically provide the space analogue of an equation involving time, and vice versa. This is not the place to enter upon a technical discussion of special relativity; it will suffice to remark that the space coordinates x, y, z bear to the corresponding momentum variables p_x, p_y, p_z the same relation that time bears to energy, and that inequality (3.3) implies the following system of inequalities for the space coordinates and momentum variables:

$$\Delta x \Delta p_x \geq h/4\pi,$$
$$\Delta y \Delta p_y \geq h/4\pi, \tag{3.4}$$
$$\Delta z \Delta p_z \geq h/4\pi;$$

these are the inequalities of the uncertainty principle that governs the accuracy of measurement of the space coordinates and their corresponding momenta.

Thus, the inherent constraints on the information that can be gained from physical measurements arise in part from properties of nature, and in part, as described by (3.1), from the mathematical—i.e., quantitative—description of the abstract process of measurement without regard to properties governed by physical law.

3.3 Theory of Measurement and the Conservation of Information

Suppose that we set out to measure some feature of a physical system, for instance, the energy of a photon of light emitted by it. An experiment is performed and our analysis shows that the error of the measurement lies within some limits, whose difference is ΔE_1. By the uncertainty principle it necessarily follows that a simultaneous measurement of the time when the particle of light was emitted must be uncertain by an amount Δt_1 such that the inequality $\Delta E_1 \Delta t_1 \geq h/4\pi$ holds. If we are clumsy experimenters (and also in certain other circumstances), the actual value of the product may be much larger than $1/4\pi$. But suppose for the moment that we are careful and that $\Delta E_1 \Delta t_1$ is as small as it can be. Now suppose that a second observation of the emitted energy is made with the purpose of refining our knowledge of its value. Then we will have a spread of error ΔE_2 which is smaller than ΔE_1, and the information gain will be, according to Chapter 2,

$$\log_2(\Delta E_1/\Delta E_2).$$

There will, of course, by a corresponding uncertainty in the moment of emission whose magnitude is Δt_2. If we have been so careful that this second experiment also achieves the minimum value possible for the error product $\Delta E_2 \Delta t_2$, then $\Delta E_1 \Delta t_1 = \Delta E_2 \Delta t_2$ whence

$$\log_2(\Delta E_1/\Delta E_2) = \log_2(\Delta t_2/\Delta t_1). \tag{3.5}$$

Since $\Delta E_2 < \Delta E_1$, we must have $\Delta t_2 > \Delta t_1$, whence the gain in information about the energy is balanced by a loss in information about the instant it was liberated. We can rewrite (3.5) in the suggestive form

$$\log_2(\Delta E_1/\Delta E_2) + \log_2(\Delta t_1/\Delta t_2) = 0, \tag{3.6}$$

which shows that an information gain for one of the two conjugate variables is exactly balanced by an information loss for the other variable.

Equation (3.6) is a kind of "law of conservation of information," which asserts that, in the best of circumstances, information gained as the result of any kind of observation whatsoever is balanced by an equal loss of information about the conjugate quantity. In situations that are suboptimal, the information gained by a measurement will be less than the information lost about the conjugate variable, and possibly about other variables as well. This latter formulation is equivalent to the *second law of thermodynamics* and sets bounds on the performance of any information-processing system.

3.4 Entropy and Information

The concept of entropy and the earliest statement of the second law of thermodynamics were introduced by Sadi CARNOT, the inventor of thermodynamics, in 1824 when he was 28 years of age. He died eight years later in a cholera epidemic in Paris, leaving behind a notebook containing other important contributions that anticipated the work of his successors which remained unpublished (with the exception of some abstracts brought to public view in 1878 under the auspices of the French Academy of Sciences into whose hands the notebook had been entrusted) until its first full presentation to the scientific public by Émile PICARD in 1927.

The concept of entropy arose in considerations connected with the theory of heat, and its relationship to information was not at first realized. Yet in retrospect it is not surprising that the analytical treatment of information should have had its origin in this area of physical science, for the subject of thermodynamics is primarily concerned with the determination of the laws which govern the conversion of mechanical and other "higher" forms of energy into "heat" energy. Heat energy is characterized by its disorder, and ultimately by the irreversibility of certain processes which involve heat transfer. Thus, an ice cube melts in a warm room, but the reverse process never is observed to occur. In this example, and in general, the tendency of things is to "run down", not "up"; structures tend with time to lose their organization to some degree, and it is this degree of organization that entropy measures. But the degree of organization of a system can also be interpreted as a measure of the quantity of information incorporated in it. This idea, which slowly began to develop during the course of the nineteenth century, was given explicit form by SZILARD in 1929. Thereafter the notion that entropy was another term for information, and that the physical measure of entropy was also a measure of the quantity of information or degree of organization of the corresponding physical system became prevalent. More precisely, the difference in the quantity of information between two states of a physical system is equal to the negative of the corresponding difference in entropy of the two states. The use of the negative sign stems from the historical accident that the inventors of the theory of thermodynamics tended to think of entropy as increasing with an increase in the degree of *dis*organization of a system, whereas from the standpoint of information, increased disorganization will correspond to a *de*crease in the quantity of information. Thinking in terms of changes in information content is the easiest and most natural way to understand entropy in the context of thermodynamical considerations, but the idea of information is of course much more natural and far-reaching.

Insofar as we are concerned with information transfers from one physical or biological system to another, there are aspects of the classical theory of thermodynamics that require our attention. Information transfers between physical or biological systems are the result of certain changes of state. Each physical change of state, such as the firing of a neuron or the chemical change induced in a photosensitive chemical pigment in the retinal receptors of the human eye, is accompanied by the release of a certain small amount of energy. But whereas there is a strict law of conservation for energy which assures us that the total amount of energy in a closed environment remains unchanged no matter how the physical circumstances may vary, this is not true for the measure of organization of the system. We conclude, along with CARNOT and other nineteenth century thinkers about thermodynamics, that the information or entropy content of a system will generally change as its energy is redistributed by a physical change of state. In nature we find that self-contained systems change from more highly organized structures to less highly organized ones; from states of higher information content (and lower entropy) to states of lower information content (and higher entropy). CARNOT was led to the concept of entropy by tracing how a physical system becomes progressively disorganized as energy is redistributed within it, and he discovered the relationship between the change in energy of a part of the system and the corresponding change in the measure of information (or entropy). This relationship will be important to us later so we will describe it here. Suppose the system of interest changes from state 1 to state 2; let us say the firing of a neuron distinguishes between them: in state 1, the neuron has not yet fired, whereas in state 2, it has. Then a certain amount of energy ΔE will be released during this process. The corresponding change in information ΔI will be given by the formula

$$\Delta E = c\Delta I,$$

where c is a constant. For physical systems this constant is proportional to the absolute temperature T of the system. If energy changes are measured in units of ergs and temperature according to the absolute (KELVIN) scale, for which T measured in Kelvins is equal to the temperature in degrees Celsius (centigrade) plus 273.16, then this equation can be written

$$\Delta E = (kT \log 2)\Delta I = (9.57 \times 10^{-17})T\Delta I \text{ ergs}, \tag{3.7}$$

where k is BOLTZMANN's constant mentioned in section 1.3 and the information I is measured in bits. We see that an enormous number of bits of information will be needed to produce an energy change which can be sensed at the macroscopic level of normal activities. This fact offers a partial explanation for why information as an independent subject of scientific interest lagged so long behind other fields, and even behind the thermodynamical approach to the subject.

3.5 Information, Energy, and Computation

There is one area of human activity where the small energy differences which correspond to information changes of only a few bits have direct and significant consequences. Contemporary computing machines are constructed from microelectronic switches (transistors). They are so small that extremely minute quantities of energy suffice to change their state, which is coded to correspond to a 1 bit change in information content. Since a change in the state of a single bit can of course be detected by a digital computing machine— this lies at the heart of its mode of operation—the corresponding energy difference will be significant although, depending on the construction of the machine, it might be so minute as to be undetectable by the direct operation of the human sensory system. The ultimate physical limits to computer performance are determined in part by the trade-offs which are implicit in equation (3.1). In principle, these limits also govern biological information processing systems although their application in this situation is obscured by the complexity of the total organism which makes it difficult to determine whether biological information processing systems actually operate anywhere near their physical limits. Nevertheless, there are some useful conclusions that can be drawn from an analysis of physical computing systems regardless of whether they are constructed from electronic "hardware" or tissue "bioware."

Equation (3.1) tells us that a change of 1 bit in information content requires the expenditure of at least $kT \log 2$ ergs of energy; this observation seems to have first been made by VON NEUMANN. One implication of this result is that it takes more energy to produce (i.e., make recognizable or measurable) a bit at high temperature than at low. This is readily understandable, for the incessant bombardment of every physical substance by the thermally agitated molecules in a room at ordinary temperatures creates a "noise" background in which the 1 bit signal may be submerged. At ordinary room temperatures T is about 293 Kelvin whence $kT \log 2$ is approximately equal to 3×10^{-14} ergs, a very small quantity indeed, much less than is released by a typical chemical reaction of even a single atom. Contemporary electronic computers dissipate about $10^{10} kT$ ergs per operation, nowhere near the fundamental physical limit, but still much too small to be directly sensed by a person.

If we could carefully construct a computing machine so that it would require about $kT \log 2$ ergs of energy to produce 1 bit of information—the best energy efficiency we can hope for—then the uncertainty principle of HEISENBERG (equation (3.1)) implies that the time required to be sure that the "switch" has changed from one state to another, and thereby yields information, is related to the energy by $kT \log 2 \Delta t \geq h/4\pi$ that is, the switching time Δt must be at least

$$\Delta t \geq \frac{h}{4\pi kT \log 2} = 5.5 \times 10^{-12}/T$$

seconds long, which is about 10^{-14} seconds at ordinary temperatures. In this time, a signal traveling at the greatest possible speed—that of light—would travel about 3 micrometers.

3.6 The Physics of Information Processing

Information processing consists of the performance of a sequence of calculations and data transmissions from one location within the processing apparatus to another.

We may think of the complexity of a computation or information processing task as measured by the number of elementary logical operations and data transfers required to perform it. Elementary logical operations are implemented in a computer by changing the state of a small number of appropriately interconnected switches. Since the change of state of a switch corresponds to the deployment of a certain amount of switching energy, E_{sw}, which must be at least $kT \log 2$ ergs in magnitude, the logical complexity of a calculation can be represented by the number of logical operations required to perform it, which in turn can be represented by the quotient E/E_{sw} of the total energy needed to complete the calculation by the switching energy E_{sw} for a single change of state of a switch such as a transistor.

Here we may follow some ideas of MEAD and CONWAY with profit. If we think of a computation as a succession of logical operations which are intended to reduce the initial data to some unique ultimate output, then each stage of logical processing reduces the inventory of possible outcome states. Put in another way, since the computation is performed by the operation of a number of interconnected logical switches (each of which we may, without loss of generality, assume to be in a definite one of two possible states at a given stage in the computation), there will be a finite number N of possible logical states in which the computing system could conceivably be found in the course of the computation. Then, following the ideas set forth in Chapter 2, the measure of the logical information in the computation will be $I_L = \log_2 N$ if the logical states are equally probable.

Every bit of logical information corresponds to an energy E_{sw} which sets a particular switch to the state required by the computation. Evidently the logical information will satisfy the relation

$$E/E_{sw} \geq I_L.$$

Similarly, in the course of the computation, data will be moved from one location to another within the information processor. In order to route data properly, switches will be appropriately set and this too will require switching energy E_{sw} per event. By estimating the number of possible communication pathways one can derive a measure of this "spatial" information I_C corresponding to communications. The total information in the computation will be $I = I_C + I_L$ and we have the relation

$$E/E_{sw} \geq I. \tag{3.8}$$

The entropy increase of the physical computing system in the course of the computation will be given by the formula

$$\Delta S = E/T$$

where T is the absolute temperature. Hence we find

$$\Delta S = E/T \geq (E_{sw}/T)I$$

so

$$\Delta S/(k \log 2) \geq (E_{sw}/kT \log 2)I, \qquad (3.9)$$

where k denotes Boltzmann's constant. Thus we see that a computing device, whether it be hardware or bioware, must dissipate sufficient heat to increase its entropy by an amount ΔS such that (3.9) holds, where $E_{sw}/(kT \log 2)$ is the ratio of the actual switching energy of the information processing system to the theoretical minimum switching energy. Mead and Conway put it thus:

> With respect to a given computation, an assemblage of data in an information processing structure possesses a certain logical and spatial entropy. Under proper conditions, we may be able to remove all of the logical and spatial entropy from a system leaving it in the state of zero logical and spatial entropy; we call this state "the right answer." A sufficient dissipation of energy and outflow of heat must occur during the computation to increase the entropy of the environment by at least an equal amount. When viewed this way, information processing engines are really seen as just "heat engines," and information as just another physical commodity, being churned about by the electronic gears and wheels of our intricate computing machinery.

3.7 What Is Life?

Life is nature's solution to the problem of preserving information despite the second law of thermodynamics. In 1944 Ernst Schrödinger wrote a little book, whose title is the heading for this section, which explored the remarkable ability of the genetic material to preserve its coded message of life and transmit it intact to future generations although it is continually buffeted by chaotic thermal bombardments that ultimately reduce and disorganize the forms of matter which do not carry the message of life. Schrödinger argued that the separation of energy levels which is a consequence of the quantization of energy ultimately provides the barrier to the disruption of the chemical bonds of the large protein molecules which carry the genetic code, and concluded that it was the peculiar properties of these exceptional molecules inherited from the universal quantum properties of matter that enable life to persist. Were the physical world governed by the laws of Newton, life would not exist.

Beyond the view of what makes life possible, one may ask what makes it different from non-living substance. This question may be approached from many different directions. Schrödinger observed that one fundamental and universal property of life is that it organizes its environment; that is, life increases the information—decreases the entropy—of the stuff from which it is composed. He claimed (and overstated his point) that we eat primarily in order to subsume within ourselves the organization of the ingested foodstuffs rather than any other of their properties. While there can be no doubt that the output side of the eating cycle is not so well organized as the input side, the food we eat does contribute some mass to replace that which is lost through metabolic processes, and of course it does supply the energy which ultimately drives metabolism. But, as all heat generating machines do, the cells of the body increase entropy—lose information—and they must find some way to regain the lost information. This, SCHRÖDINGER claimed, is the principal purpose of consuming highly organized plant and even more highly organized animal foodstuffs; we siphon off their information content as well as some of their energy and a small portion of their bulk.

In 1953 CRICK and WATSON discovered how the information which controls the genesis of new life is coded in the deoxyribonucleic acid (DNA) molecule. At its most fundamental level, the secret of life is contained in the message this code carries rather than in the medium that carries it. Thus, the student of the science of life must necessarily be in part a student of information science.

3.8 The Theory of Measurement and Information

We have seen that the amount of energy that is needed in order to obtain one bit of information is at least as great as $kT \log 2$. Since the binary expansion of a real number consists of an infinite number of bits (binary digits), it follows that it is physically impossible to know a real number with complete precision. This is the viewpoint expressed in Chapter 2 and in the previous sections of the present chapter. One consequence of the inability to observe physical phenomena with complete accuracy is an unexpungable uncertainty in the values of observables. As we have seen, the intrinsic uncertainties in observation are interrelated according to the uncertainty relations of HEISENBERG. Although they are generally considered to be an essential constituent of the physical theory known as quantum mechanics, it is also possible to adopt the position that the uncertainty principle is primarily the quantitative expression of a limitation on the accuracy of observations which does not depend on the details of a particular physical theory. Indeed, if the uncertainty relations are accepted within the framework of classical physics, then they imply certain restrictions on the predictive capabilities of that theory which express the consequences of the inherent limitations on real observations as a physical system evolves through time.

One of the simplest physical systems already illustrates how observational uncertainties and the limitation on the amount of information that can be obtained from observations conspire to dissipate information with the passage of time.

Consider an elastic ball which moves along the x-axis between two rigid walls at $x = 0$ and $x = 1$ in the absence of forces. According to NEWTONian mechanics, the ball will move with constant velocity v, which we assume to be different from 0. When the ball meets one of the rigid walls, its velocity will reverse direction but maintain the same magnitude. According to classical physics, the motion of the ball is completely specified if its position and momentum are known at any given time. If the mass of the ball is m, then its momentum p and velocity v are related by $p = mv$, so it amounts to the same thing to say that simultaneous knowledge of the ball's position and velocity completely determine its present state and future course.

If x and p are used as coordinates on a graph, then at each instant (x, p) corresponds to a point and, as time passes, the point will trace a curve in the x–p plane which describes the motion of the ball. This form of graphical representation is particularly suitable when considering problems related to the HEISENBERG uncertainty relation $\Delta x \Delta p \geq h/4\pi$.

Suppose the ball is observed at some time t_0. According to classical physics, it will have a well-defined position and a well-defined momentum. Perhaps so, but no measurement will confirm this. The best that can be done is to observe that the position of the ball lies in some small interval, say between x_0 and $x_0 + \Delta x$, and that its momentum lies between, say, p_0 and $p_0 + \Delta p$. The uncertainty relation of HEISENBERG requires that $\Delta x \Delta p \geq h/4\pi$ which means that the position in the x–p plane of the point which corresponds to the ball cannot be known; the best that can be done is to confine that point to a rectangle or other region R of uncertainty having an area no smaller than $h/4\pi$. Figure 3.8.1 illustrates this situation. (That p is positive in the diagram means that the ball is moving to the right; negative momentum corresponds to leftward motion).

As time passes the ball will move, first to the right and, after rebounding from the rigid wall at $x = 1$, to the left and still later, after rebounding from

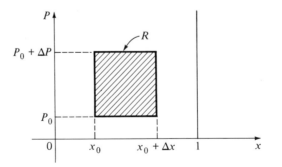

Figure 3.8.1   Limits to observation.

the rigid wall at $x = 0$, to the right again. How will the region of uncertainty R behave as the ball moves?

Since the velocity of the ball is constant $v = p/m$ (constant apart from changes in direction upon rebounding from one of the walls), the coordinate x increases according to the formula

$$x = x_0 + (t - t_0)(p/m)$$

for points which have momentum p. But for points which have momentum $(p + \Delta p)$, and hence greater speed, the position changes according to

$$x = x_0 + (t - t_0)\left(\frac{p + \Delta p}{m}\right); \tag{3.10}$$

intermediate values of the momentum correspond to proportionally intermediate positions. This means that as time passes, the rectangular region of uncertainty R in Figure 3.8.1 becomes deformed, the points corresponding to greater values of p moving more rapidly than those corresponding to lesser values of p. Figure 3.8.2 displays the shape of R as the ball moves forward from its initial position between x_0 and $x_0 + \Delta x$ toward the wall at $x = 1$. Notice that R is deformed into an increasingly elongated parallelogram whose base Δx and height Δp remain unchanged; hence the area of R remains constant as it is deformed, in agreement with the uncertainty relation.

After the elapse of some time the top of the parallelogram R will meet the rigid wall—that is, if the momentum of the ball were so large as to correspond to points at the parallelogram's top, the ball would meet the wall—and rebound toward the left. Figure 3.8.3 illustrates the first occurrence of a rebound.

After contact with the wall, the ball's velocity (and hence momentum) changes direction and therefore algebraic sign, so the rebounding part of region R is depicted in the region of the x–p plane where p is negative.

From eq(3.10) we discover that the region R becomes an increasingly elongated parallelogram until finally, when $(t - t_0)(\Delta p/m) = 1$, i.e., when $t - t_0 = m/\Delta p$, the elongated parallelogram covers the entire line segment from 0

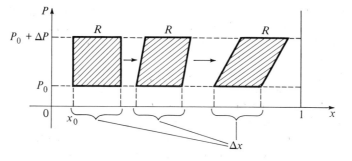

Figure 3.8.2   Deformation of R as the ball moves.

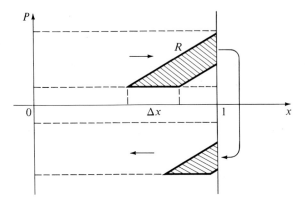

Figure 3.8.3   Rebounding from a rigid wall.

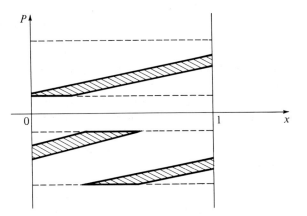

Figure 3.8.4   *R* is elongated beyond the width of the interval.

to 1. From this instant on, it is no longer possible to say more about the position of the ball than that it is *somewhere* between the two walls unless we *also* know something more about its momentum than we originally knew. Moreover, as time passes, it will be necessary to know the momentum with greater and greater accuracy if we wish to be able to specify the position of the ball between the walls.

As the parallelogram elongates further, it can no longer fit between the walls as a single region: part of it is rebounding to the right from the wall at $x = 0$ while another part is rebounding to the left from the wall at $x = 1$, and the region of uncertainty will split into two smaller parallelograms whose sides are parallel. This situation is illustrated in Figure 3.8.4. And after a still longer time has elapsed, the original rectangle *R* will have become deformed and separated into a large number of parallelograms whose corresponding sides are parallel.

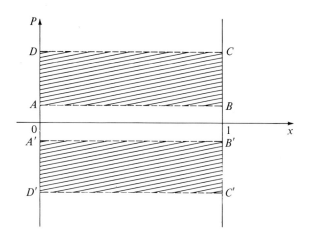

Figure 3.8.5 *R* becomes increasingly dense.

These parallelograms become increasingly horizontal, increasingly thin, and increasingly close to one another as time passes although the sum of their areas remains constant and equal to the area of the original region *R*. This situation is illustrated by Figure 3.8.5 which suggests that *R* becomes increasingly densely distributed in the rectangles *ABCD* and *A'B'C'D'*. Since any practical measurement can be made with only limited accuracy, after some time it will no longer be possible to distinguish the different mini-parallelograms according to the different values of *p* to which they correspond and then all knowledge about the position of the ball is lost except that it is confined between the walls.

Thus, although the information which was initially obtained from observation is still coded in the physical system which has evolved according to the equations of NEWTONian mechanics, it has become so diffused throughout the *x*–*p* plane that it is no longer accessible to observers using finite instruments.

As another example, consider a particle that traverses a circular orbit at constant speed, e.g., a planet revolving about a star. Let *r* denote the distance of the particle from the origin of a polar coordinate system centered on the axis of revolution, and let the central angle θ correspond to a point on the circle.

The uncertainty principle asserts that $\Delta s \Delta p \geq h/4\pi$ where *s* denotes distance along the perimeter of the circle and *p* denotes the tangential momentum. The angular momentum corresponding to *p* is $p_0 = mvr$ where *m* denotes the mass of the particle and *v* its tangential velocity. From $s = r\theta$ we obtain

$$\Delta s \Delta p = \Delta\theta\Delta(mvr) = \Delta\theta\Delta p_\theta \geq \frac{h}{4\pi}.$$

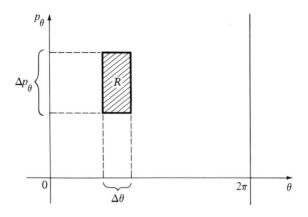

Figure 3.8.6   Phase plane representation.

Figure 3.8.6 displays the phase plane that corresponds to this situation. Simultaneous measurement of θ and p_θ at time t_1 leads to a rectangle of uncertainty R in the phase plane as shown in the figure. From the point of view of physics, the uncertainty is an inescapable consequence of HEISENBERG's principle, but from the point of view of information science, it is an inescapable consequence of the inability to measure the infinitely many digits in the decimal expansions of the observed quantities.

As time passes the particle will continue its motion along the circle at a constant speed but its predicted future position will not be known with certainty because of the uncertain knowledge of the initial values of θ and p_θ. Points along the lower horizontal edge of the uncertainty rectangle R in the phase plane correspond to a smaller speed than do points along the upper edge. It follows that, as time passes, R will be distorted into a parallelogram whose base and height remain unchanged.

Values of θ that differ by 2π correspond to the same point on the orbit. By limiting θ to lie in the range $0 \leq \theta < 2\pi$, we are able to represent R as a vertical rectangular strip on a section of a cylinder, as exhibited in Figure 3.8.7.

With the passage of time the points along the upper bounding circle of the cylindrical section move at constant speed relative to points on the base of the cylinder. If the region R is thought of as an infinitely flexible sheet of rubber, then, as the upper circular boundary of the cylinder rotates, R will be stretched thinner and thinner as it is wound around the cylinder until, in the limit, it will be dense in the cylinder: every point on the cylindrical section will lie arbitrarily close to the region of uncertainty. Ultimately it will not be possible to distinguish the different local branches of the phase space orbit of R whence the the location of the particle in phase space will be totally undetermined; see Figure 3.8.8. Chaos will have emerged from order as a result of the fundamental limitations on the processes of measurement.

The reader should *not* believe that the direction of time can be defined this

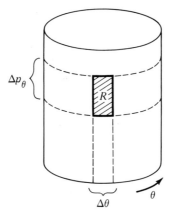

Figure 3.8.7   Cylindrical representation of phase space.

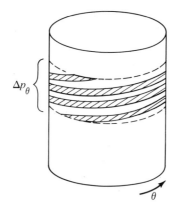

Figure 3.8.8   Dense evolution of the region R.

way, that is, as the direction from a single (connected) region R to the multiple-component and diffused R into which it changes, for if the equations are followed *backward* in time from t_0, the same phenomenon occurs. The reader will find it profitable to consult P.C.W. DAVIES's interesting work *The Physics of Time Asymmetry* on this question.

3.9 Time's Arrow

> All the king's horses and all the king's men
> Cannot put Humpty Dumpty together again.

The apparently unidirectional flow of time has puzzled philosophers and scientists from the start. There is nothing in the laws of physics, save one exception, that either dictates the direction of the passage of time or contradicts a flow in the opposite direction, from future to past. The dynamical equations that govern motion in NEWTON's classical mechanics, in MAXWELL's theory of electromagnetism, in EINSTEIN's general relativity, and in the quantum mechanics of HEISENBERG and SCHRÖDINGER are indifferent to the sign of the variable that denotes time and, consequently, to whether events proceed from the past to the future or contrariwise. In these models of nature, the past and the future are symmetrically linked to each other and (subject to certain technical reservations implicit in the probabilistic nature of quantum phenomena) any "initial" physical configuration can be restored from a "final" configuration:

> The famous Duke of York
> With twenty thousand men,
> He marched them up to the top of the hill
> And marched them down again.

The one exception is the *second law of thermodynamics*, which asserts that as time passes the entropy of an isolated system cannot decrease. In terms of information, as an isolated physical system evolves with the passage of time, information about the state of the system is sometimes conserved but more generally it is lost.

The second law of thermodynamics prescribes a direction for time's arrow: the direction of the future is the direction of information loss and entropy gain. Since the discovery of thermodynamics in the nineteenth century, this "heat death" or "running down" of the universe as information is lost through the dissolution of organized structures and the homogenization of the contents of the cosmos has captured the imagination of poets, philosophers, and theologians, and has provided a profound paradox for scientists.

Common sense and everyday observation appear to contradict the progressive decrease in organization implied by the second law of thermodynamics. In Steven FRAUTSCHI's words,

> A look at our present picture of the history of the universe reveals a remarkably different and more interesting situation. In the beginning there is a hot gas, nearly homogeneous and in thermal equilibrium. As it expands it breaks into clumps of matter—galaxies, stars, planets, rocks, dust, and gas—with a wide range of temperatures. Some of these objects develop highly organized structures and, on at least one planet, self-replicating structure called "life" develop. Finally, a form of life emerges with the capability to ask questions about these systems.

It has been argued that althougth the second law of thermodynamics calls for a general decrease in information with the passage of time, pockets of the universe might exist where local fluctuations could lead to increases in information and the development of increasingly organized material structures.

But this would occur at the expense of a compensating more rapid loss of information elsewhere to balance the account and insure that the overall average change in information is negative. Proponents of this view find it difficult to explain what caused the fluctuations and why they have endured so long, or why we have been unable to to observe regions of the universe that are not highly structured. Everywhere we turn we see stars and star clusters organized on ever increasing scales of regularity.

Within the past decade a new approach to this fundamental question was made possible by the discovery of black holes and by J.D. BEKENSTEIN and S. HAWKING's deduction of a formula that measures the amount of information of a black hole, that is, a measure of the degree of structure of a black hole in terms of the parameters that describe its physical state.

Black holes are creatures of the gravitational force. Thus it was recognized that the role of universal gravitation might prove crucial to resolving the paradox of how organization appears to increase as time passes while information necessarily decreases in obeisance to the second law of thermodynamics.

Our description of these new ideas about information and gravitation follows the discussion given by FRAUTSCHI in 1982.

Traditional theories of the consequences of the second law of thermodynamics do not take the effects of gravitation into account. Their assumptions are, in essence, equivalent to positing a closed nongravitating box in which all matter and energy approach thermodynamic equilibrium, the state of minimum information that is accessible to the system. The real universe differs from this idealized model in at least three respects: (1) it is expanding; (2) the effects of gravitation have a long range; and (3) the interplay between the rate of expansion of the universe and the rate at which its material and energetic contents can adapt themselves to new conditions of equilibrium.

As the universe expands, the number of states accessible to the matter-energy in it increases. If W_1 denotes the number of accessible states at one instant and W_2 denotes the number of accessible states at a later instant, then $W_2 > W_1$ because of the expansion, and the corresponding information difference is measured by

$$\Delta I = \log_2(W_1/W_2) < 0; \tag{3.11}$$

information is lost as a consequence of the expansion.

It is possible to estimate the number of accessible states by modeling the matter-energy contents of the universe as a gas of N free particles with given temperature in a given volume, accounting for the contribution of black holes by means of the formula due to BEKENSTEIN and HAWKING. This leads to a model very different from the classical "running down" of the universe. Indeed, in the initial stages of the "big bang," the universe was probably a homogeneous hot gas, which is a state of maximum entropy, i.e. minimum information. According to the new theory, as the universe expands, the number of accessible states grows more rapidly than the matter-energy can accomodate itself to fill them. Since the actual number of instantaneous accessible states continues to increase with time, there is a corresponding absolute loss of information. But because the total number of accessible states increases more rapidly than the number that is instantaneously accessible as a result of the expansion of the universe and the finite speed of propagation of signals, the ratio of instantaneously accessible states to the total number of accessible states can decrease with the passage of time. Let us suppose that this is so.

Denoting the total number of accessible states at a time t_1 by W_1^{total} and the number that are instantaneously accessible to W_1^{inst}, and the corresponding numbers at a later time t_2 by W_2^{total} and W_2^{inst}, these relations can be expressed by the following inequalities:

$$W_1^{total} < W_2^{total},$$

$$W_1^{inst} < W_2^{inst}, \tag{3.12}$$

$$W_1^{total}/W_1^{inst} < W_2^{total}/W_2^{inst}.$$

These inequalities have the following interpretation: Considering the total number of states, there is a loss of information that is measured by the quantity $\log_2(W_1^{total}/W_2^{total}) < 0$ as time passes from t_1 to t_2. Similarly, there is a loss of information when only the instantaneously accessible states are considered, amounting to $\log_2(W_1^{inst}/W_2^{inst}) < 0$. The third inequality implies that

$$\log_2\left(\frac{W_2^{total}}{W_2^{inst}}\right) - \log_2\left(\frac{W_1^{total}}{W_1^{inst}}\right) > 0. \tag{3.13}$$

If we interpret

$$\log_2\left(\frac{W_i^{total}}{W_i^{inst}}\right)$$

as the information gained by confining the accessible states to those that are instantaneously accessible, then this inequality shows that *there is a relative net gain of information as time passes*. In order to understand why this could be true, interpret W_i^{total} as a volume in an appropriate space of states and W_i^{inst} as a smaller volume contained within the first. Then the expression $\log_2(W_i^{total}/W_i^{inst})$ is the usual measure of information for observation W_i^{total} followed by W_i^{inst}. If the ratio (W_i^{total}/W_i^{inst}) increases as time passes from t_1 to t_2, then the relative accuracy of the observation increases and therefore the information gained does too.

Thus, the expansion of the universe provides a source of net relative information gain. As FRAUTSCHI has put it,

> We have come to a conclusion which stands the closed 19th-century model on its head.
>
> Far from approaching equilibrium, the expanding universe as viewed as a succession of causal regions falls farther and farther behind achieving equilibrium. This gives ample scope for interesting nonequilibrium structures to develop out of initial chaos, as has occurred in nature.

3.10 Information Gain and Special Relativity

We have already observed that the theory of special relativity destroyed the classical concept of simultaneity. Neither space nor time has an absolute meaning from which it follows that separate measurements of space variables or of the time variable cannot yield results that are independent of the observer's relative motion. Distinct observers in constant relative motion will record different measurements of intervals of length and of time when they conduct identical experimental observations of a phenomenon.

What does this mean for the *information* yielded by an experimental observation? We shall see that the principle of special relativity is consistent with our previous demonstration that a single measurement of an intrinsically unbounded physical variable (such as length or duration) cannot yield information: information only results from the comparison of two measurements of a variable quantity.

Indeed, let O and O' denote observers who are in relative motion with constant velocity v in the direction of the x-axis of a Cartesian coordinate system, and let a physical length be measured by O and by O' by the same experimental means. If the measured results are l_1 and l'_1 respectively, then the theory of Special Relativity shows that

$$l'_1 = \frac{l_1}{\sqrt{1 - v^2/c^2}}$$

where c denotes the velocity of light.

Suppose that O and O' perform a second, more refined, measurement of the physical length, again using the same experimental means, with respective result l_2 and l'_2. According to the results of Chapter 2, the information gained by O will be $I = \log_2(l_1/l_2)$. Making use of the relationship between primed and unprimed quantities and coordinate systems, we find that $I = I'$: the gain in information that results from a pair of measurements of a physical length is independent of the relative state of unaccelerated motion of the observer.

Similarly, measurements t and t' of a time interval in frames of references that are moving with constant relative velocity v leads to the relation

$$t' = \frac{t}{\sqrt{1 - v^2/c^2}}$$

from which we conclude that the information gained by making a pair of measurements of a temporal interval is also independent of the relative state of unaccelerated motion of the observer.

More generally, if the space-time coordinate quadruples $q = (x, y, z, ict)$, resp. $q' = (x', y', z', ict')$ of a point are observed by O, resp. O', from their different frames of reference, then the vectors q and q' are related by a LORENTZ transformation whose matrix M satisfies $q' = Mq$ and $M^t M = 1$, where M^t

denotes the transpose of the matrix M and $\mathbf{1}$ denotes the unit matrix. The matrix equation implies that LORENTZ transformations preserve the so-called *space-time interval* $ds^2 = dx^2 + dy^2 + dz^2 - c^2\, dt^2$ as well as the space-time volume invariant derived from this metric form. Observations of these space-time invariants will be independent of the state of unaccelerated motion of the observer. But these quantities are composite and cannot be directly observed: their measurement is ultimately thrown back upon direct measurement of the fundamental physical variables of space and time.

The quantity of information gained by a measurement with respect to some prior measurement made under the same conditions of motion should be independent of the state of motion. We may adopt this as a fundamental principle of the theory of measurement. Our analysis of the special theory of relativity shows that it is consistent with this principle.

CHAPTER 4

Principles of Information-Processing Systems and Signal Detection

What we call reality consists of a few iron posts of observation between which we fill in by an elaborate papier-mâché construction of imagination and theory.

—John Archibald WHEELER

4.1 Purpose of This Chapter

The quotation which heads this chapter can be interpreted from at least three different points of view. First, that reality consists of a continuum of phenomena of which we observe only a few, and upon those few observations we build a conceptual theoretical structure which enables us to imagine those aspects of reality which have not been observed. This is a philosophical position adopted by many physicists. A second interpretation is less concerned with objective reality and more concerned with subjective, that is, with psychological, perception. Even when we do observe the physical world, our subjective impressions of that objective reality do not bear a one-to-one relationship to it: the cognitive interpretations of our sensory impressions differ from the latter in many ways (some of which will be discussed for the modality of vision in Chapter 5). Thus we may say that our subjective theoretical models of objective reality do not even agree in all cases with the "iron posts of observation"; these iron posts are often systematically bent to conform with the seemingly more flimsy papier-mâché constructions of imagination—with the illusion of reality. Yet it is surprising how well our conceptual models of reality, both the individual's personal subjective description and the theoretical description of science, appear to agree with observations and to accurately forecast what one would observe if the effort of observation were made. This leads to the third interpretation: that although objective phenomena vary through continua of values, they are already completely determined by their values for some discrete set of observational conditions. This implies that the iron posts of observation are in principle sufficient to specify the state of objective reality and deficient only in so far as real observers can make only finitely many observations. But this need not be a critical barrier if the state of a system is primarily determined by observations that are localized in space and time, i.e., observations which refer to a confined spatial region and to the present and recent past.

Although this interpretation may appear to constrain the potential variability of objective phenomena, it does correspond to circumstances that occur not infrequently and that are summed up by the so-called *sampling theorems* of signal detection that precisely specify the conditions under which a signal which varies continuously as a function of some parameter (such as time or position) will nevertheless be completely determined by values sampled at uniform intervals of the independent variable. The famous sampling theorem of SHANNON is of this sort: it states that if the signal is thought of as a superposition of periodic functions of various frequencies (as any sound can be considered as a combination of tones of various pitches) and if the range of frequencies is limited, then the signal can be exactly reconstructed from sample observations which are made sufficiently frequently; we will discuss this important result in section 4.5. There is some evidence that the signals received by the human sensory apparatus satisfy, or nearly satisfy, the conditions which assure that "a few iron posts of observation" will provide all that there is to know about the observed phenomena. For image processing and human vision these ideas have been considered by MARR and POGGIO and their coworkers in the context of the detection and identification of edges. We will return to this question in section 6.4. In the present chapter we will concentrate on general questions concerned with how information-bearing signals can be detected and analyzed, and on the types of structures which appear to be most efficient for information processing.

The word *signal* refers to many different kinds of communication techniques. A signal can be a quantity which varies with time: such a signal will possess a value for every time t within some prescribed range of times; the value of the signal at t may be denoted by $f(t)$. Sound patterns provide one example of signals of this kind, and amongst sound signals, acoustic speech waveforms provide an important particular example. Figure 4.1.1 displays an acoustic speech waveform which corresponds to an utterance of the linguistic signal "speakers." (Each column describes the distribution of sound energy in the speech waveform during a 12.8 millisecond interval and successive columns correspond to time intervals separated by 6.4 milliseconds; each column therefore overlaps 6.4 milliseconds of those on either side of it. The vertical scale measures frequency, ranging from 0 hertz to 5000 hertz. The

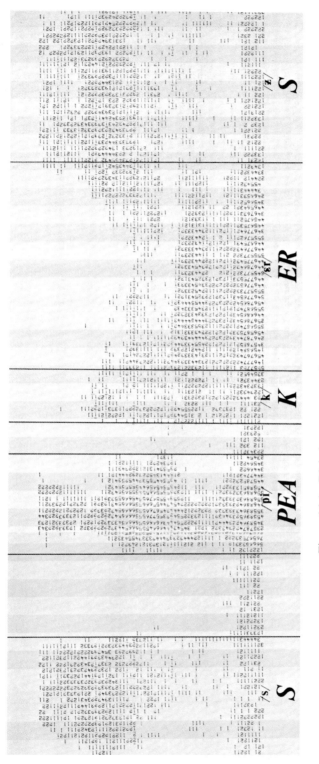

Figure 4.1.1 Digital spectrogram of an acoustic speech waveform.

numeric entries are proportional to the logarithm of the quantity of energy in the speech waveform which fell into the different frequency and time intervals). Or the signal may consist of a single image, which associates with each point in the Euclidean plane (or in some confined region of it, such as the visible surface of a television screen) a quantity. If we consider a "gray level" image, then to the point with (Cartesian) coordinates (x, y) there is associated a nonnegative number which measures the magnitude of the brightness of that image point or, expressed in terms of an objective physical quantity, it measures the intensity of illumination at that point. Sometimes measurements are recorded relative to the maximum intensity of illumination for the image under consideration and the corresponding numerical measure varies along a black-gray-white scale with 0 corresponding to black, meaning the absence of illumination, and the maximum numerical value corresponding to the maximum illumination. It is often convenient to choose the unit of measurement of grayness so that the maximum illumination—called "white"—corresponds to the measure 1. In this case the levels of gray will be measured by numbers between 0 and 1. In real situations only a finite number of distinct gray levels will be distinguishable. Typically, $256 = 2^8$ categories are adequate for displaying the variation of grayness in a given image (although the eye can sense a much greater range of variation) and this is convenient for computer implementations since the eight bits which constitute a computer byte of information suffice to distinguish precisely 256 categories. Thus the entire range of gray levels can be conveniently represented by a single byte, and the signal which describes such an image can be thought of as associating a byte with each distinct point of the image.

Signals which represent color images are slightly more complicated because each point of the image requires more than the measure of brightness (to use subjective terminology) to characterize it. Isaac NEWTON was one of the first to realize that colors can be specified by triples of numbers corresponding to measures of the subjective variables called hue, saturation, and brightness. The three-color theory of visual perception proposed by Thomas YOUNG implied that every subjective color could be represented by a superposition of three primary colors, and that the original color could therefore be identified with the triple of nonnegative numbers which specifies the quantity of each primary which enters into the superposition. His theory is realized by the familiar red-green-blue commercial color television signal. Although it must be suitably modified and augmented to provide an adequate theory of human color vision, it evidently does correspond to normal perception with a generally satisfactory degree of verisimilitude.

In any event, color image signals require the association of a triple of numbers with each image point. If the coordinates of the point are (x, y), then we may say that the color at (x, y) is described by the triple of numbers $(f_1(x, y), f_2(x, y), f_3(x, y))$.

A triple of numbers is often called a *3-dimensional vector*. Thus we conclude that a color image signal is described by associating a vector in 3-dimensional space with each point of the image; a color image signal is a vector-valued function of the two coordinate variables x and y.

Another kind of signal is the sequence of alphabetic characters which constitutes some natural language text, such as this sentence itself. In this case, the signal consists of a sequence of symbols drawn from the fixed inventory consisting of the letters of the alphabet (more precisely, from the characters available on the keyboard of a typewriter). Thus the signal can be represented by a sequence $S(1), S(2), \ldots, S(n), \ldots$ where each of the symbols $S(n)$ is drawn from the fixed inventory.

Information is transmitted from its source to its destination. Whatever means, or channel, is used for transmission, there will be a maximum rate of transmission, called the *capacity* of the channel, which is determined by the intrinsic properties of the channel but is independent of the content of the transmitted information or the way it is encoded. Channel capacity is measured in bits per second. The channel capacity of a telephone line is about 56,000 bits per second, whereas the channel capacity of neurons varies but is generally of the order of 1,000 bits per second. A channel having a capacity of about 150 million bits per second is required in order to transmit a color television signal in the United States; greater capacity is required for European and Russian television signals, which have somewhat higher resolution and carry greater information.

We can obtain a rough estimate of the channel capacity for the eye engaged in normal daylight vision. Under these circumstances the 6 million photosensitive pigments of the cones, which are responsible for color vision, are active. Restricting attention to them alone, the rate at which they receive information is approximately the product of the number of bits of information used to discriminate light intensity (about 8), the refresh rate required to maintain the illusion of continuous motion (about 50 images per second), and the number of receptor cones in the retina (about 6×10^6). The product, 2.4×10^9 bits per second, exceeds the channel capacity (*data transfer rate*) of even a large computer. Indeed, it exceeds the channel capacity to the higher cortical centers of the brain, which implies that much of the input information must be omitted by the higher processing centers due to this overload of information.

Studies have been made of the channel capacity of various animal sensory and neural subsystems, and some consideration has been given to how complex organisms or other systems can arrange to combine numbers of information transmitting components which have a low channel capacity, such as neurons, in a way that increases the overall ability of the system to process information. Since the channel capacity of the individual components of the system cannot be increased, this question really concerns the *structure of the system*: how a communication network should be organized to increase its capacity. This problem will be examined in the next section, with particular

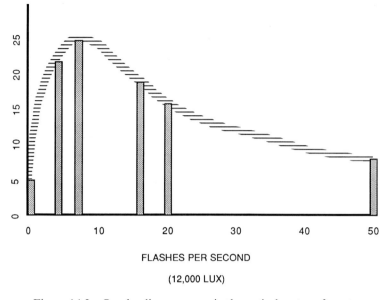

FLASHES PER SECOND

(12,000 LUX)

Figure 4.1.2   Overloading a neuron in the optical cortex of a cat.

emphasis on the properties of hierarchically organized information processing structures.

The problem of information overload is intimately related to the principle of selective omission of information: the former is the reason why the latter is necessary. As the rate of information which is inserted into a communication channel increases beyond the channel's capacity, only a portion of the input can be transmitted and the accuracy of transmission declines. In the case of biological information processing systems, the output of the channel typically falls below the channel capacity. This form of omission of information acts to reduce the overload on the system, but it is not selective and, consequently, it does not favorably affect the organism's ability to detect a meaningful signal embedded within a background of noise. Figures 4.1.2 to 4.1.4 display examples of this kind of omission at the cellular level. Similar phenomena occur at the aggregated level of complete systems. Figure 4.1.5 illustrates the decline in human typing performance as the rate of symbol input is increased for random sequences of equiprobable symbols. The channel capacity for this type of information transfer varies with the number of keys used, ranging from about 9.6 bits per second up to about 14.5 bits per second.

At the decision-making level similar problems of overload occur, but they can be partially ameliorated by the substitution of parallel processing for sequential processing. The benefits are suggested by CHRISTIE and LUCE's calculation of the theoretical channel capacities for processing information in these two ways, illustrated in Figure 4.1.6.

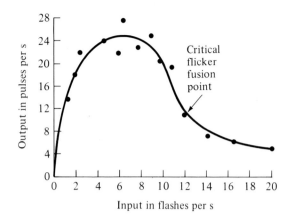

Figure 4.1.3   Overloading a single cell in the lateral geniculate of a cat. (From Bartley, *Archives of Opthamology*. Volume 60, pp. 784–785. Copyright 1958, American Medical Association. Reprinted with permission.)

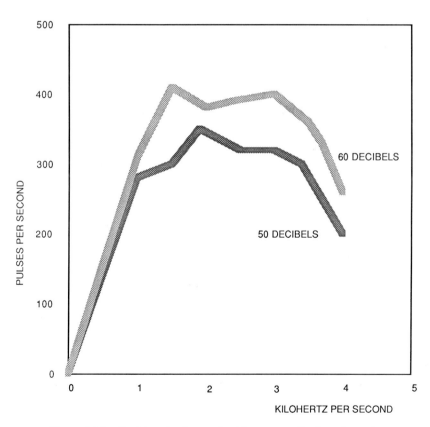

Figure 4.1.4   Cochlear nucleus output from one cell of a guinea pig.

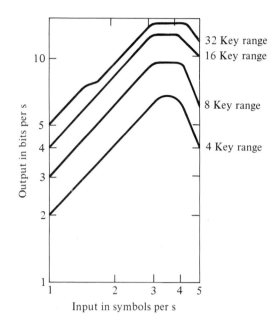

Figure 4.1.5   Typing performance curves.

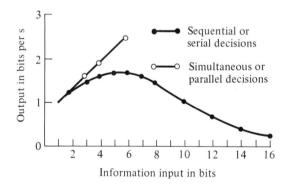

Figure 4.1.6   Sequential vs. parallel decision making channel capacity.

As information input rates rise, there are only two ways to cope with the increase: improve technology so that the channel capacity of the components of the information system is increased, or organize large numbers of components in a novel way so that the system is able, through parallel processing and interaction, to process information at a greater rate than one component can. The first approach is often used for electronic information processing

Figure 4.1.7   Ant and chip. (Photo courtesy of North American Philips Corporation.)

systems where advances in technology have often been able to substitute for organizational improvements in the architecture of the system. The replacement of coaxial copper telephone cable by fiber optic cable, and the successive replacement of the vacuum tube by the transistor, and the transistor by the integrated circuit, offer examples. But this method is not accessible to organisms. Biological nature must rest content with information processing components that cannot be rapidly changed and for which the range of potential improvement is quite limited. In these circumstances system design becomes the most effective way to improve information processing performance. These issues are strikingly illustrated by Figure 4.1.7, which carries other important messages as well.

These remarks highlight two of the three main topics considered in this chapter. The first concerns the organization of information processing systems to perform complex tasks. We will examine hierarchical systems in some detail because they appear to provide a particularly powerful and general way of achieving this objective. The second topic concerns the acquisition of signals and their reconstruction. Here the principal ideas come from communication engineering, and the method of FOURIER analysis plays a central role. We will discuss the various ways in which signals can be transmitted by *modulation*, that is, by varying the physical properties of some carrier, and how they can be reconstructed under appropriate conditions from sample observations. This discussion will form the basis for our consideration of edge detection and texture analysis algorithms in Chapter 6. Throughout all our discussions, the idea that natural information processing systems evolve with time so as to more closely approach a system that maximizes or minimizes a quantitative measure of information relative to some constraints which arise from the particular problem is a recurrent theme which lies just beneath the surface if it is not always explicitly stated. This is the third main topic of the chapter.

The *quantity* of information that is gained from an observation or calculation cannot be property considered without also considering the *cost* or *effort* required to perform the measurement or computation. The relationship between the gain in information and the cost to obtain it determines which information processing algorithms and implementations will be practical. This remark applies to information processing systems that have been designed by people for human needs but it also reflects a fully general feature of nature: physical systems are so organized that they cannot "communicate" and change their state unless, in the absence of external intervention, they preserve the time, energy, action, or other natural constraints of their initial configuration, and biological organisms, through the process of evolution, have taken the further step of selecting those information processing systems that are most "cost-effective" among the alternatives that are compatible with the natural constraints.

FERMAT's classical variational principle of "least time" and MAUPERTUIS and HAMILTON's principle of "least action" express this parsimony of nature in a mathematical form: the evolution of a physical system follows that path amongst all conceivable alternatives that *extremizes*, i.e., *maximizes* or *minimizes* a suitable *cost function*, such as *time, action, or energy*. Thus, the path of a ray of light through an optically inhomogeneous medium *minimizes* the *time* required to pass from the initial position to its emergent point.

The second law of thermodynamics is a dynamic version of this general physical principle. It can be interpreted as asserting that an isolated physical system will evolve toward an equilibrium configuration for which the entropy of the system is as large as possible. Once the system has reached a state of maximum entropy, it will remain in it, or in some state of equal entropy, thereafter. From the informational point of view, the system evolves to a state about which least is known *a priori*, and therefore about which a measurement will yield as much information as possible. In the evolution toward the state of minimum information it is essential that the system be physically isolated. This means, in particular, that the energy of the system does not change with the passage of time as the system evolves. Thus the various configurations of the system as it evolves toward the stable state of minimum information occur by repartitioning the fixed total energy of the system amongst its constituent parts. The available energy constrains the evolution of the system and limits its possible intermediate configurations as well as the final state.

The quantity of energy available to the physical system can be thought of as a kind of measure of the "cost" or "effort" that the system can "pay" to modify its state, for the greater the available energy, the greater will be the variety of states into which the system can be configured and the correspondingly greater will be the amount of information that could be gained from an observation.

It has been said that there is no 'free lunch"; even nature must pay for information, but her currency is *energy*.

Information systems created by people, including computing machines, also must pay for the information they produce. One system will be clearly better than another if the first yields more information per unit effort or cost than the second. "Cost" can be measured in units of time, or expended energy, or number of elementary operations, or some surrogate measure such as money. But in general, natural as well as manufactured information structures that have a stable organization or configuration are designed to extremize the amount of information per unit cost that they deliver, and isolated natural dynamically varying structures evolve toward that stable state.

The study of the organization of information processing structures would be misleading as well as incomplete without an examination of the relationship between the cost of information and the amount of information a system yields. One of the grave defects of contemporary economic theory is its inability to account for either the commodity aspects of information or its role in economic decision making in an environment in which each is capable of influencing the destiny of national economies in significant ways. But in other disciplines the picture is clearer.

In section 4.2 we consider hierarchically organized information structures. Implicit throughout the discussion is the constraint of the cost of processing the information measured by the number of operations performed or, if all operations are idealized so that they are peformed serially and are of equal duration, by the *time* that elapses until completion of the processing. Thus the *parameter values* that select one particular hierarchical information processing structure from amongst a parameterized family of structures are determined by the condition that the selected system maximize the quantity of information per unit time.

In section 4.3 we take up the more general situation of a statistical ensemble of systems that consists of a number, possibly infinite, of discrete states each of which occurs with some probability. The SHANNON measure of the average information gained from an observation can be constrained by some cost function. We show how the constraint determines the probability distribution and express it in terms of the cost function for systems that extremize information. We consider applications to the transmission of information by systems such as telegraph codes and natural languages (including the so-called ZIPF's law); to statistical mechanics; to quantum mechanics; and to the distribution of the size of cities.

4.2 Hierarchical Organization of Information-Processing Systems

Divide and conquer.

Although the variations in human physical and mental abilities may seem to be quite considerable, an impartial observer would probably judge their general comparability more remarkable than their range of differences. Apart from the exceptional intellects, athletes, and other talents, we all perform in roughly similar ways and with roughly similar degrees of effectiveness. The tasks individuals and societies face are not, however, roughly comparable in their complexity. This raises the question of how it is possible to organize the activity of individuals or other "components" of complex systems which are all approximately equivalent in their capabilities so that the combination will perform many times more effectively than its constituent parts can perform independently. Over the course of thousands of years, social structures have been developed which, however imperfect they may otherwise be, have made it possible to bring enormously complex projects to a successful conclusion within a human lifespan. We will see that the organizational relationships which societies have evolved to make this possible are but special instances of much more general and far-reaching principles for organizing information-bearing structures.

The problems that beset the leaders of governments or other large organizations provide an interesting example. No one can doubt that the president of the United States faces difficulties that are many times more numerous and more complex than those which confront the ordinary citizen. Yet problems and responsibilities fill a large part of each of our lives, which implies that a president cannot hope to perform an adequate job merely by working longer hours or performing more efficiently than the average person, for the scope of variation of individual performance is far too limited for such methods of attack to make a significant difference. One reads with amusement news reports which compare the number of hours one president spends at his desk and the number of memoranda he reads with his predecessor's corresponding statistics. Compared with the number of memoranda available to be read or the hours that would be required for a direct personal attack on issues such as preserving domestic tranquility, resolving economic problems, allocating national resources, and assuring general improvement in the health, education, and well-being of the citizenry, the fluctuations of effort from one president to another are irrelevant.

What then enables a society to undertake complex tasks, such as placing a footprint on the moon and bringing home the foot that placed it there?

Every collection of people that has been charged with the completion of a complex task has evolved a largely *hierarchical structure* to organize the activities of its members. It is no accident that governments, military organizations, and large business enterprises all have a pyramidal, or hierarchical, structure, for analysis shows that such organizational structures are greatly superior in effectiveness to forms of organization in which individuals bear a more nearly equal relationship to one another in the performance of tasks which serve the goals of the enterprise. Here of course we are restricting our attention to just those aspects of the individual's activities which involve the complex task initially prescribed. A hierarchical structure may have undesirable "side effects" in other areas of personal life, but it is not our intention to explore the desirability of one or another form of social structure here; rather, we merely seek to identify the general information-processing principle which has led to the prevalence of hierarchical structures and to explain why they are effective.

A concrete example will help to make plain the technical issue. Suppose that we wish to create a telephone system which will enable each telephone to communicate with every other one. Suppose that there are a total of N telephones. For the United States at the present time, N is approximately equal to $10^8 = 100$ million. How shall these telephones be connected to one another? The communications engineer will ask what *topology* the *telephone network* should possess.

We might try to connect each telephone directly to every other one. N telephones will require $N(N - 1)/2$ connections. For $N = 10^8$ telephones, there would be about 5×10^{15} connections, so many as to be far beyond our capacity to construct or afford them. This network configuration can be represented by N points in general position joined by straight line segments in every possible way. The vertices correspond to the telephones; the line segments to direct connections between them. Figure 4.2.1 displays several such networks for small values of N.

At the opposite extreme, we could connect each telephone to a single central office which would switch a call from its source to the proper destination. For N telephones this would require only N connections—the least possible number if each telephone is able to access all others—but the switching load on the central office would be enormous, potentially resulting in long waiting times, and the total amount of connecting wire required (if wires were used to connect telephones to the central office) would be very large if the telephones

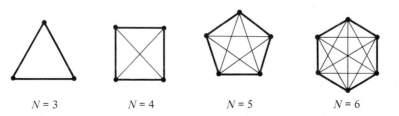

$N = 3$ $N = 4$ $N = 5$ $N = 6$

Figure 4.2.1   Totally connected network topology.

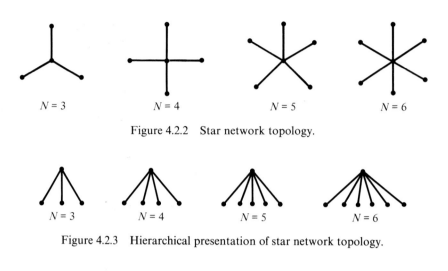

Figure 4.2.2   Star network topology.

Figure 4.2.3   Hierarchical presentation of star network topology.

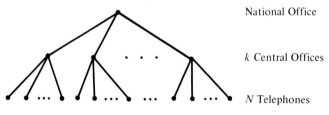

Figure 4.2.4   3-Level hierarchical network topology.

were widely distributed geographically. This network configuration can be diagrammed as a *star* with each of N points directly connected to an additional distinguished point which corresponds to the central office. Figure 4.2.2 displays some star networks for small values of N. The information represented by this diagram can be organized in another way which we will find more useful. If the "central office" is denoted by a node at the top of a diagram and the N telephones are denoted by nodes arranged below it in a row, then the star networks of Figure 4.2.2 appear as shown in Figure 4.2.3. This presentation emphasizes the hierarchical arrangement of the network and shows that it consists of 2 "levels."

Suppose that k central offices each serve N/k telephones directly, and that the k central offices are themselves connected to a single "national" office. This network configuration is diagrammed in Figure 4.2.4. We will refer to it as a 3-level hierarchical network topology.

This arrangement of the network has a total of $(N + k)$ connections. The central offices are chosen so that each one switches calls directly for telephones to which it is connected, but if communication between telephones connected to different central offices is desired, the request is routed through the national office. If we agree to use a network like this, we can still inquire how many central offices would be needed in order to minimize the waiting time. That is, what is the optimal value for k? Starting from a given telephone, the waiting time required to obtain service from its central office will be proportional to the number of telephones the central office serves, namely N/k. Similarly, the time required to obtain service from the national office will vary with the number k of central offices it serves. For simplicity let us assume that the constants of proportionality (which correspond to the rates at which the offices process information) are the same; if not, the general form of the result remains valid but the numerical aspects will be different. Then the total time required will be proportional to $k + (N/k)$. Straightforward analysis shows that this will be a minimum when $k = \sqrt{N}$, so the average total time required to communicate will be roughly proportional to $2\sqrt{N}$, whereas the star network shown in Figure 4.2.2 requires a time proportional to N in order to serve N telephones. For $N = 10^8$, a 3-level hierarchy reduces the waiting time by a factor of $\frac{1}{2}\sqrt{N} = 5{,}000$ compared with the star network, which can be thought of as a 2-level hierarchy.

The advantages of the additional level of "central offices" in the 3-level hierarchy raises the questions of whether the interposition of additional levels would further improve performance, and if so, how the optimum number of levels could be determined. In fact, there is an optimum number of levels. This problem will be solved in section 4.6.

Hierarchical structures occur throughout civilization. As Herbert SIMON has pointed out, complex manufactured items consist of a collection of systems which are themselves composed of subsystems, asemblies, sub-assemblies and so forth, and that it is this ordered decomposition into a hierarchy of simpler parts that makes it possible to manufacture the whole. An electrical power distribution system for a large city, the space shuttle system, and an ordinary automobile are examples. By repetition of the process of partitioning the system into disjoint and independent sub-units, a level of simplicity is finally reached for which individual expertise suffices to design and produce these elementary components. Each of the other levels involves design of an interface and interconnection of units belonging to the next simpler level. If the hierarchy is properly arranged so that the difference in complexity from one level to the next remains about the same and well within the span of individual human competence, then the entire complex system can be successfully constructed. The same is true of complex decision-making problems. By recursively decomposing them into simpler constituents such that the difference in complexity from one level to the next is kept within the limited range of human capabilities, very complex decisions can correctly be made by teams of people organized into a hierarchical structure.

The advantages of hierarchical organization for the performance of complex tasks extend beyond the artificial organizational structures and manufactured objects created by people to natural biological and physical systems as well. James G. MILLER has stressed the hierarchical and "systems" nature of organisms in his interesting book, *Living Systems*. At the most fundamental level of life, the complex chemical combinations which carry the genetic code could hardly, in the limited time since the cooling of the earth, have come into direct existence from an atomic "soup" wherein atoms of each type found their proper location in what would ultimately become the DNA molecule as the result of random thermal motion. Rather, DNA consists of nucleotide building blocks each of which is in turn constructed from simpler standardized chemical combinations. These combinations have a much higher probability of coming together in the proper circumstances and in the time available for evolution to form a DNA molecule as a succession of hierarchically ordered chemical events. Thus, we may think of our own existence as a consequence of the efficiency of hierarchical information processing structures.

It will be worthwhile to recall some of the details of how the code which carries genetic information is constructed so that we can estimate how much information it carries. The information in deoxyribonucleic acid (DNA) is coded in terms of ordered triples of four chemical substances called *nucleotide bases*. The ordered triples are called *codons*. The bases, and a shorthand abbreviation for each, are

A: Adenine $\Big\}$ Purines
G: Guanine

C: Cytosine $\Big\}$ Pyrimidines
T: Thymine

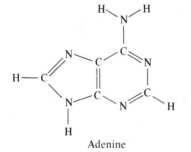

Adenine

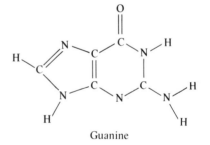

Guanine

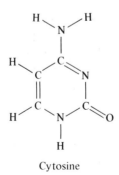

Cytosine

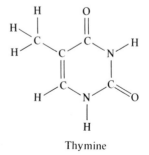

Thymine

Figure 4.2.5   Nucleotide bases for genetic code.

Each of these compounds is a relatively small organic molecule; their chemical structural formulae are displayed in Figure 4.2.5. Ordered triples of nucleotide bases correspond to the 20 primary amino acids which are the building blocks for proteins, and the purpose of the DNA code is to guide the synthesis of proteins by prescribing the appropriate sequence of amino acids. Since there are 64 possible permutations of the 4 bases taken 3 at a time, it follows that a given amino acid may correspond to more than one codon. The complete correspondence is displayed in Table 4.2.1.

According to George BEADLE, there are about 5 billion nucleotide bases in the genetic material of a human cell. The volume occupied by this material is

Table 4.2.1 The Codon to Amino Acid
Correspondence of the DNA Code

Codon	Amino Acid	Codon	Amino Acid
AAA	lysine	CAA	glutamine
AAG	lysine	CAG	glutamine
AAC	asparagine	CAC	histidine
AAU	asparagine	CAU	histidine
AGA	arginine	CGA	arginine
AGG	arginine	CGG	arginine
AGC	serine	CGC	arginine
AGU	serine	CGU	arginine
ACA	threonine	CCA	proline
ACG	threonine	CCG	proline
ACC	threonine	CCC	proline
ACU	threonine	CCU	proline
AUA	isoleucine	CUA	leucine
AUG	methionine	CUG	leucine
AUC	isoleucine	CUC	leucine
AUU	isoleucine	CUU	leucine
GAA	glutamic acid	UAA	gap (comma)
GAG	glutamic acid	UAG	gap (comma)
GAC	aspartic acid	UAC	tyrosine
GAU	aspartic acid	UAU	tyrosine
GGA	glycine	UGA	tryptophane
GGG	glycine	UGG	tryptophane
GGC	glycine	UGC	cysteine
GGU	glycine	UGU	cysteine
GCA	alanine	UCA	serine
GCG	alanine	UCG	serine
GCC	alanine	UCC	serine
GCU	alanine	UCU	serine
GUA	valine	UUA	leucine
GUG	valine	UUG	leucine
GUC	valine	UUC	phenylalanine
GUU	valine	UUU	phenylalanine

very small: were the bases strung end to end, they would form a strand about five feet long, but because the diameter of the molecules is very small, the strand would be invisible to the unaided human eye and only barely visible with the aid of a modern microscope. The 5 billion bases correspond to about 1.7 billion codons. In order to gain some intuitive understanding of the amount of information that can be stored in this long sequence of codons, we may compare it to a long sequence of alphabetic characters forming some linguistic message. Since $64 = 2^6$, each codon corresponds to 6 bits of information. When considered as a sequence of letters of the alphabet, a typical book contains about 5 million bits; hence, the information content of DNA is roughly comparable (excluding comparisons of redundancies in both the genetic and in the linguistic codes) to the information content of 2,000 books.

The density of information storage in the DNA molecule far surpasses anything that contemporary information technology can do, for although one videodisk can hold 10 billion bits—the blueprint for a human life—the 5 foot long DNA strand is normally folded upon itself in such a compact way that it fits into the nucleus of a cell. As BEADLE points out, if it "were wound back and forth, one layer thick, on the head of a pin, it would cover only about one-half of one percent of the head of the pin." Thus, 50 angels can dance on the head of a pin.

At a still more fundamental level, the structure of matter itself is hierarchically organized. We speak of the phase of a material substance (whether it is in a solid or fluid state), of chemical combinations, of atoms, of atomic nuclei, and of elementary particles. These conventional distinctions describe a hierarchy whose levels are distinguished by the amount of energy which is needed to break the corresponding structure apart into its constituents of the next level of "simplicity."

The thermal energy associated with the motion of an atom or molecule is about kT, where k is BOLTZMANN's constant to which we have previously referred and T denotes the absolute temperature. For ordinary temperatures this is about 0.04 eV, where eV stands for *electron-volts*; 1 eV $= 10^{-12}$ erg, a very small amount of energy but a convenient unit for the comparisons we wish to make. The thermal background energy of 0.04 eV can be thought of as typical of the energy that a constituent atom or molecule of some macroscopic object will pick up or lose at each encounter as the result of the incessant random collisions it undergoes at ordinary temperatures (for substances such as water, whether it is in a solid, liquid, or gaseous state will depend on the ambient level of thermal energy; as the latter varies from about 0.037 to 0.043 eV, the state of water varies from ice to steam). The stability of chemical combinations and of atoms themselves demonstrate that much more energy must be necessary if the bonds which hold them together are to be disrupted. Indeed, molecular binding energies lie between about 0.25 and 5 eV: on the average, about 30 times as great as the thermal background disturbances. But the binding forces between the molecules of liquids like water are so small that the thermal energy overcomes them above their "freezing" point.

Atoms are bound together in chemical combination by sharing one or more of their electrons. Since atoms generally have a greater degree of permanence than their chemical combinations, one should suspect that electrons are bound more tightly by the nucleus of their parent atom than by the chemical binding force which shares electrons amongst several atoms, and indeed this is generally true. The binding energy of an electron, called its *ionization potential*, varies from 3 to 25 eV, on the average perhaps 5 times as great as chemical binding forces. These quantities of energy are the ones which are normally available per molecule at the human scale of events, and they govern all of the biochemical processes which support life as well as the chemical processes which lie at the foundation of industrial civilization.

When atoms are stripped of their electronic sheath, the atomic nucleus remains. The energy necessary to disrupt the nucleus of an atom ranges from 2 to 20 million electron-volts (i.e., 2–20 Mev). This tremendous jump from the level of chemical structure to the level of nuclear structure is the reason that nuclear phenomena do not play a direct role in human affairs in ordinary circumstances, and why they cause so much trouble when they do. The drama of ordinary life is played on the chemical level of the hierarchy of physical structures, not on the nuclear level.

Nuclear particles such as protons and neutrons are not the ultimate constituents of matter, but their binding energies are beyond the range accessible to current experimental equipment, which can produce energies as high as 4×10^{11} eV.

The hierarchy of material structures is diagrammed in Figure 4.2.6. In this figure, a physical structure corresponding to one node in the diagram is composed of constituents which belong to the next lower level; the links are intended to designate the corresponding constituents. The indicated energies decompose, or "split" the structure on their corresponding level into constituents belonging to the next lower level.

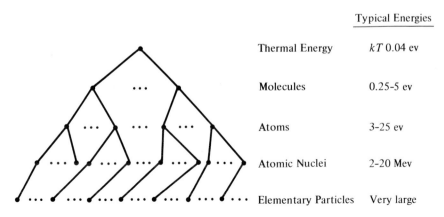

	Typical Energies
Thermal Energy	kT 0.04 ev
Molecules	0.25–5 ev
Atoms	3–25 ev
Atomic Nuclei	2–20 Mev
Elementary Particles	Very large

Figure 4.2.6   Hierarchical structure of matter.

There is no need to dwell upon the prevalence of hierarchically structured systems. Language itself is hierarchically organized in terms of phrase grammar structures. The sentence of English decomposes into a noun phrase and a verb phrase, each of which is further decomposable until the ultimate phonemic or graphemic primitives are reached. Military commands are organized into squads, platoons, divisions, corps, and armies. Automobile roadways are hierarchically organized according to the number of their lanes, varying from narrow one lane alleys to eight and ten lane interstate superhighways with median separators and breakdown shoulders. Airline route structures consist of hierarchical "spoke and hub" arrangements and the larger airports themselves reinforce and extend the hierarchy with their collections of separate terminals connected by ground transportation links. In all these examples, the driving force which motivates the design of the network is efficiency. And in all these examples the main purpose of the organizational design is to increase the efficiency of information transfer.

Earlier we mentioned that a videodisk, which can store 30 minutes of a color television program, holds about as much information as a strand of human DNA. Videodisk and magnetic disk storage architectures are themselves examples of the use of hierarchical organization in order to provide rapid random access to their contents. In both cases, the surface of the disk is organized, in effect, as a series of concentric circular annuli. Information is recorded, either magnetically or optically as the case may be, along the annular tracks. A sensing element can move from track to track along a radius of the disk. By simultaneously spinning the disk about its axis and moving the sensing element along a radius, any storage location on the disk can be brought into position under the sensing element so that its contents can be read.

If there are k tracks on the disk, and if each track contains the same number of bits of stored information, say N/k bits per track, then the disk will hold a total of N bits. Current versions of the videodisk will hold $N = 10^{10}$ bits distributed amongst 54,000 tracks, and rotate at 30 revolutions per second. The process of searching for the contents of a particular location on the disk is organized as a 3-level hierarchy. The topmost level represents the disk itself; the next level consists of 54,000 nodes representing the tracks, each of which is connected to the topmost "disk" node; and the bottom level consists of 10^{10} nodes, each of which represents a storage location on disk, with $10^{10}/54,000 = 185,000$ of these joined to the node which represents the track on which the storage locations are located. The order of nodes at each level reflects the order of the tracks and storage locations on the disk.

The process of reading the contents of a disk location proceeds by moving the sensing element until it is positioned above the track containing the storage location and then waiting until the rotation of the disk brings the desired location into position under the sensing element. If the sensing element moves at a rate of a tracks per second and the rate of rotation of the disk covers b storage locations per second, then, for a disk which has a total of N

storage locations equally distributed amongst k tracks, the average time required to position a particular location so that it can be read will be

$$t_{av} = \frac{1}{2}\left(\frac{k}{a} + \frac{1}{b}\frac{N}{k}\right); \tag{4.1}$$

the factor $1/2$ appears because it will be necessary, on the average, to search through half the tracks, and half the locations on the selected track, before the desired storage location is found. We have previously considered the case of a 3-level hierarchy where the "search rates" are equal ($a = b$); in that case the average search time is minimized if the number of tracks k is related to the total quantity of information stored N by the equation $k^2 = N$. In the present case, the search rates may be different but a similar analysis (which will be carried through in a more general context in section 4.6) shows that the average search time is minimized if

$$k = \sqrt{(a/b)N}. \tag{4.2}$$

For videodisks available in 1984, $N = 10^{10}$, $k = 54,000$, and $b = (N/k)/30 = 617$ locations per second. If the engineering parameters of the disk were selected so that for the given search rates a and b the choice of the number k of tracks is such as to minimize average search time, then it would follow that a satisfied (4.2); substitution of that value of a into (4.1) will yield the minimum average search time corresponding to systems with search rates a and b. Carrying out the calculations yields $a = 1800$ tracks per second and $t_{av} = 30$ seconds.

The advantages of hierarchical structures for information processing systems have deep roots. They are related to the way in which numbers can be represented and to how the fundamental arithmetic operations can be performed.

Let us examine how hierarchical structures are employed in elementary arithmetic to simplify it, thereby making accessible to every educated person, a capability that otherwise, because of its complexity, would be limited only to specialists.

A primitive and ancient way to denote a positive whole number N is to arrange N similar strokes, as shown in Figure 4.2.7; we may refer to this as a "counting" notation. Counting notations take up a large amount of space; the number 10^6 would require about 300 pages for its expression if each stroke occupied the space of one typed character, whereas the base 10 positional notation for 10^6 consists of just seven symbols: 1,000,000. If two numbers M

$$N = 3 \qquad N = 6 \qquad N = 10$$

Figure 4.2.7   "Counting" notation for numbers.

and N were represented by strokes, their product would consist of MN strokes and the effort to "count" the product in order to determine its value would be proportional to the number of strokes, that is, to MN.

One of the greatest intellectual advances in human history was the Babylonian invention of the more efficient positional notation for numbers more than 4,000 years ago. Positional notation is itself a hierarchically organized structure. Suppose we wish to express the whole number N in positional notation relative to base B. If we write $N = (n_k n_{k-1} \ldots n_1 n_0)_B$, we must determine the base B digits $n_0, \ldots n_k$. It will be convenient to think of the familiar case $B = 10$ in what follows. The first step is to determine the largest integer k such that $B^k \leq N$; this will specify how many digits will occur in the base B notation for the number N, and it also tells us how many levels will appear in the hierarchy. The next step is to determine the leading digit n_k of the base B representation of N. n_k will be one of the digits 0, 1, ..., $B - 1$. It is the largest integer such that $n_k B^k \leq N$. The next digit n_{k-1} is the largest integer such that

$$n_{k-1} B^{k-1} \leq (N - n_k B^k),$$

and so forth; the details are of no importance for our interests. These steps are illustrated in Figure 4.2.8 which shows that the process of finding the positional notation expression for a number corresponds to selecting a path through the hierarchical structure which corresponds to the particular base selected.

There is a tremendous gain in computational efficiency (which generally goes unnoticed by most people who use arithmetic, although it is what makes its use possible) when two numbers expressed in positional notation are multiplied. Let M and N be expressed in base B (B might be 10, corresponding to the usual decimal notation, or 2, corresponding to binary expansions). That is,

$$M = (m_k \ldots m_1 m_0)_B,$$

$$N = (n_1 \ldots n_1 n_0)_B,$$

which is the positional notation for the expressions

$$M = m_k B^k + \cdots + m_1 B + m_0,$$

$$N = n_l B^l + \cdots + n_l B + n_0.$$

The product

$$M \times N = m_k \ldots m_1 m_0$$

$$n_k \ldots n_1 \; n_0$$

is calculated by multiplying the single digits and adding the resulting columns according to the usual familiar prescription for $B = 10$. This will involve a total of $(k + 1)(l + 1)$ elementary (i.e., one-digit) multiplications followed by

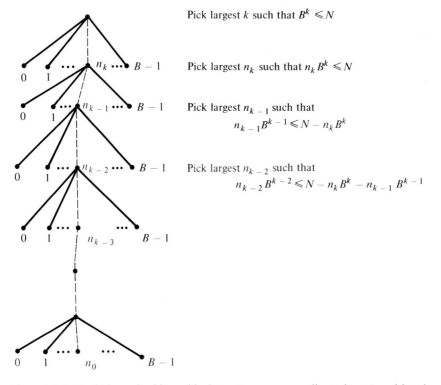

Pick largest k such that $B^k \leqslant N$

Pick largest n_k such that $n_k B^k \leqslant N$

Pick largest n_{k-1} such that
$$n_{k-1} B^{k-1} \leqslant N - n_k B^k$$

Pick largest n_{k-2} such that
$$n_{k-2} B^{k-2} \leqslant N - n_k B^k - n_{k-1} B^{k-1}$$

Figure 4.2.8　Path through a hierarchical structure corresponding to base B positional notation.

about the same number of additions to calculate the product MN. Thus the effort will be roughly proportional to $(k + 1)(l + 1)$, which is very nearly the same as $(\log_B M)(\log_B N)$. This is much smaller than MN, and the improvement becomes more striking the larger are the factors M and N.

The traditional method for multiplying numbers is so old and familiar that it came as a shock when, in 1963, the mathematicians KARATSUBA and TOOM independently discovered a still more efficient way to multiply large numbers which also depends upon an application of hierarchical organization to simplify the calculation. Their method is characteristic of the way in which complex problems can often be partitioned, subdivided, or re-organized to obtain a larger number of interrelated and potentially simpler problems which yields an information processing advantage. It is this powerful and general principle of dividing to conquer upon which attention should be focused, rather than on the particular application to the mathematics of multiplication. But since this application is very simple, and it provides a good illustration of the main point, we will briefly describe it.

Suppose that the numbers M and N are expressed in base B positional notation. Let each of them consist of n digits. If n is even, say $n = 2m$, then we can write

$$M = x_0 + x_1 B^m,$$

$$N = y_0 + y_1 B^m,$$

where x_0, x_1, y_0, y_1, are m-digit numbers in base B notation. Then the product is

$$MN = x_0 y_0 + (x_0 y_1 + x_1 y_0)B^m + x_0 y_0 B^{2m}; \qquad (4.3)$$

it appears to require 4 multiplications of m-digit numbers, for which the effort required will be proportional to $4(m^2) = (2m)^2 = n^2$ if each of these smaller products is calculated by the traditional method. But following KARATSUBA and TOOM, we can rewrite the term $x_0 y_1 + x_1 y_0$ in (4.3) as

$$x_0 y_1 + x_1 y_0 = x_0 y_0 + x_1 y_1 - (x_1 - x_0)(y_1 - y_0),$$

which requires only 3 multiplications, so the effort will be proportional to $3(m^2) = \frac{3}{4}(2m)^2 = \frac{3}{4}n^2$, which is only 75% the effort required by the traditional method.

We can introduce a hierarchical decomposition by repeating this process (at least if m is even; but it is easy to show that whether n, or m, is even really doesn't matter). The final result is that two n-digit numbers can be multiplied with an effort proportional to $n^{\log_2 3}$ rather than n^2 required by the ordinary method. This makes a great difference because $\log_2 3 = 1.58\ldots$. For instance, the effort required to multiply two 100-digit numbers by the traditional method is proportional to $n^2 = 100^2 = 10{,}000$, whereas for the new method it is only $100^{\log_2 3} = 1475$; for 1,000-digit numbers the corresponding efforts are 1,000,000 and 18,957, respectively. It is conceivable that calculating prodigies are "wired" to perform multiplication by using this more efficient method in place of the traditional one. Perhaps certain natural computations performed by the neural net as it processes sensory inputs and other information utilize this and other exceptionally efficient hierarchically organized procedures in order to compensate for the inherent slow speed and unreliability of biological information-processing components.

It is worthwhile to note that these ideas can be carried substantially further. By decomposing the multiplicands into more than two parts, it is possible to reduce the computational cost of multiplication of two n digit numbers so that it is $O(n^{1+\varepsilon})$ for any positive ε. The coefficient of the order of growth increases as ε decreases. This asymptotic result has little meaning for n of practical sizes. The interested reader should consult KNUTH for more detailed information.

The use of hierarchical organization to increase efficiency is intimately related to the notion of *parallel processing* of information. Whereas most contemporary computers operate in a largely sequential mode wherein elementary tasks follow one another in the control stream with data called up from storage as the need for it arises, it has been recognized that substantial future increases in processing speed will become more dependent upon processing information concurrently—in parallel—rather than sequentially. The higher biological organisms rely on intensive use of parallel processing to compensate for the low data transfer rate of neurons. Little is known about how organisms manage this activity, nor is the mathematical theory of parallelism in computation and other types of information processing yet out of its infancy.

The idea of parallel processing is based on the successive subdivision of a problem into simpler constituents which can be solved independently and simultaneously. Each subproblem may admit further subdivision. This process replaces the original task by a hierarchically organized collection of interrelated subproblems. In principle, those which lie at the lowest level, i.e., at the end of the subdivision process, will be simpler than those which lie above them in the hierarchy, and those which belong to a given level can be processed simultaneously; when combined, their solutions provide the solution to the original problem.

For some simple and frequently occurring problems, such as the multiplication of numbers, a decomposition into subproblems is possible. Indeed, both the traditional and the newly discovered methods for multiplication make it possible to process the digits of the factors in parallel, which thereby still further reduces the time that would be required by a sequential method. Other special problems can be partitioned into subproblems in a natural way, which makes them good candidates for parallel processing techniques. Calculations which depend on vector analysis often have this character because many vectorial operations can be performed independently on the components of the vectors. But it is not yet known to what extent general information processing tasks can be cast into a parallel, or hierarchical, form.

Among the uses of parallel processing by organisms, its role in human vision stands out. There are several stages of processing between the initial detection of incident light by the retina and the ultimate cognitive response in the higher cortical centers where parallel processing of information plays a crucial role. The processes of lateral inhibition and excitation, which aggregate sensed data from neighboring retinal receptors, are perhaps the best understood. We will return to this topic in Chapter 5.

It is evident that hierarchical decomposition is a powerful tool for organizing information processing systems. Since biological organisms in effect perform certain "natural computations" when they process sensory data, it will not be surprising to find that evolution has arranged their information processing structures so that they can take advantage of the efficiencies offered by hierarchical organization.

Hierarchical structures have been freely discussed throughout this section although they have not yet been precisely defined. We will fill this gap now and, as a consequence, we will be able to state a general design property for certain multilevel hierarchically organized systems which process information in an optimally efficient way.

A hierarchical structure is often called a *tree* structure, especially in computer science. A *graph* consists of a finite set of points, called *nodes*, and a set of line segments, called *edges*, which join some pairs of nodes. A tree is an acyclic graph. The nodes can be thought of as corresponding to decision states for the system which the tree is intended to describe; an edge which joins two nodes indicates that the corresponding decision states are connected in the sense that one transmits information to the other or, more generally, that one precedes the other in the sense that the former decision state must act in order that the latter have the opportunity to act. When joined nodes are ordered as to precedence in this way, we may think of the edge which connects them as having a direction, which may be indicated by an arrowhead, to specify the pecedence ordering. A tree whose nodes are joined by directed edge is called a *directed tree*. Most of the examples discussed earlier in this section were of directed trees. When the directions are obvious, it will be convenient to let them remain implicit and not clutter the tree diagram with arrows.

The nodes of a tree are organized in *levels*, which are labeled from 0 to the maximum level number in a particular tree, say *l*. There is only one node belonging to level 0; it is called the *root* of the tree.

Each node other than the root of the tree is joined to precisely one node belonging to the previous level; the latter node is called the *parent* of the former one, and the former is a *child* of the latter. A node other than the root has exactly one parent, but each node may have many children.

A node which does not have any children is called a *leaf* of the tree. All of the nodes belonging to the greatest level are leaves, although nodes belonging to other levels may also be leaves.

For some perverse reason, it has become conventional to draw trees with their roots located at the top of the diagram and their leaves (belonging to the maximal level) at the bottom. Figure 4.2.9 illustrates the definitions.

The nodes belonging to a given level can be *ordered* (say from left to right). We say that the tree is ordered if all of its levels are ordered, and if the ordering on successive levels is compatible; that is, if *m* and *m'* are nodes belonging to

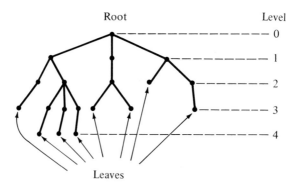

Figure 4.2.9   A tree with 5 levels.

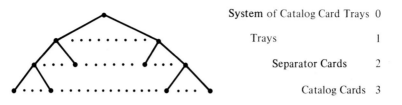

System of Catalog Card Trays 0

Trays   1

Separator Cards   2

Catalog Cards   3

Figure 4.2.10   Library catalog card system tree.

the same level and m is prior to m' in the ordering (we can write this as $m < m'$), and if n is a child of m, n' is a child of m', then n is prior to n' in the ordering on their level.

The number of nodes that belong to level k is denoted $N(k)$. Thus, $N(0) = 1$ because the sole node belonging to level 0 is the root of the tree.

As we have already seen, hierarchical structures, i.e., trees, correspond to information processing structures such as the retrieval structure of a magnetic or videodisk computer memory. A library card catalog system provides another helpful illustration. In this case, the root of the tree corresponds to the collection of card catalog trays; each tray corresponds to a node which belongs to level 1; within a tray, the "3 × 5" "tabbed" separator cards which group the catalog cards into various categories are the nodes of level 2 which are the children of the node corresponding to their particular tray; and the catalog cards which are grouped behind a tabbed separator card correspond to the nodes of level 3 which are also the leaves of the tree (Figure 4.2.10). This tree is evidently ordered alphabetically and it is also directed.

The process of locating or retrieving information from a magnetic disk storage device or from a library card catalog corresponds to a process of *sequential search* through the tree. A sequential search through a tree consists of an ordered sequence of nodes which begins with the root and ends at a leaf. This ordered sequence determines a polygonal path. The path proceeds from the root to the first node of level 1, and thence along the level 1 nodes until some desired level 1 node $n(1)$ is reached. Then the search continues by proceeding through those level 2 nodes which are the children of $n(1)$, beginning with the first of these (in the ordering of the level) until the desired level 2 node $n(2)$ is reached. This search process is continued indefinitely until a leaf is reached; this is the sought for data object.

A sequential search is illustrated by the dotted line in Figure 4.2.11 which proceeds from the root, labeled a, to the data object b, which is a leaf of this tree.

Suppose that each node belonging to a given level has the same number of children. How many nodes should each level contain in order that the average time required to complete a sequential search is as small as possible? This is a generalized version of the problem of retrieving information from a magnetic disk storage device that we considered earlier in this section. In order to be

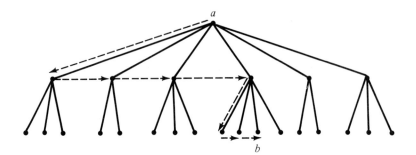

Figure 4.2.11   Sequential search.

able to state the solution to this information system design problem, let us suppose that the rate of searching data objects represented by the level k nodes is denoted by v_k. Then the time required to search $N(k)$ nodes belonging to level k will be $N(k)/v_k$.

We will show that the average search time through such a tree will be minimized if the number of nodes $N(k)$ which belong to level k is related to the number $N = N(l)$ of leaves—which is just the number of data objects on the magnetic disk or in the file—by the formula

$$N(k) = (v_1 \ldots v_k)\left(\frac{N}{v_1 \ldots v_l}\right)^{k/l}, \quad k = 1, \ldots, l. \tag{4.4}$$

For example, if $l = 2$, then $N(1) = v_1(N/v_1v_2)^{1/2} = \sqrt{(v_1/v_2)N}$, which agrees with formula (4.2) if we write $v_1 = a$ and $v_2 = b$ and recognize that the number of nodes belonging to level 1 was denoted k in that equation but is denoted $N(1)$ here.

In order to derive formula (4.4) observe that, whatever the value of l, each subtree of a tree which minimizes average sequential search time must be so organized that sequential searches through that subtree must also minimize average sequential search time for otherwise it would be possible to replace the subtree by a more efficient one, thereby decreasing the average search time for the full tree. But this contradicts the optimality of the latter.

Based on this observation, formula (4.4) can be proved by induction.* The original tree consists of $l + 1$ levels. Suppose that each subtree which has the same root and consists of l levels is optimal in the sense that the average time required for a sequential search is minimized. As the induction hypothesis, we will assume the validity of equation (4.4) for $k < l$ and will prove that it must be valid for $k = l$ as well. From that equation we find

$$N(1) = v_1\left(\frac{N}{v_1 \ldots v_l}\right)^{1/l}$$

---

*The reader should recognize that this proof utilizes a divide and conquer strategy.

so

$$(v_1 \ldots v_k)\left(\frac{N(1)}{v_1}\right)^k = (v_1 \ldots v_k)\left(\frac{N}{v_1 \ldots v_l}\right)^{k/l} = N(k);$$

this provides the following formula for $N(k)$ from which $N = N(l)$ is absent:

$$N(k) = (v_1 \ldots v_k)\left(\frac{N(1)}{v_1}\right)^k. \tag{4.5}$$

The average time required for a sequential search of the complete tree is easily seen to be

$$2t_{av} = \frac{1}{v_1}\frac{N(1)}{N(0)} + \frac{1}{v_2}\frac{N(2)}{N(1)} + \cdots + \frac{1}{v_l}\frac{N(l)}{N(l-1)} \tag{4.6}$$

where we recall that $N(0) = 1$; the factor 2 on the left side appears because, on the average, only half the nodes at each level are encountered in a sequential search. Now suppose that the subtree through level (l) is optimal. Then the numbers $N(k)$, $k < l$, are related to $N(1)$ by equation (4.5); substitution in (4.6) yields

$$2t_{av} = \frac{N(1)}{v_1} + \frac{1}{v_2} \times v_2 \frac{N(1)}{v_1} + \cdots + \frac{1}{v_{l-1}} \times v_{l-1}\frac{N(1)}{v_1}$$

$$+ \frac{1}{v_l}\frac{N}{\left\{(v_1 \ldots v_{l-1})\left(\frac{N(1)}{v_1}\right)^{l-1}\right\}}, \tag{4.7}$$

where we have written N in place of $N(l)$. Simplification produces

$$2t_{av} = (l - 1)\frac{N(1)}{v_1} + \left(\frac{N}{v_1 \ldots v_l}\right)\frac{1}{(N(1)/v_1)^{l-1}}. \tag{4.8}$$

The total number of nodes N is given; our task is to determine $N(1)$ so that t is minimized. This can be accomplished by solving the equation $dt/dN(1) = 0$. Differentiation of (4.7) leads to the equation

$$0 = \frac{dt}{dN(1)} = \frac{1}{2}\left\{\frac{(l-1)}{v_1} - \frac{(l-1)}{v_1}\left(\frac{N}{v_1 \ldots v_l}\right)\frac{1}{(N(1)/v_1)^l}\right\}$$

or

$$N = N(l) = (v_1 \ldots v_l)\left(\frac{N(1)}{v_1}\right)^l,$$

which is formula (4.5) for $k = l$. This completes the induction step.

It remains to check that the formula is correct for $l = 2$. But this is the case of a 3-level hierarchy. For this case the average sequential search time is

$$2t_{av} = \frac{1}{v_1}\frac{N(1)}{N(0)} + \frac{1}{v_2}\frac{N(2)}{N(1)} = \frac{1}{v_1}N(1) + \frac{N}{v_2} \times \frac{1}{N(1)}.$$

Then

$$0 = \frac{dt}{dN(1)}$$

implies

$$0 = \frac{1}{v_1} - \left(\frac{N}{v_2}\right)\frac{1}{N(1)^2}$$

whence

$$N = \frac{v_2}{v_1} \times N(1)^2 = (v_1 v_2)\left(\frac{N(1)}{v_1}\right)^2,$$

which is formula (4.5) for $k = 2$. This completes the proof.

Now suppose that a hierarchical structure which has N leaves and $l + 1$ levels is so arranged that it minimizes the average time required for a sequential search. How long is the search time? And how will it vary if the number of levels is changed? Finally, how many levels should there be so that the average sequential search time is not only minimized among hierarchies of a given number of levels, but among all hierarchies which have the given number N of terminal nodes? These questions are easily answered.

Twice the average search time is given by (4.7). Substitution using the formula that relates the sizes of the levels for a sequential search time minimizing $(l + 1)$-level hierarchy yields

$$t_{av} = \frac{1}{2}\left\{\frac{N(1)}{v_1} + \cdots + \frac{N(1)}{v_1}\right\} = \frac{l}{2}\frac{N(1)}{v_1} = \frac{l}{2}\left(\frac{N}{v_1 \ldots v_l}\right)^{1/l} \tag{4.9}$$

where (4.5) has been used in the last equality. Thus the minimum average sequential search time grows as the $(1/l)$-th power of the number of terminal nodes (which correspond to the inventory in a data base, etc.). The dependence on the number of levels is more complicated, for as l increases, $(N/v_1 \ldots v_l)^{1/l}$ decreases to 1, but the factor $l/2$ grows. This suggests that there should be an optimum number of levels which will yield the smallest possible sequential search time t_{av}. In order to determine l—or at least to find a good approximation to it—one can treat l as a continuous variable and then determine the value which minimizes t_{av} as a function of l. If this value is not an integer, then the integer which provides the optimum hierarchical structure will be either the smallest integer greater than l, or the largest integer less than l. Let us determine the value of l, considered as a continuous variable, which minimizes the average search time given by formula (4.9). Express the relationship between t_{av} and given by (4.9) in the form

$$t_{av} = \frac{1}{2}\left(\frac{N}{v_1 \ldots v_l}\right)^{1/l} = \frac{l}{2}\exp\left(\frac{1}{l}\log\left(\frac{N}{v_1 \ldots v_l}\right)\right), \tag{4.10}$$

where exp denotes the exponential function and log is the natural (base e)

logarithm. Suppose that the search rates $v_1, \ldots, v_l$ are all equal and choose the unit of time so that their common value is 1. Then differentiation of (4.10) with respect to l yields

$$\frac{dt}{dl} = \frac{1}{2} \left\{ 1 - \frac{\log N}{l^2} \right\} \exp\left(\frac{1}{l} \log N \right)$$

and the equation $dt/dl = 0$ has the unique solution

$$l = \frac{1}{2} \log N = \frac{1}{2} (\log 2) \log_2 N \simeq 0.35 \log_2 N. \tag{4.11}$$

It is easy to verify that this corresponds to the minimum search time, and substitution in (4.10) yields the result that the minimum average sequential search time as a function of the number N of terminal nodes in a hierarchical structure is

$$t_{min} = \frac{l}{2} \exp\left(\frac{1}{l} \log N \right) = \frac{\log N}{2} \exp\left(\frac{\log N}{2 \log N} \right)$$

$$= \frac{\sqrt{e}}{2} \log N = \left(\frac{\sqrt{e} \log 2}{2} \right) \log_2 N, \tag{4.12}$$

i.e.,

$$t_{min} \simeq 0.94 \log_2 N. \tag{4.13}$$

As an example, consider a library catalog card tray, which typically contains about $N = 800$ cards. According to (4.11), $l = 0.5 \log_2 800 \simeq 3.5$ will minimize average sequential search time, and that time will be approximately 5.5 units. Traditional library card catalog trays seldom have an indexing hierarchy for which l is greater than 3.

4.3 Extremal Systems and the Cost of Information

The practical effectiveness of an information processing system, whether it be designed by humans for human use or is the result of millenia of biological evolution, depends upon a combination of the amount of information it produces and the effort, or cost, required to obtain the information. Generally, the purpose of the system is to provide rather than to suppress information, and in this case maximization of the information gained for a given effort invested, or minimization of the effort required to yield a given amount of information, will be the natural formulation of this objective. In instances wherein the information-processing system is deluged by input not all of which is relevant, a good strategy will be to minimize information of the unwanted type. In both cases, the quantity of information provided by the information processing system and the amount of effort required to obtain it are inextricably interwoven: the information gain per unit effort for efficient and stable systems generally assumes an extreme, i.e., maximum or minimum, value.

In the previous section we studied hierarchical information processing systems and a particular way of using them. The systems belong to a family of hierarchical structures that are characterized by the values of certain structural parameters. If one assumes that the time required for each elementary operation performed by the hierarchically organized information-processing system is independent of the operation, then the time required to provide a unit of information will be a good measure of effort or cost. If the parameters of the system are selected so that the corresponding structure minimizes the processing time, the system will be optimal for its intended purpose *within the family of structures considered*. There may, of course, be more suitable structures than those belonging to the given family.

More generally, suppose that $\mathscr{S}$ denotes a family of information-processing structures whose members are characterized by certain real valued numerical parameters, or variables, $x_1, \ldots, x_n$. Let $S(x_1, \ldots, x_n) \in \mathscr{S}$ denote the system whose parameters have the indicated values. Suppose that it is possible to assign a cost or effort $E(x_1, \ldots, x_n)$ that measures the cost required for the given structure to produce the quantity I of information. If $I = I(x_1, \ldots, x_n)$, that is, if the quantity of information produced can also be expressed in terms of the structural parameters of the system, then the problem of determining the optimal information-processing structure belonging to the family $\mathscr{S}$ is equivalent to the problem of extremizing the ratio I/E, i.e., maximizing or minimizing this ratio; usually one is interested in the *maximum*. This in turn is equivalent to extremizing I if E is prescribed, or extremizing E if I is prescribed. These are problems of the calculus of variations. In the previous section we saw that one type of problem of this class reduces to an ordinary maximization problem in the differential calculus. In this section we will investigate the more general situation in the context of the properties of statistical information systems for which SHANNON's formula expressed in terms of the probabilities of occurrence of the various states of the system provides a measure of the average information gain per observation.

Consider an information-processing system that has a possibly infinite but countable number of states $S_1, \ldots, S_n, \ldots$ and let p_k denote the probability of occurrence of the state S_k. Then the average information provided by a single observation of the system is

$$I = -\sum_1^\infty p_k \log_2 p_k. \tag{4.14}$$

When applying the methods of the calculus we will often want to use the corresponding expression where the base 2 logarithm is replaced by the natural logarithm in order to simplify differentiation. Information expressed in this form is measured in *nats* instead of bits, and we have the relation

$$I_{nat} = -\sum_1^\infty p_k \log p_k = I \log 2. \tag{4.15}$$

Since the p_k are a probability distribution,

$$1 = \sum_1^\infty p_k. \tag{4.16}$$

Let the cost or effort of using state S_k be measured by E_k. Then the average cost per observation of using the information system will be E, where

$$E = \sum_1^\infty p_k E_k. \tag{4.17}$$

Suppose that the average effort per observation E is given. Let us find that probability distribution that will extremize, i.e., maximize or minimize, the information. We refer the reader to a textbook on the calculus of variations for a discussion of problems of this type and their solution by the method of LAGRANGE *multipliers*. The solution proceeds by introducing the LAGRANGE multipliers α and β and the expression

$$L = I_{nat} + \alpha \sum p_k - \beta \sum p_k E_k$$

which incorporates the information and the constraints expressed by the previous two equations. The numerical value of the multipliers will be determined later. Now consider each of the probabilities p_k to be an independent variable and seek the extrema of L considered as a function of the variables p_k. Differentiation of L with respect to p_k yields

$$0 = -1 - \log p_k + \alpha - \beta E_k$$

whence

$$p_k = e^{(\alpha-1)-\beta E_k}.$$

From eq(4.16) find

$$e^{1-\alpha} = 1/e^{\alpha-1} = \sum_1^\infty e^{-\beta E_k}. \tag{4.18}$$

This expresses the LAGRANGE multiplier α in terms of the known quantities E_k and the parameter β whose value is still to be determined. It follows that the extremizing probability distribution is

$$p_k = \frac{e^{-\beta E_k}}{\sum e^{-\beta E_k}}. \tag{4.19}$$

The parameter β can be found from the effort constraint (4.17):

$$E = \sum_1^\infty p_k E_k = \frac{\sum E_k e^{-\beta E_k}}{\sum e^{-\beta E_k}}. \tag{4.20}$$

Since β appears interwoven through the terms of the equation, it may be a difficult numerical task to obtain an accurate approximation to its value.

With the values of the LAGRANGE multipliers given implicitly by eqs(4.18, 4.20), the probability distribution (4.18) is called the *canonical distribution*. Calculate I_{nat} for the canonical distribution to obtain:

$$I_{nat} = -\sum p_k \log p_k = -\sum p_k((\alpha - 1) - \beta E_k) = (1 - \alpha) + \beta E. \quad (4.21)$$

Hence, if ΔI_{nat} and ΔE, respectively, denote differences in corresponding values of I_{nat} and E, then

$$\Delta I_{nat} = \beta \Delta E,$$

or, expressed in differential form, and recalling the relationship between I_{nat} and I,

$$dI_{nat} = \beta dE, \qquad dI = \frac{\beta}{\log 2} dE;$$

equivalently,

$$\frac{dE}{dI} = \frac{\log 2}{\beta}. \qquad (4.22)$$

This relation shows that $\frac{\log 2}{\beta}$ measures the infinitesimal incremental effort required to obtain an infinitesimal increment of information: more precisely, it is the rate of change of effort as a function of information gained. The reader should remember that (4.22) is valid only for the canonical distribution which extremizes information gain as a function of effort.

Those acquainted with statistical thermodynamics will find all this familiar: I of course measures the entropy and changes in I correspond to the negative of changes in entropy; the effort function E can be interpreted as the energy of a physical system, and in this case $\beta = 1/kT$, where k is BOLTZMANN's constant and T denotes temperature in degrees Kelvin. It follows that a physical system for which this interpretation is valid will extremize information for a given value of the energy by evolving toward a canonical distribution of its states which *maximizes the entropy*, that is, *minimizes the information*. Thus, an observation of the system will yield the maximal amount of information about the systems's state since that state will have been as disorganized as possible prior to the observation.

Let us consider several applications of these ideas.

EXAMPLE 4.3.1. Suppose that there are a finite number N of states and that observation of each of the states requires the same effort: $E_k = \text{constant} = c$. Then

$$p_k = \frac{e^{-\beta c}}{\sum_1^N e^{-\beta c}} = \frac{1}{N}$$

and $I = \log_2 N$. This example shows that our earlier analysis of the average information per observation, which did not explicitly take the cost of acquiring the information into account, *implicitly* assumed that the cost was equal for all states under the hypothesis that the distribution was canonical. Observe that the probability distribution is independent of the specific magnitude of the effort required by the various states; it is only the relative effort that matters.

EXAMPLE 4.3.2. Let $E_k = \log k$, where k runs through the positive integers 1, 2, ... This effort function is an approximation of the effort required to express the integer k in positional notation with respect to some base, for it requires approximately $\log_B k$ digits selected from the base B inventory of B symbols in order to represent k. Similarly, if one assumes that languages have evolved in a parsimonious way so that the more frequently used words are shorter, then one would expect as a crude approximation that the number of letters in the kth most frequent word in the lexicon should be roughly proportional to $\log k$. In fact, natural languages have too few words for such a relation to persist for large values of k, but the degree to which it appears to hold is surprising.

The canonical distribution belonging to the logarithmic effort function is easily calculated. We have

$$e^{-\beta E_k} = e^{-\beta \log k} = k^{-\beta}$$

from which it follows that

$$p_k = \frac{k^{-\beta}}{\sum_1^\infty k^{-\beta}} = \frac{k^{-\beta}}{\zeta(\beta)}, \tag{4.23}$$

where we have written $\zeta(\beta) = \sum k^{-\beta}$ for the well-known RIEMANN *zeta function*. Substitution of this expression into the formula for the corresponding average information yields, after some simplification,

$$I_{nat} = \frac{d}{d\beta}(\beta \log(\zeta(\beta))). \tag{4.24}$$

This relation shows that the information is determined by β, i.e., by the cost per bit.

The so-called ZIPF's law asserts that the relative frequency of occurence of the kth most frequent word is approximately proportional to $1/k$, or perhaps k^{-s}, where the exponent s is approximately equal to 1 (cp. Figure 4.3.1). It is evident that this "law" cannot be correct with $s = 1$ because the harmonic series $1 + \frac{1}{2} + \cdots + \frac{1}{k} + \cdots$ diverges whereas the sum of the terms of a discrete probability distribution must equal 1.

EXAMPLE 4.3.3. Suppose that the effort function is $E_k = Ak + B$, where $k = 0$, 1, 2, ... and A and B are positive constants. For this case,

$$p_k = \frac{e^{-\beta A k - B}}{\sum_0^\infty e^{-\beta A k - B}}.$$

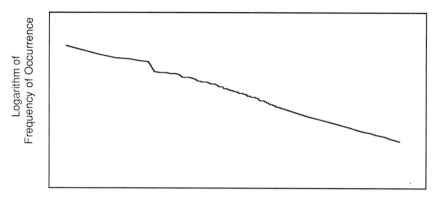

Logarithm of Rank

Figure 4.3.1   Relative frequency of occurrence of English words in text illustrating "Zipf's law." (From data in Kucera and Francis, *Computational Analysis of Present-Day Edited American English*, Brown University Press, 1967.)

Set $e^{-\beta A} = x$. Then

$$\sum_0^\infty e^{-\beta A k} = 1 + x + \cdots + x^n + \cdots = \frac{1}{1-x}$$

so

$$p_k = (1 - e^{-\beta A})e^{-\beta A k} \tag{4.25}$$

and the effort constraint equation expresses the average effort as

$$E = \sum p_k E_k = \sum_0^\infty p_k(Ak + B) = B + \frac{A}{e^{\beta A} - 1}, \tag{4.26}$$

after some algebraic simplification.

In the theory of quantum mechanics, perhaps the simplest example is the harmonic oscillator (that is, the quantized spring). In this case, E_k denotes the energy of the kth quantized energy level of the oscillator. The energy of the kth level is $E_k = hv(k + \frac{1}{2})$, where h denotes PLANCK's constant, and v denotes the fundamental frequency of oscillation of the oscillator. Thus, substituting

$$A = hv, \qquad B = \tfrac{1}{2}hv$$

in the previous equation, and recalling that $\beta = 1/kT$ for applications in statistical physics, yields the expression

$$E = \frac{hv}{2} + \frac{hv}{e^{hv/kT} - 1}$$

for the average energy of the harmonic oscillator as a function of its fundamental frequency v. Physical systems that emit light radiation can be considered as collections of charged harmonic oscillators. As a function of the frequency v, the preceding formula exhibits the variation of average energy as a function of fundamental frequency and is the key step in obtaining PLANCK's law for the distribution of equilibrium thermal (*black body*) radiation. Thus we see that the distribution of black body radiation has an "informational" as well as a "physical" aspect: the informational structure evolves toward a steady state in which the distribution of states of the various oscillators is canonical and minimizes information inherent in the structure.

4.4 Signals, Modulation, and Fourier Analysis

There are many different types of information-bearing signals. Communication by smell consists of the perception of molecules carried from one organism to another through air or water by diffusion processes (BOSSÉRT and WILSON); communication by sound involves the perception of pressure differences in an ambient fluid; vision is mediated by variations of an electromagnetic field which occur within a certain range of frequencies. There are other means by which organisms and machines sense their surroundings and communicate with one another. Some nocturnal tropical spiders communicate by means of low frequency mechanical vibrations which are transmitted over relatively large distances through the banana plants on which they live (ROVNER and BARTH); the rattlesnake combines visible light and infrared radiation to form its view of the world (NEWMAN and HARTLINE); and the eel has evolved a nonjammable electrolocation system (MATSUBARA).

Other indirect modes, based upon a command of technology, have been adapted to human communication needs. Portions of the electromagnetic spectrum ranging from the 10^4 meter-long AM radio waves to the 10^{-9}–10^{-12} meter long x-rays used for medical purposes have been harnessed as communication media. Electrical currents transmit telephone signals and alter the internal states of computing machines as they process information. And finally, biochemical changes in the brain create action potentials and electrical currents which are the physical embodiment of its knowledge and reasoning powers.

Although information-bearing signals have greatly varying physical realizations, there are but a few principles which govern their construction and their accurate transmission through the communication channel. The signals themselves consist of a systematic variation of some physical quantity in time and/or space. The variations constitute a code which, if it is accurately transmitted, can be deciphered by a properly prepared receiver. In all practical circumstances the signal will be corrupted by *noise*, that is, by unwanted and uncontrollable perturbations of the source or transmission system or of the receiver itself which modify the systematic variation of the physical communication channel in an unpredictable way. In this section we will discuss the elementary characteristics of information bearing signals from a general standpoint which is applicable to all particular realizations.

All signals require that energy be transferred from the source to the receiver. If energy is transferred only when a signal is transmitted, then the communication system is generally said to be *unmodulated*; otherwise it is said to be *modulated*. In the latter case, the energy transferred in the absence of signal transmission is referred to as the *carrier*. Thus, speech is an example of an unmodulated system whereas radio is an example of a modulated system because in the former case sound energy is transmitted to the hearer only when signals are being transmitted; whereas in the case of radio, a carrier wave is continually broadcast without regard to whether speech or music or another signal is being transmitted.

The term *modulation* refers to the systematic variation of the carrier to encode a given signal upon it which can be decoded by the receiver. Unmodulated communication can be viewed as the limiting case in which the carrier is absent and the "coded" signal is identical to its given form. This is a consistent way to think of unmodulated communication, but since useful communication systems tend to have either no carrier or a relatively large carrier, it is not often useful to combine the concepts.

Signals are impressed on the systematic transfer of energy which constitutes the communication process by means of a code which enables the receiver to identify which symbol from some prearranged inventory has been transmitted. The symbol inventory and the code that expresses it may be artificially created and established by agreement, as for the telegraphic MORSE code or the secret codes of governments. It may have developed in a less structured way, as spoken and written language codes did, or it may have evolved over long periods of time as did the vision system or the sense of smell in various organisms. In the latter cases, neither the inventory of symbols nor the coding algorithms need be consciously known by the organism; indeed, only since the 1960's has vision research begun to identify some of the "symbols" that the human vision system employs when it uses reflected light as a means for obtaining information about the environment.

Modulated communication relies on a physical carrier whose characteristics are systematically varied. Although speech may be considered as an unmodulated communication mode because energy is transferred only when the signal is transmitted, within the periods of time when the source is speaking it can be regarded as modulating the stream of air expelled from the lungs. Similarly, the detection and analysis of sunlight reflected from objects can be considered as the receiving end of a (one-way) communication system which consists of a carrier—sunlight—which is modulated by the processes of reflection and absorption by physical objects. In this instance, the variation of the carrier which constitutes the signal will correspond to changes in the spatial composition of the scene rather than in time variation; although if the constituents of the scene move or otherwise change with time, then the carrier light wave will be modulated in time as well as in space.

The carrier sound wave produced by expelling air past the vocal cords in singing or speaking vowels consists of a very nearly periodic variation of air pressure. Similarly, the spectral hues which are the components of sunlight consist of a periodically varying electromagnetic field. The carrier waves used for radio and television broadcasting also consist of a periodically varying electromagnetic field, but the frequency of variation is different in each case. Periodic variation plays an important direct role in sensory perception. HELMHOLTZ proposed that the organ of CORTI in the ear peformed an analysis of sound waves into their sinusoidal periodic constituents. About a century later, in 1968, the role of physiological "channels" tuned to relatively narrow spatial frequencies was discovered by CAMPBELL and ROBSON. They and their followers have shown that higher mammalian vision systems contain detectors that respond to sinusoidal gratings of various spatial frequencies. Thus, it appears that the decomposition of physical signal stimuli into constituents which vary periodically (in time or space) plays an important role in the processing of sensory information.

The branch of mathematics that is devoted to the study of periodically varying function is called *FOURIER analysis* after the great nineteenth century French mathematician, Joseph FOURIER. The purpose of the remainder of this section is to summarize the main aspects of FOURIER analysis to the extent that they will be needed in what follows, and then, in section 4.5, to apply them to provide a derivation of the uncertainty relation for measurement. This will complete the discussion of section 2.6 and provide a derivation of the SHANNON sampling theorem, which will be used in section 6.4.

FOURIER analysis can be applied to a great variety of signals and there is a correspondingly rich inventory of examples which illustrate its properties to a greater or lesser extent and with varying degrees concentration on specific aspects of the method. Although no illustrative example can convey the theory with precision and completeness—only a mathematical treatment can approach that goal—some examples of its application are particularly effective for developing the intuition of the mathematically inexperienced reader and for drawing connections between the abstract concepts of the mathematical theory and familiar physical circumstances. For these reasons, we will motivate our discussion by considering the application of FOURIER analysis to the diffraction of light. This is just the situation that prevails in psychophysical experiments concerned with the visual perception of periodic gratings and the trade-offs inherent between the advantages of localization of information in space and the advantages of localization in spatial frequency.

Consider a screen S with apertures in it through which parallel beams of linearly polarized monochromatic light are transmitted. Figure 4.4.1 displays an example of such a screen which has a single point aperture. The light which issues forth from this point aperture radiates as if from a point source. If we follow the light wave emitted by the aperture at a given instant of time, we find that it spreads throughout a hemisphere whose radius grows at the speed of light.

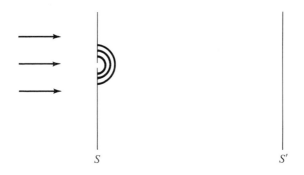

Figure 4.4.1   Screen with point aperture.

If a screen S' is set up parallel to but at some distance from S, the spherical wave front will bathe S' with some distribution of light. This distribution is said to be the *diffraction pattern* produced on S' by light passing through the aperture in S. This pattern depends on the distance between S and S'. As S' is moved farther away from S, the spherical wave fronts which lap on S' will correspond to spheres that are ever increasing in radius and the wave fronts themselves will become increasingly planar in the neighborhood of each of their points. In the limit where S' is at infinite distance from S, the wavefronts will be plane waves, so we conclude that the diffraction image at infinity of a point aperture is a plane wave.

In practice, diffraction images cannot be observed at infinity although one would like to be able to do so because planar wave fronts are simpler than spherical ones and the trajectories of light waves, which are orthogonal to the wave fronts, become parallel rays rather than divergent radii of spheres. At infinity, calculations are easier to make and the optical characteristics of the diffraction image are less complicated. We can obtain the advantages of observations at infinity without requiring that the screen S' be at infinite distance from the source screen S by means of a *lens*. A double convex lens placed so that S and S' are its focal planes has the property that the spherical wave fronts of light emitted from a point source on S will be transformed into planar wave fronts after passing through the lens, and the radial rays of light emanating from the point aperture will be transformed into rays parallel to each other and perpendicular to the screens (see Figure 4.4.2). Thus we may say that the lens is a convenient substitute for placing S' at infinity. Here we may think of the lens of the eye as well as the lens of an artificial optical instrument, although the former focuses the image on the nearly spherical surface of the retina rather than on a plane.

It is evident from what has already been said that observation of the diffraction pattern on S' using the lens or, equivalently, on the plane at infinity, transforms a point-source of light P on S into a uniform image on S'. Suppose the roles of S and S' were interchanged: let S' act as the source and observe

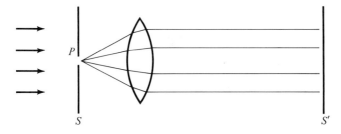

Figure 4.4.2   Effect of a double convex lens.

the resulting distribution of light on S. If S' were the source of a uniform bundle of parallel rays of light, then the lens would focus the light onto the point P; everyone knows this because a double convex lens will concentrate the (nearly) parallel rays of sunlight.

Now suppose once again that the screen S is the source of light, but let the point aperture P be stopped and another point aperture Q be opened. Then Q would emit spherical waves of light which would also be transformed into planar waves after they pass through the lens. Although this situation appears to lead to the same distribution of light on the screen S as before, we know that cannot be true because reflecting the distribution of light back to S from S' leads to the point P, not to Q. This shows that something important is missing from our model. What has been left out is the description of the electromagnetic oscillation which is the light wave. All that we have described thus far are the *geometry* and intensity of the wave motion.

The most convenient description of the periodic spatial variation of the electromagnetic field that constitutes a monochromatic light impulse involves the use of complex valued functions. Recall that if i denotes the positive square root of -1, then

$$e^{ix} = \cos x + i \sin x,$$

(where x is measured in radians). Hence the fundamental periodic functions $\cos x$ and $\sin x$ can be expressed in terms of the complex exponential functions by the familiar formulae:

$$\cos x = (e^{ix} + e^{-ix})/2,$$

$$\sin x = (e^{ix} - e^{-ix})/2i.$$

The distribution of light on the screen S' due to a point source of light on S can be expressed in terms of the exponential function; in fact, the exponential function e^{ivx} corresponds to the spatial variation of a planar light wave with spatial frequency v at some instant of time. In terms of real quantities, we may think of the plane wave as characterized by the sine and cosine functions. Introduce coordinate systems on S and S' as shown in Figure 4.4.3. For simplicity of exposition we have assumed that the screens S and S' are

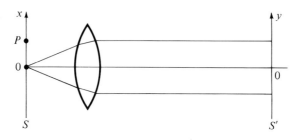

Figure 4.4.3   Distribution of light due to a point source.

1-dimensional (in effect, we are looking at a lateral view of a realistic experimental arrangement). If P and S correspond to the point with coordinate x, then the distribution of light at y on S' which results from a point source at P will be proportional to $e^{-2\pi ixy}/2$. The constant of proportionality depends on the distance between S and S' and other fixed physical parameters; for the sake of convenience, and for other reasons which will become clear later, we will choose our units so that this constant is equal to 1.

If P is at the origin of the coordinate system, $x = 0$, then the distribution of light on S' at some instant will be described by $e^{-2\pi ixy} = e^0 = 1$; that is, the image is *constant* on S'. So we see that the difference between the light distributions due to point sources at different places on the screen S is expressed by the parameter x in $e^{-2\pi ixy}$.

The screen S may have more than one aperture in it, and an aperture may be translucent rather than transparent and so transmit only some fraction of the light that falls upon it. Both properties can be considered together by introducing the notion of a *transmittance function*. The transmittance function varies from 0 to 1; if the transmittance at x is 0, then light is not transmitted at x; if the transmittance is 1, then the entire amount of light incident at x is transmitted; and if the transmittance lies between 0 and 1 at x, the corresponding fraction of the incident light is transmitted.

The light emitted through the (translucent) apertures in S will consequently be described by a function $x \mapsto f(x)$ which is the product of the magnitude of the incident light distribution (which we have assumed is uniform) and the transmittance function that defines the apertures on the screen.

We may think of this distribution f of emitted light as the superposition of light from point source apertures at every point on S, each point aperture weighted by the value $f(x)$ to specify "how much" light it emits. The resulting image on the screen S' will be the superposition of all the corresponding wave fronts. Since the wave front corresponding to a single point at x weighted by the source magnitude $f(x)$ is

$$e^{-2\pi ixy}f(x),$$

the superposition of all these plane waves on S' will be given by the formula

$$\int_{-\infty}^{\infty} e^{-2\pi i x y} f(x)\, dx.$$

That is, if $x \mapsto f(x)$ describes the distribution of emitted light on S, then the expression above will describe the distribution of observed light on S'.

This expression defines a function of y which depends on the function f. This new function will be denoted by $\mathscr{F}f$ and its value at y will be written $(\mathscr{F}f)(y)$; that is,

$$(\mathscr{F}f)(y) = \int_{-\infty}^{\infty} e^{-2\pi i x y} f(x)\, dx. \tag{4.27}$$

$\mathscr{F}f$ is the FOURIER *transform* of f. We will see that the FOURIER transform provides an analytical link between an observable and its dual observable in the sense of section 2.6, and that the uncertainty relation (2.32) is a consequence of its properties. But of more general and immediate significance, the FOURIER transform of a function f describes its decomposition into a superposition of "plane waves" or periodic exponential functions (or, what amounts to the same thing, its decomposition as a sum of sines and cosines).

If f denotes a distribution of light on S, then the total intensity of the distribution is measured by

$$\int_{-\infty}^{\infty} f(x)\overline{f(x)}\, dx,$$

where $\overline{f}$ denotes the complex conjugate of f. On physical grounds we would expect that the total intensity of the image on S' would be the same. Indeed, this relationship is expressed by a fundamental property of the FOURIER transform:

$$\int_{-\infty}^{\infty} f(x)\overline{f(x)}\, dx = \int_{-\infty}^{\infty} (\mathscr{F}f)(y)\overline{(\mathscr{F}f)(y)}\, dy. \tag{4.28}$$

We have already come to expect that if a distribution of light on S' is "reflected", one should observe the original source distribution on S. Mathematically, this means that if $g(y)$ is a distribution on S', there must be some source distribution f on S such that $g(y) = (\mathscr{F}f)(y)$. It turns out that f can be found from g from the formula

$$f(x) = \int_{-\infty}^{\infty} e^{2\pi i x y} g(y)\, dy; \tag{4.29}$$

note the plus sign which appears in the exponent. Since $g(y) = (\mathscr{F}f)(y)$, equation (4.29) is the formula for calculating the *inverse* FOURIER transform.

The FOURIER transform decomposes a function f into sinusoidal or exponential periodic "vibrations." Formula (4.29) states that the "vibration" $e^{2\pi i x y}$ occurs in the function f weighted by the magnitude $g(y)$; if $g(y)$ is 0, then the "frequency" which corresponds to y is absent from f.

Table 4.4.1 FOURIER Transforms

$f(x) = \int_{-\infty}^{\infty} e^{2\pi ixy}(\mathscr{F}f)(y)dy$	$(\mathscr{F}f)(y) = \int_{-\infty}^{\infty} e^{-2\pi ixy}f(x)dx$
(1) $f(x)$	$(\mathscr{F}f)(y)$
(2) $af(x) + bg(x),$	$a(\mathscr{F}f)(y) + b(\mathscr{F}g)(y),$
(3) $f(x)g(x)$	$((\mathscr{F}f) * (\mathscr{F}g))(y) = \int_{-\infty}^{\infty}(\mathscr{F}f)(u)(\mathscr{F}g)(y-u)du$
(4) $(f * g)(x) = \int_{-\infty}^{\infty}(u)f(x-u)du$	$(\mathscr{F}f)(y)(\mathscr{F}g)(y)$
(5) $f(ax), \quad a > 0$	$(1/a)(\mathscr{F}f)(y/a)$
(6) $f(-x)$	$(\mathscr{F}f)(-y)$
(7) $e^{2\pi iax}f(x)$	$(\mathscr{F}f)(y-a)$
(8) $f(x-a)$	$e^{-2\pi iay}(\mathscr{F}f)(y)$
(9) $1/\sigma\sqrt{2\pi}\exp(-(1/2)(t^2/\sigma^2))$	$\exp(-\frac{1}{2}(2\pi\sigma)^2 y^2)$
(10) $\sin(2\pi ax)/2\pi ax, \quad a > 0$	$\begin{cases} \frac{1}{2a} & \text{if} \quad \|y\| < a \\ 0 & \text{if} \quad \|y\| > a \end{cases}$
(11) $\begin{cases} \frac{1}{2a} & \text{if} \quad \|y\| < a, \\ 0 & \text{if} \quad \|y\| > a, \quad \text{with} \quad a > 0 \end{cases}$	$\sin(2\pi ay)/2\pi ay, \quad a > 0.$

In the formulae, a and b are constants.

Some important general relationships between functions and their FOURIER transforms are listed in the first part of Table 4.4.1.

Although it is difficult to calculate the FOURIER transform of a function analytically, one can rely on widely available computer algorithms to calculate an accurate numerical approximation. (It is interesting to note that the most efficient of these, the "Fast Fourier Transform" algorithm, is based on a divide-and-conquer strategy.). There are some important special cases for which the FOURIER transform can be explicitly evaluated. Suppose, for instance, that f is the function

$$f(x) = \begin{cases} 1 & \text{if} \ -a < x < a \\ 0 & \text{otherwise.} \end{cases} \tag{4.30}$$

This function corresponds to an aperture of length $2a$ centered at $x = 0$ on the plane S in Figure 4.4.3. From equation (4.27) the FOURIER transform of f is

$$(\mathscr{F}f)(y) = \int_{-\infty}^{\infty} e^{-2\pi ixy}f(x)dx$$

$$= \int_{-a}^{a} e^{-2\pi ixy}f(x)dx$$

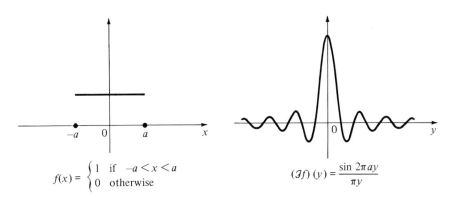

$$f(x) = \begin{cases} 1 & \text{if } -a < x < a \\ 0 & \text{otherwise} \end{cases} \qquad (\mathcal{F}f)(y) = \frac{\sin 2\pi ay}{\pi y}$$

Figure 4.4.4   A function and its Fourier transform.

since $f(x) = 0$ outside the interval, and $f(x) = 1$ inside it. The last integral is easily evaluated:

$$= \frac{e^{-2\pi ixy}}{-2\pi iy}\bigg|_{-a}^{a}$$

$$= \frac{e^{-2\pi iay} - e^{2\pi iay}}{-2\pi iy}$$

$$= \frac{\sin(2\pi ay)}{\pi y}.$$

Sketches of the graphs of f and $(\mathcal{F}f)$ are shown in Figure 4.4.4. This example will play an important role when we consider the SHANNON sampling theorem.

Another valuable example of a function and its FOURIER transform is provided by the GAUSSIAN or normal probability density function. If

$$f(x) = \frac{1}{\sigma\sqrt{2\pi}}e^{-x^2/2\sigma^2},$$

then the FOURIER transform of f is

$$(\mathcal{F}f)(y) = e^{-(2\pi\sigma y)^2/2},$$

another GAUSSIAN. Whereas the standard deviation of f is σ, the standard deviation of its FOURIER transform is $1/(2\pi\sigma)$; hence, a narrowly peaked GAUSSIAN has a FOURIER transform with a broad peak, and conversely. This relationship is illustrated in Figure 4.4.5. It plays an important role in computational models of edge detection.

In general, if the intensity in the source distribution f is narrowly concentrated near some point, then its FOURIER transform $\mathcal{F}f$ will be broadly spread over a large interval, whereas if the image is narrowly concentrated, then the source must be broadly spread. As will be seen later, it is not possible for both

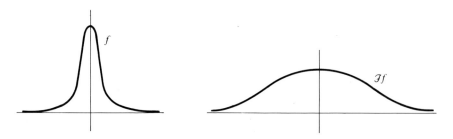

Figure 4.4.5   Gaussian Fourier transform pair.

source and image—for a function and its FOURIER transform—to be concentrated on small intervals.

If we think of x as having the dimensions of length, then y must have the dimensions (1/length) in order that the exponent in $e^{-2\pi i x y}$ be a pure number. Thus far we have acted as if x and y both have dimensions of length, but in reality our choice of coordinates suppresses the inverse linear dependence of the exponent on the *wavelength* of the light and on the *distance* from the origin on S to the point of observation on S' and treats x and y as pure numbers. If x were correctly treated as a position coordinate, then y would denote a position coordinate on S' *divided by* the product of the wavelength and the distance from the origin on S to the point of observation on S'. Hence y would have the dimensions of (1/length): it would measure *spatial frequencies*.

Without any loss of generality, we can think of the FOURIER transform of the light distribution emitted by a source scene as providing a spatial frequency analysis of the light distribution. From this point of view, it is easy to understand why spatially concentrated source distributions have FOURIER transforms that are broadly spread throughout a wide band of *spatial frequencies*. Similarly, if the variable x were interpreted as time, the FOURIER transform variable would have dimensions (1/time), i.e., frequency, and the impossibility of simultaneously localizing a function in time and in frequency is an expression of the uncertainty relation for these dual variables. Let us now turn to consider the uncertainty relation and the SHANNON sampling theorem.

4.5 Shannon's Sampling Theorem and the Uncertainty Relation

In this section we will discuss two important applications of FOURIER analysis. The first, SHANNON's *sampling theorem*, states sufficient conditions that a continuous function (a *signal*) can be exactly reconstructed from periodic samples of its values and provides the formula that accomplishes the reconstruction. This result is one of a large family of "interpolation" formulae. Its particular significance lies in its close connection to communication engineering design techniques, but it is of special interest for us because there is evidence that the human vision processing system employs a related method to capture light signals that vary continuously throughout the visual plane from values sampled by the discrete collection of photosensitive retinal receptors. This application of the sampling theorem will be discussed in section 6.4.

The second application of FOURIER analysis considered in this section is a derivation of the uncertainty relation for measurement introduced in Chapter 2. This relation plays a fundamental role in all of our considerations, although it frequently acts "behind the scenes" rather than in a directly perceptible way. We will find, for instance, that the trade-off between localization of light energy in a visual scene and detection of the details of the scene is the same as the trade-off between localization of *spatial position* variables and simultaneous localization of the dual *spatial frequency* variables. The extent to which such simultaneous localization is possible is described by the uncertainty relation.

Let us make these remarks more precise. Suppose that a function $x \mapsto f(x)$ is given, where x is measured in units of length. Then the FOURIER transform of f is

$$(\mathscr{F}f)(y) = \int_{-\infty}^{\infty} e^{-2\pi i x y} f(x) dx.$$

Since the exponent xy must be a pure number, y must have units of $(1/\text{length})$: y is a *spatial frequency* just as a variable whose units are $(1/\text{time})$ is an ordinary (i.e., chronological) frequency.

Elements of a scene that correspond to high spatial frequencies have small extent: thus, the fine details of a scene may be said to be represented by the high spatial frequency components of the light signal whereas the more gross characteristics correspond to low spatial frequencies.

The consequences of removing high spatial frequencies from an image are illustrated by the interesting pair of images in Figure 4.5.1.

A physical measuring instrument will necessarily have limited resolution. Some of the consequences of this unavoidable limitation were discussed in Chapter 2. From our present standpoint, this is just another way of saying that there is some maximal spatial frequency beyond which the instrument is insensitive. In this case, a remarkable interpolation theorem discovered many times, perhaps first by E.T. WHITTAKER, but later by Claude SHANNON, who made the most profound use of it, asserts that it is possible to exactly reconstruct a function from knowledge of its values at a discrete set of uniformly spaced points if the "spatial bandwidth" of the function is limited. This suggests, as was mentioned, that it may be possible for the discrete collection of retinal photoreceptors in the eye to provide enough information for the essentially perfect reconstruction of the incident image if the spatial frequencies of the image are suitably limited. And this, in turn, would constitute a very powerful form of selective omission of information at the signal gathering stage without a corresponding loss at the signal interpretation stage.

Shannon's sampling theorem asserts the following: If the spatial frequencies of the function $x \mapsto f(x)$ are limited by W, then f can be written in the form

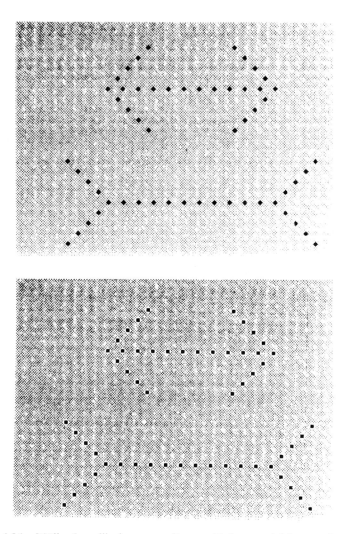

Figure 4.5.1   Müller-Lyer illusion: upper figure: with low spatial frequencies present; lower figure: with low spatial frequencies removed (image disappears when viewed from a distance). (From Julesz and Schumer, *Annual Review of Psychology*, Volume 32. © 1981 by Annual Reviews Inc. Reproduced with permission.)

$$f(x) = \sum_{n=-\infty}^{\infty} f(n/2W) \frac{\sin \pi(2Wx - n)}{\pi(2Wx - n)}. \qquad (4.31)$$

The condition that the spatial frequencies are limited by W means that the FOURIER transform of f satisfies the condition

$$(\mathscr{F}f)(y) = 0 \quad \text{unless} \quad -W < y < W. \qquad (4.32)$$

The infinite series (4.31) expresses f in terms of its sample values at the points $x = n/2W$, where n runs through the integers. Thus the formula tells us how to interpolate between the sampled values $f(n/2W)$ in order to reconstruct the function.

We will sketch a proof of the sampling theorem for those readers who are familiar with FOURIER series.

If $\mathscr{F}f$ satisfies (4.32), then it can be expressed as a FOURIER series on the interval $-W < y < W$, namely

$$(\mathscr{F}f)(y) = \sum_{n=-\infty}^{\infty} a_n e^{-2\pi i n y/2W}. \tag{4.33}$$

The coefficients a_n are constants whose value will be calculated later.

Substitute the series expression (4.33) for the FOURIER transform $\mathscr{F}f$ in equation (4.29) to obtain

$$f(x) = \int_{-W}^{W} e^{2\pi i x y} \left(\sum_{n=-\infty}^{\infty} a_n e^{-2\pi i n y/2W} \right) dy; \tag{4.34}$$

the limits of the integral are $-W$ and W rather than $-\infty$ and ∞ because $(\mathscr{F}f)(y) = 0$ unless $-W < y < W$. Interchanging the order of integration and summation in (4.34) yields

$$f(x) = \sum_{n=-\infty}^{\infty} a_n \int_{-W}^{W} e^{2\pi i (x - n/2W) y} dy$$

$$= \sum_{n=-\infty}^{\infty} a_n \frac{e^{2\pi i (x - n/2W) y}}{2\pi i (x - n/2W)} \bigg|_{-W}^{W}$$

$$= \sum_{n=-\infty}^{\infty} a_n \left\{ \frac{e^{2\pi i (x - n/2W) W} - e^{-2\pi i (x - n/2W) W}}{2\pi i (x - n/2W)} \right\}$$

$$= \sum_{n=-\infty}^{\infty} 2W a_n \frac{\sin \pi (2W - n)}{\pi (2W - n)}.$$

Now calculate the constants a_n. If we let x approach $m/2W$ for some integer m, then

$$\frac{\sin \pi (2Wx - n)}{\pi (2Wx - n)} \quad \text{approaches} \quad \frac{\sin \pi (m - n)}{\pi (m - n)};$$

this is 0 unless $m = n$, and in the latter case the quotient approaches the value 1 as x approaches $m/2W$. Hence, only one term in the infinite series survives so, upon comparison of left and right sides, we find

$$f\left(\frac{m}{2W}\right) = 2W a_m.$$

Substitution of this value for $2W a_m$ yields the sampling formula, equation (4.31).

If f is not band-limited as called for by the hypothesis of the sampling theorem, then it cannot be reconstructed from its sample values because high frequency components of f are mixed in with low frequency components to create *alias* or *moiré* pattern artifacts in the reconstructed function. This

Figure 4.5.2   Sine waves of different frequencies with the same set of equally spaced sample values.

situation is illustrated in Figure 4.5.2: it is impossible to distinguish between the two functions whose graphs are shown by the values at the indicated sampling points. If frequencies that are too great for the sampling rate are excluded, then the sampled function can be reconstructed.

Recall the graph of the function $x \mapsto (\sin x/x)$ displayed in Figure 4.4.4. SHANNON's sampling theorem tells us that we can reconstruct f by superposing translates of this curve (appropriately scaled to take the value of W into account) multiplied by the sampled values of f. Hence, if the values of f decrease rapidly outside a neighborhood of some point, only relatively few terms in the expansion will contribute much and the approximation provided by those sample values corresponding to points within the neighborhood will be a good one.

Now let us turn to the *uncertainty relation*, which describes the relationship between the *width* of the function $\bar{f}f$ and the width of the corresponding expression formed from the FOURIER transform of f, $(\mathscr{F}f)(\overline{\mathscr{F}f})$. Here the bar "$\bar{\;}$" denotes the complex conjugate.

Suppose for simplicity that the average value of x relative to the function $\bar{f}f$ is 0, that is,

$$x_{av} = \int_{-\infty}^{\infty} xf(x)\overline{f(x)}\,dx = 0,$$

and similarly that

$$y_{av} = \int_{-\infty}^{\infty} y(\mathscr{F}f)(y)\overline{(\mathscr{F}f)(y)}\,dy = 0.$$

These conditions can always be arranged by translating the graph of the function f to a suitable position.

We want to derive the uncertainty relation $\Delta x \Delta y \geq 1/4\pi$, so first we must define Δx and Δy. Set

$$(\Delta x)^2 = \int_{-\infty}^{\infty} x^2 f(x)\overline{f(x)}\,dx \bigg/ \int_{-\infty}^{\infty} f(x)\overline{f(x)}\,dx \qquad (4.35)$$

and

$$(\Delta y)^2 = \int_{-\infty}^{\infty} y^2(\mathscr{F}f)(y)\overline{(\mathscr{F}f)(y)}\,dy \bigg/ \int_{-\infty}^{\infty} (\mathscr{F}f)(y)\overline{(\mathscr{F}f)(y)}\,dy. \qquad (4.36)$$

Δx and Δy are obtained by extracting the positive square root of $(\Delta x)^2$ and $(\Delta y)^2$, respectively.

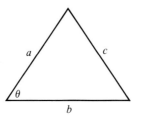

Figure 4.5.3   Law of cosines.

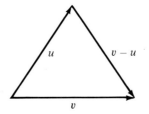

Figure 4.5.4   Vectorial representation of law of cosines.

The proof of the uncertainty relation relies on *Schwarz's inequality*, which is a far-reaching generalization of a well-known fact from elementary trigonometry. In the triangle (Fig. 4.5.3), let a, b, c denote the lengths of the indicated sides. Then the *law of cosines* states

$$c^2 = a^2 + b^2 - 2ab \cos \theta,$$

or equivalently,

$$\cos \theta = \frac{a^2 + b^2 - c^2}{2ab}. \tag{4.37}$$

If the sides of the triangle are represented by vectors as shown in Figure 4.5.4, then $a^2 = u \cdot u$, $b^2 = v \cdot v$, $c^2 = (v - u) \cdot (v - u)$, where the dot denotes the "scalar product," and (4.37) can be written in the symmetrical form

$$\cos \theta = \frac{u \cdot v}{\sqrt{(u \cdot u)(v \cdot v)}}.$$

Squaring both sides yields

$$\cos^2 \theta = \frac{(u \cdot v)^2}{(u \cdot u)(v \cdot v)}.$$

Since $0 \le \cos^2 \theta \le 1$, this implies the inequality

$$(u \cdot v)^2 \le (u \cdot u)(v \cdot v),$$

which becomes, when written out in coordinates,

$$(u_1 v_1 + u_2 v_2)^2 \leq (u_1^2 + u_2^2)(v_1^2 + v_2^2).$$

An analogous inequality is satisfied by functions and can be derived in a similar way:

$$\left| \int (f_1 g_1 + f_2 g_2) dx \right|^2 \leq \left(\int (f_1 \bar{f}_1 + f_2 \bar{f}_2) dx \right) \left(\int (g_1 \bar{g}_1 + g_2 \bar{g}_2) dy \right). \quad (4.38)$$

We will apply this to our circumstances.

Set

$$f_1 = xf, \qquad f_2 = x\bar{f}$$
$$g_1 = \overline{df/dx}, \qquad g_2 = df/dx.$$

The integral on the left side of (4.38) reduces to

$$\int \left(xf \frac{\overline{df}}{dx} + x\bar{f} \frac{df}{dx} \right) dx$$

which we recognize is equal to

$$\int x \frac{d}{dx}(f\bar{f}) dx.$$

Integrate by parts to obtain

$$\int_{-\infty}^{\infty} x \frac{d}{dx}(f\bar{f}) dx = xf\bar{f} \Big|_{-\infty}^{\infty} - \int_{-\infty}^{\infty} f\bar{f} dx.$$

The first term on the right side will be 0 because the signal f rapidly decreases to 0 at ∞ (this is a consequence of the hidden assumption that the signal must have "finite energy" or, in the mathematician's terms, be *square integrable*). Now apply the general relation (which can be proved by integration by parts)

$$\mathscr{F}\left(\frac{df}{dx} \right)(y) = \int_{-\infty}^{\infty} e^{-2\pi ixy} \frac{df}{dx} dx$$
$$= -2\pi iy (\mathscr{F}f)(y)$$

and obtain from SCHWARZ's inequality (4.38):

$$\left| \int f\bar{f} dx \right|^2 \leq 4(2\pi)^2 \int x^2 f\bar{f} dx \int y^2 (\mathscr{F}f)(\overline{\mathscr{F}f}) dy.$$

Using the definitions (4.35) and (4.36), this inequality can be rewritten as

$$(\Delta x \Delta y)^2 \geq \frac{1}{(4\pi)^2},$$

whence extraction of the square root yields the *uncertainty relation for measurement*,

$$\Delta x \Delta y \geq \frac{1}{4\pi}. \tag{4.39}$$

If the variable x is interpreted as *time* (we will write t in place of x), then y has units (1/time), that is, *frequency*, and we will write Δv in place of Δy. Then (4.39) becomes $\Delta t \Delta v \geq 1/4\pi$. If we are concerned with a quantum mechanical system, then the energy ΔE which corresponds to the oscillation of frequency Δv is $\Delta E = h \Delta v$, where h denotes PLANCK's constant, so we have derived the *HEISENBERG uncertainty principle* for time and energy,

$$\Delta t \Delta E \geq \frac{h}{4\pi},$$

from a universal mathematical property of the FOURIER transform.

CHAPTER 5

Biological Signal Detection and Information Processing

The present state of Artificial (Machine) Intelligence has deep within it a strange paradox. It is just those aspects of control and the selection of relevant from irrelevant data which are the most difficult to mechanize—though they were the first problems to be solved by organisms.

—R.L. GREGORY

5.1 Purpose of This Chapter

At the present time, the most complex and capable information-processing systems are the biological organisms. In this chapter we will investigate how the general principles of information science are embodied in particular biological information processing systems in order to gain an appreciation for the variety of realizations that they may have.

The principles that will be studied are

- *The selective omission of information,*
- *The organization of systems to maximize or minimize information,*
- *The hierarchical organization of information processing systems,*
- *The role of group invariant functions.*

Each of these principles plays an important role in restructuring the information provided by an organism's sensory signal detection systems, and all the evidence suggests that they play an equally central role at the level of cognitive information processing in the higher animals.

We will focus on the study of particular biological information processing systems rather than attempt to survey them comprehensively. There are theoretical reasons for doing this in addition to the obvious practical limitations of time and space that forestall an attempt to treat the subject comprehensively. For the moment, let us limit our attention to human information processing. Since one of our purposes is to study the nature of abstract patterns of information and how they are organized, we can restrict our considerations to one or another of the sensory modalities without loss of generality because there is compelling evidence that all sensory data are transformed into common knowledge representations at the highest levels of cognition. This point and its implications will be briefly discussed in section 5.2. If it be granted, then we can study one of the high channel capacity sensory modalities without loss of generality insofar as our concerns are limited to the abstractions that represent information as knowledge at higher cognitive levels. We will concentrate on the modality of vision. This choice has several advantages. It is easier to illustrate properties of the vision system as they are discussed than it would be to provide examples for the auditory, somatic, or smell senses. Embryologically, the retina of the eye is a displaced part of the brain whence its capabilities may be more representative of the processes used by the higher cognitive centers to create and act upon percepts than are those of other sensory transduction systems. The eye is also an instrument of remarkable sensitivity; this makes it possible to perform experiments in which the parameters of the input signal are varied in precise and refined ways in order to reveal the "fine structure" of human image processing. The reknowned work of JULESZ on stereopsis, which made use of computer generated random masks, provides an excellent example of the accessibility to experiment of the vision system. Visual illusions are better known, more numerous, and have been a subject of study for far longer than illusions of the other sensory modalities and are in general far easier to create and modify than are other kinds of illusions. They constitute a special kind of experimental laboratory that tests the capabilities of the vision system at its performance limits. By understanding the mechanisms that are responsible for visual illusions, one can hope to learn something about the way the vision system encodes and analyzes image data: the analysis of illusions may even provide some insight into the *intent* of human image processing, by which we mean the general principles that govern the conversion of sensed data into perceptual and cognitive entities.

For these reasons we will concentrate on the modality of human vision in our study of biological information processing, although our purpose is to try to penetrate beyond the level of sensory transduction and early image processing to the higher level cognitive strategies that may be common to all forms of biological information processing.

Section 5.3 summarizes the neurophysiology of the human eye in order to establish terminology and provide a quantitative basis for the later discussions. The following section considers the information processing implications of the physical presence but psychological absence of the "blind spot." Simple experiments, directly accessible to a sighted person, demonstrate the extent to which the higher cognitive centers continuously execute a program of "pattern continuation" which artificially completes the retinal image so that it is perceived as a simply connected topological continuum whereas it physically is a (discrete approximation to a) manifold rent by a large "hole." Section 5.5 examines the implications of the stabilized vision experiments, which demonstrate that only spatio-temporal *differences* in an image are perceived. This is a mechanism which selectively omits information in the original signal and also provides means for adapting the image transduction system to a wide range of stimulus magnitudes. Sections 5.4 and 5.5 provide an indication of the extent to which the human vision system selectively omits information, and they also suggest some of the means that are employed. Sections 5.6 and 5.7 explore this theme further by studying the perceived distribution of information along a contour. The underlying objective is to identify where information is concentrated, for if information is not uniformly distributed, then it will be unnecessary (or at least less necessary) for the cognitive centers to retain and process those parts of a contour that are low in information content. But we know that contours appear to be seen in their continuous entirety, which suggests that either the low information parts are retained in memory despite their relative lack of value or, that they are regenerated as needed by an algorithm. The well known and amusing examples of subjective contours suggest that the latter is the case. As a consequence of the recent discovery by VON HEYDT, PETERHANS, and BAUMGARTNER of cortical cells that respond (sic!) to subjective contours, a new light has been cast on the crucial role of "hardwired" pattern continuation in higher cognitive processing of sensory stimuli. Section 5.7 is devoted to determining the algorithm which generates the cognitive contour by applying the principle that information is extremized by the human information processing system. This result provides further evidence that much of what we perceive as a faithful image of a physical scene is a mental construction created by a pattern continuation process which fills in the scene between relatively few and carefully selected observational data. This procedure may be closely related to the ability to recognize distinct images as instantiations of a single object, as when views of a face seen from various angles are correlated and identified with a single face-object.

Section 5.8 explores human color vision as an example of the role of group invarience and perceptual "symmetries" in reducing information requirements. By representing the space of perceived colors as a homogeneous space of a certain LIE group whose action corresponds to changing the background illumination of the scene, the classical experimental observations of NEWTON and GRASSMANN and the color metric ideas of HELMHOLTZ find their most natural expression, the latter arising automatically from its characterization as a metric that leaves the perceptual i.e. psychological, distance between colors in a scene unchanged under changes of background illumination in normal daylight vision.

The eye consists of many different vision subsystems integrated into a unit which is effective across a broad range of intensities and frequencies of light and degrees of resolution and aperture. The monochrome night vision ("scotopic") subsystem complements the ("photopic") color subsystem used for normal daylight vision; the 3-color scheme is sensitive to stimulation by electromagnetic radiation whose wavelength varies between about 0.4 micrometers and 0.7 micrometers, and various adaptive processes enable the vision system to accommodate intensity variations whose extremes are in the ratio $10^{12}:1$. As regards resolution, the vision system employs a hierarchical organization. The bulk of the photosensitive cone receptors produce a relatively low resolution color image in normal daylight vision which is supplemented by the relatively high resolution capabilities of the foveal region of the retina. The latter comes into action when we direct our gaze upon an object of interest; thus the foveal gaze acts like a flying-spot scanner. This use of a hierarchical information processing system to increase effectiveness is discussed in section 5.9.

The principles that govern human image processing are just those that were proposed in Chapter 1 as general principles of information science. They find various realizations and embodiments in both animate and inanimate information processing systems. One striking example of how different the embodiments can be even when the principles are directed toward a common goal is provided by the sound-based echolocation system that certain bats have evolved to provide them with information about their surroundings that animals which are active in daylight normally obtain from light-based vision systems. The echolocation system of the horseshoe bat is described in some detail in section 5.10; its signal processing capabilities and the hierarchical structure of this sensory system are emphasized in the hope that the reader will recognize the general principles that this particular embodiment illustrates.

Section 5.11 contains a glossary of visual illusions that can be used to test some of the general principles of information processing for their explanatory power. We propose novel explanations for several classical illusions based on these principles and trust that the reader will be able to apply the principles to understand others that we do not discuss.

The effects of undersampling signals have only recently been examined in the domain of vision. Undersampled images generally exhibit *aliases*, visual artifacts that are not present in original signal. It is evidently of some importance for an organism to be able to distinguish between image artifacts and faithful representations of sensory stimuli, but insufficient data is a constraint that no amount of ingenuity can overcome. Nevertheless, just as the HEISENBERG uncertainty principle establishes and governs the trade-off between accurate knowledge of one variable and accurate knowledge of its dual, so too (and for essentially the same reason) is it possible to trade aliases for another undesirable consequence of limited knowledge—noise. Section 5.12 explores this particularly subtle manifestation of the principle of selective omission of information and its consequences for human vision.

5.2 Interconvertibility of Information Representations

It is well known that information which is encoded in one form can be recoded into another. This is routinely done whenever one of the electrical-based communication systems is used. For instance, use of the telephone involves coding a mentally generated message into neural signals that drive the muscles of the articulatory tract to produce pulses of air pressure which, when directed to the mouthpiece of a telephone, are in turn recoded into a time varying electrical signal. If this signal is transmitted over a long distance, it may be recoded again for microwave or satellite transmission. At the receiving end, the reverse process occurs until the stage is reached where the signal, once again realized as a train of air pressure variations, can be detected by the ear. At that point it is converted into neural impulses which travel to the higher brain centers. Similarly, information can be represented by the symbols of written languages which are sensed by the vision system as characteristic patterns. The path from creation of a mental message by one person to its reception by the same or another person via the intermediate medium of the written or typed page involves a similar variety of representations.

All this is well known and therefore hardly ever given thought. But these examples and other similar ones suggest that information must be stored in the brain in an abstract form that is compatible with all sensory input and output modalities. The different sensory modalities must be to the corresponding cognitive concepts what the representation of a vector in coordinates is to its realization in the underlying basis-free abstract vector space. This comparison is intended, of course, metaphorically. Nevertheless, the evidence for an abstract canonical cognitive form into which all sensory information is transformed is compelling. Such a means for processing sensory information would also be efficient, for there would be no need to replicate stored knowledge or processing algorithms for each modality. Indeed, the most efficient method for combining the results of sensory inputs would probably be to convert the input data—after selective omission—into the canonical form at as early a stage as possible in the processing structure.

If this is in fact the strategy that biological information processing systems use, then it may be possible to learn a great deal about the abstract canonical forms and the algorithms that act upon them by tracing the various transformations that are experienced by data acquired by one—any one—of the senses.

Some of the evidence for the interconvertibility of information representations by biological systems is obvious. A person can draw a picture of an object following instructions that are spoken. It is possible to "hear" a musical composition by reading its score. Other examples are more striking. GREGORY reported the case of S.B., which is of particular interest. It appears that S.B. became blind at an early age, perhaps prior to his first birthday. His vision was restored by surgical procedures in his fifty-second year. He had been educated in the use of braille and led an unusually active life for a person with his handicap. S.B. had been taught to recognize relief capital letters of the Roman font, such as those that appear affixed to institutional buildings, by touch at the school for the blind. He was able, upon regaining his vision, to directly transfer his tactile knowledge to the vision modality, and could recognize capital (but *not* lowercase!) letters by sight. This remarkable example supports the assumption that the various sensory modalities provide inputs that are processed and stored in the higher brain centers in a form which is independent of the sensory channel through which they were perceived.

Research by BELLUGI and her coworkers has provided increasing evidence that American Sign Language (ASL) is a natural language that possesses a full range of syntactic and semantic processes. Because ASL relies on the visual rather than the acoustic sensory channel, it provides one means for studying the extent to which the cognitive component of language information processing is independent of the sensory input and output coding of cognitive representations. Indications that ASL communication is primarily processed by the left cerebral hemisphere despite its visual-spatial input coding suggests that left-hemisphere language processing functions may be specialized at a level of language representation that is more abstract than, or at least independent of, speech-coded realizations.

The significance of such speculations would be greatly tempered were it to be shown that the expressive power of ASL is markedly inferior to the expressive power of spoken natural languages, for then special, and possibly quite simple, cognitive structures might suffice for processing the limited ASL signal inventory whereas more general, and probably more complex, mechanisms would be necessary for spoken language. For this reason it is worthwhile to estimate the rate of transmission of information (which we take as a measure of the "expressive power" of a language) in ASL, and to compare it to spoken English.

A standard version of spoken English consists of the sequential production of phonemes selected from an inventory of 42 possibilities, at a rate of about 10 per second. Were there no contextual constraints on the production of grammatically and semantically correct speech sounds, the transmission rate would be $10 \log_2 42 = 54$ bits per second. The actual rate of transmission of semantic information, as distinguished from this rate of unconstrained signalling, is significantly smaller. It is difficult to measure the transmission rate precisely. The analogous problem for written English was considered by SHANNON in 1951; he estimated that printed English is about 75% redundant. The redundancy of spoken English is probably comparable; if so, the rate of transmission of semantic information in spoken English is probably about 10 to 15 bits per second.

The lexical elements of ASL are *signs*. Each sign can be decomposed into constituents called *primes* which are drawn from four distinct categories. The instantiation of a sign consists of the simultaneous display of one prime from each category, although not all combinations of primes so selected correspond to a currently valid sign, just as not every combination of English phonemes, or of alphabetic letters, corresponds to a word.

The categories into which primes are grouped correspond to

1. *Configuration* of the hands,
2. *Location* of the hands in space,
3. *Orientation* of the palm,
4. *Movement* of the hands.

Orientation of the palm has only recently been recognized as a distinctive feature in ASL; it was not, for instance, included in STOKE's pioneering dictionary. Signs are formed by placement and motion of the hands, which may touch the face but are normally confined to an imaginary vertical index plane in front of the signer. This plane should be thought of as having an independent existence for the signer can "store" a sequence of signs (and the information they represent) for future reference by assigning to them a position in the index plane which stands as a marker for the sign sequence. Pointing to the assigned position "calls up" the referenced sequence.

It is currently estimated that there are 41 primes of *configuration*, 27 of *location*, 15 of *orientation*, and 11 of *movement*. Since primes that belong to distinct categories are independent, and an ASL sign consists of one prime drawn from each category, there are a maximum of $(41)(27)(15)(11) = 182,655$ lexical signs. This is roughly comparable to the number of one vowel string words of written English (or one syllable words of spoken English) that would be expected were all admissible initial consonant, vowel, and final consonant strings of letters (or phonemes) that are consistent with the gross structural linguistic constraints were to actually occur. In reality, the largest dictionaries suggest that only about 7,000 one vowel string (or one syllable) words have been systematically used in English. Although the question has not yet been studied, we would anticipate that a similarly small fraction of the possible lexical signs are actual ASL signs. Notwithstanding the correctness of this conjecture, the inventory of potential ASL lexical signs is very much smaller than the inventory of words in English or other widely used spoken languages.

ASL is a highly inflected language, more comparable in this aspect to ancient Akkadian or Latin than to English or Chinese, and it uses inflection as a mechanism for increasing expressiveness. A large number of morphological processes can be syntactically applied to any lexical sign, and may be applied if the result is compatible with the semantics. Thus, referential indexing may be applied to designate a pronominal object. Six categories of inflectional processes and three categories of derivational processes have been indentified thus far. The application of these various processes to a lexical sign is accomplished within the interval of time normally required to produce the lexical sign itself by speeding up or omitting nondistinctive features of the signing process. This results in a substantial increase in expressiveness.

Recent studies indicate that if pause time is included in continuous signalling, the rate of transmission is approximately 2 signs per second. Although more detailed studies of a statistically representative sample of native signers may modify this estimate, it is unlikely that it will be very greatly changed. It follows that the rate of transmission of information by lexical signs unconstrained by context is approximately $2(\log_2 41 + \log_2 27 + \log_2 15 + \log_2 11)$ bits per second $= (2)(17.48) = 34.96$ bits per second. Recall that there are approximately 42 phonemes in English, and that they are transmitted at the rate of approximately 10 per second: it follows that the rate of transmission of information by English phonemes unconstrained by context is $10 \log_2 42 \simeq 54$ bits/second. Thus it appears that information carried by ASL lexical signs is transmitted at about 2/3 the rate it is transmitted in English phonemes, with both cases independent of context constraints.

The morphological processes of ASL can be applied to lexical signs without reducing the rate of transmission of the lexical elements. This increases the rate of transmission of information, an upper bound for which can be determined by the following considerations. Let there be n categories of morphological processes $M_1, \ldots, M_n$ such that but one process from each category can be applied to a lexical sign in a given instantiation, although representative processes from each of the n categories can be simultaneously applied (i.e., applied within a common instantiation of the lexical sign. Note that morphological processes applied to a given lexical sign may have a meaning which depends on the order of application within the instantiation of the sign, and that there are morphological processes which can be repeatedly applied within a single lexical instantiation. These very interesting, but nonetheless very special cases will not change the main features of our conclusion). Finally, suppose that the set of morphological categories is complete, i.e., that each morphological process is accounted for. Suppose that the ith category M_i consists of m_i morphological processes. Then the contribution of the processes to the information transfer rate of ASL— to its "expressiveness"—is $2(\log_2 m_1 + \log_2 m_2 + \cdots + \log_2 m_n)$ bits per second.

There may be morphological processes of ASL, as well as other means for increasing its expressiveness, which have not yet been identified by linguistic researchers. But the analysis presented by BELLUGI and KLIMA provides an indication of the extent to which morphological processes extend the capabilities of ASL to transmit information.

Consider *referential indexing*. Since lexical signs do not carry intrinsic nominal or verbal marker, we may assume that referential indexing can be applied to most, if not all, lexical signs. There are four instances of the referential process, including the "identity" instance which leaves the lexical sign unchanged. Thus the option of referential indexing adds two bits of information per sign, or $(2)(2) = 4$ bits transmitted per second. This is a special case of the more general ability to use the index plane to "define" subjects for later indirect reference, a technique of importance in computer languages as well.

The process of *reciprocal inflection* operates on verbs to indicate mutual relations or actions. It therefore contributes not more than 1 bit per sign, or 2 bits per second.

Inflections that involve *grammatical number* and certain *distributional aspects* appear to form another morphological category consisting of perhaps 9 processes. These yield $2 \log_2 9 = (2)(3.17) = 6.34$ bits per second in the absence of constraints.

There are 8 inflections for *temporal aspect* (including the identity process which leaves the lexical sign unaltered), which contribute $2 \log_2 8 = 6$ bits per second.

Inflections for *manner* and *degree* express 4 and 5 possible distinctions respectively, and thereby add $2 \log_2 4 = 4$ and $2 \log_2 5 = 4.64$ bits per second, respectively.

Nouns and verbs can be transformed one to another by a process which adds $2 \log_2 2 = 2$ bits per second.

Other morphological processes are known but their description is not yet sufficiently detailed to permit an estimate of their contribution to expressiveness.

Combining the above estimates, morphological processes may add approximately $(4 + 2 + 6.34 + 6 + 4 + 4.64 + 2) = 28.98$ bits per second to the lexical expressiveness of 34.96 bits per second, for a total of $34.96 + 28.98 = 63.94$ bits per second.

Once again we stress that this estimate does *not* take into account syntactic or semantic constraints of context other than those which have been explicitly noted. Nevertheless, it appears reasonable to assert that ASL and English probably have comparable measures of expressiveness, which is, of course, what one would anticipate if ASL and English are comparably efficient in communicating information of comparable complexity.

Although the rates of transmission of information by ASL and spoken English are comparable, and although both systems of communication appear to be very general in terms of the complexity and level of abstraction of the information they can communicate, these similarities do not prove that their constituent information bearing units are converted into an independent and common abstract form for processing by the higher cognitive centers. Yet, the ability of a person trained in both ASL and spoken English to fluently translate from one to the other in "real time" and in either direction strongly suggests that translation must be a peripheral process, accomplished by "switching" to the appropriate modality on input and output, rather than a process which deals with two distinct categories of cognitive representations.

The reader should bear these ideas in mind as we investigate human image processing with an eye to seeing in this particular case the universal properties of biological information processing systems.

5.3 Human Vision

He believes that he sees his retinal image: he would have a shock if he could.

—George HUMPHRY

The eye is the input transducer for the human vision system. Its physical properties reflect surprising and complex aspects of the information processing characteristics of the vision system and provide some of the constraints that limit the nature of the mental representation of images. Figure 5.3.1 illustrates the major components of the vision system.

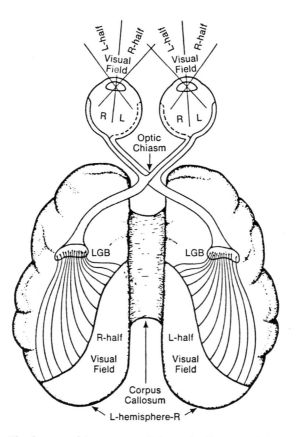

Figure 5.3.1   The human vision system. Schematic diagram of the central visual pathway in the human, as viewed from above. "LGB" denotes the lateral geniculate body in the thalmus—which projects onto the primary visual cortex in the occipital lobe of the brain. (From Popper and Eccles, *The Self and Its Brain,* © Springer-Verlag, 1977. Reproduced with permission.)

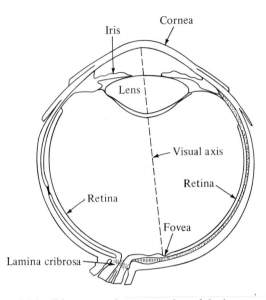

Figure 5.3.2   Diagrammatic cross section of the human eye.

The human eye is generally similar to the eyes of other vertebrates although it is more complicated than most. It is a spheroid of radius approximately 12 millimeters. Figure 5.3.2 exhibits a cross section through an axis perpendicular to the lens. It schematically highlights the retina (considered embryologically to be a portion of the brain) which contains the photosensitive receptors and their neural connections to the nerve sheath that, exiting the eyeball through the *lamina cribrosa*, the "blind spot," conducts image information to the higher cortical centers through the optic nerve.

The area of the human retina is approximately 1,000 square millimeters (mm^2) or, in units that will be more convenient, 1 billion (10^9) square micrometers (recall that a micrometer, 1 μm, is one-millionth of a meter). The retina of the human eye contains about 125 million photosensitive receptors of two types: *rods* and *cones*. The rods appear to be responsible for vision in low levels of illumination, whereas the cones mediate normal daylight vision and the perception of color. The rods are of a single type but there are three types of cones, called *red, green, and blue,* whose maximum sensitivity depends upon the frequency of the incident light. The rods are the dominant type of receptor: there are only about 6.5 million cones. On the average, there is one receptor for every 8 square micrometers of retinal surface. The fovea is the retinal region of greatest visual acuity in normal daylight vision because the density of cones in this small circular region (whose diameter is approximately 1,500 μm) abruptly increases from its average value of about 0.006 cones per square micrometer throughout the remainder of the retina to a maximum value of about 0.147 cones per square micrometer. A circular subregion, the *foveola*

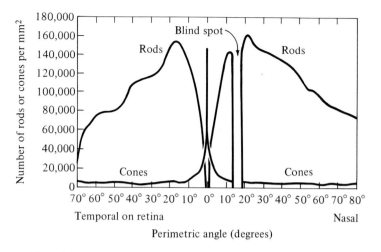

Figure 5.3.3   Distribution of rods and cones in the human retina. (From Pirenne, *Vision and the Eye*, © 1967 Chapman & Hall, Ltd. Reproduced with permission.)

(whose radius is approximately 400 μm), consists of tightly packed color sensitive cones; rods and blood vessels are absent from this "avascular" region of highest acuity. The rods, which respond to the intensity of light but not to its spectral composition, are spread throughout the remainder of the retina. Figure 5.3.3 displays the distribution of rods and cones as a function of distance from the center of the foveola, which is the intersection of the retina with the visual axis of the eye. A distance of 290 μm on the retinal surface corresponds to a visual angle of 1 degree.

The radius of the outer segment of a cone depends upon its location in the retina, but for cones in the foveola, it is about 0.5 μm, which is the approximate size of minimum features currently attainable in advanced microelectronic fabrication technology. Center-to-center distance between cones in this region is about 3 μm. Were the cones packed in a hexagonal lattice, there would be some 16,000 of them in the foveola with a density of about 0.128 per square micrometer; the experimental density measurements indicate a packing having about 0.147 cones per square micrometer, which is consistent with the assumption of nearly optimal packing; we will return to this subject in section 5.12. The high density of receptors in the foveola provides spatial resolution close to the limits of diffraction limited imagery in an ideal optical instrument. As the eye sweeps across a scene in a combination of tracking and saccadic motion, the foveal cones act like a flying spot scanner, redirecting the gaze about 20 times per second. Whereas the cone receptors of the fovea are sensitive to color, the rods which occupy the rest of the retina are achromatic and have relatively poor spatial resolving power although they are of high sensitivity.

The 125 million photosensitive receptors are connected to horizontal,

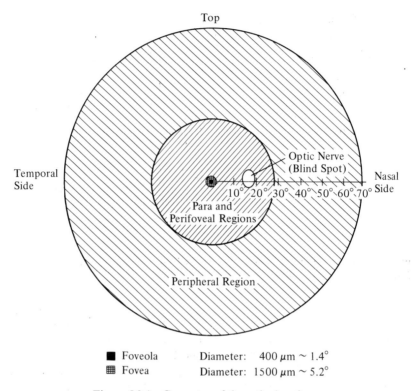

Figure 5.3.4   Geometry of the retinal surface.

bipolar and amacrine cells which in turn interact with ganglion cells whose axons are the optic nerve fibers. The optic nerve, which transmits image data to the higher cortical centers, consists of only about 1 million fibers, so there is a considerable selective omission of information as the signals gathered by the receptors are aggregated and restructured for transmission. For the highly sensitive fovea, however, the ratio of receptors to nerve fibers is about 1:1. The optic disk, where the fibers of the optic nerve gather in the retina to exit the eye, forms an elliptical region which is insensitive to incident light. The major axis of the ellipse is aligned vertically. The major and minor diameters of the optic disk are between 2,000 and 2,400 μm and 1,500 and 1,800 μm, respectively. Figure 5.3.4 illustrates the geometry of the retinal surface.

Figure 5.3.5 schematically illustrates a cross section of a portion of the retina. Notice that light entering the eyeball falls on the retina from below in the figure. From an information-processing point of view, one of the most interesting features of the retina is that the photosensitive chemical pigments that generate neural signals are located as far as possible from the incident light source, which must pass through numerous layers of signal processing cells, optic nerve fibers, and blood vessels in order to reach the receptors. It

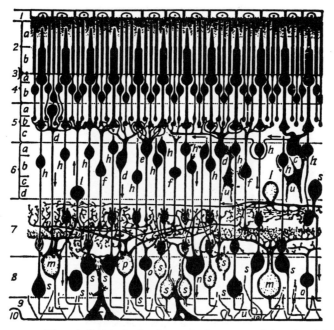

Schematic diagram showing the neural microstructure of the retina of the monkey, based on Golgi impregnations. The layers are:

   1. Pigment epithelium
   2a Outer segment of rods and cones
   2b Inner segment of rods and cones
   3. Outer limiting membrane
   4. Outer nuclear layer
   5. Outer plexiform layer (cone-pedicles and rod spherules)
   6. Inner nuclear layer
   7. Inner plexiform layers
   8. Ganglion cells (origin of primary visual projections)
   9. Layer of optic nerve fibres
  10. Inner limiting membrane
        c, horizontal cells; d, e, f, g, bipolar cells; i, l, amacrine cells; m, n, o, p, r, s, ganglion cells.

Figure 5.3.5   A portion of the retina. (From F.L. Polyak, *The Retina.* University of Chicago Press, 1941. Reproduced with permission from the author's estate.)

is remarkable that the shadows of the neural fibers, blood vessels, and various detritus floating in the pressurized gel which fills the eyeball are never seen by an observer in normal vision.

5.4 Continuity of the Visual Manifold

In the small region of the retina where the optic nerve exits, the eyeball is blind. This blind spot is a gap in the manifold of visual receptors in the retina (cp. Figure 5.3.4). Information conducted from the receptors is used by the

Figure 5.4.1   Location of the blind spot.

higher cortical centers to establish a cognitive manifold of visual perception. It is a fact of everyday experience that this perceptual manifold does not have gaps or holes; it is *simply connected*. By fixating the right eye on the cross in Figure 5.4.1 (with the left eye closed) and viewing the figure from an appropriate distance (about 25 centimeters for the author), the field of vision to the right of the cross will appear to be uniformly black. This simple experiment shows that the higher cognitive centers have filled in the region corresponding to the blind spot in such a way as to preserve the pattern of the surrounding region, which in the present case is entirely homogeneous. Since the area of the blind spot is about 23 times as great as the area of the foveola and more than one and one-half times as great as the area of the fovea, the "filling in" is assuredly a nontrivial information processing problem. This is the first and

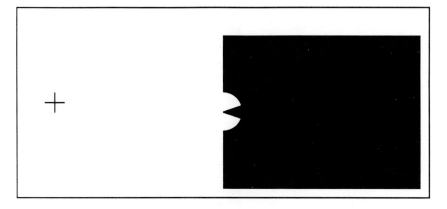

Figure 5.4.2   Half-planes.

simplest example of a cognitive process that may be called the *principle of pattern continuation,* which is an important mechanism for selective omission of information.

Before we attempt to define pattern continuation of visual images, let us consider several classes of examples. Figure 5.4.2 displays two differently colored half-planes meeting along a line except for a peculiar small indentation. If the right eye is focused on the cross at a distance such that the line runs through the center of the blind spot, the perceived image remains unchanged; indeed, no matter how the figure may vary in the region covered by the blind spot, it will still remain the same. This shows that the pattern continuation process does not fill out the image in the blind spot by a relaxation process which computes image intensities at one point by iteratively averaging the putative values of the nearest neighbor intensities subject to the constraint that the boundary values for points in visible regions just outside the blind spot be preserved. This latter procedure, which is well known in pattern recognition studies, is the discrete analogue of solving LAPLACE's equation for the region covered by the blind spot subject to the boundary values given by the visible boundary. The solution agrees with perception for the homogeneous background of Figure 5.4.1, but for Figure 5.4.2, the solution of LAPLACE's equation on the blind spot disk will continuously vary in a complicated way from white on the left through gray to black on the right. This is not, however, what the normal observer perceives.

A still more complex example shows that there are image configurations whose pattern continuation is not uniquely determined, i.e., more than one continuation is compatible with the visible boundary conditions, so that the corresponding light intensity functions are multi-valued. Consider, for instance, the configuration of Figure 5.4.3, taken from HELMHOLTZ who remarks that most observers "thought they sometimes saw one arm of the cross,

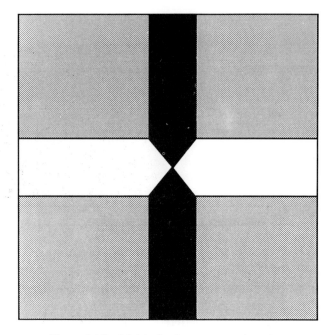

Figure 5.4.3   Multivalued pattern continuation.

sometimes the other, lying above, but more frequently the horizontal arm, probably because the horizontal diameter of the gap is less than the vertical." The "gap" is the blind spot. In this example, the two natural pattern continuations which are suggested by the visible boundary conditions are not mutually compatible and therefore require a cognitive choice for which no basis in evidence exists. Hence the perceived oscillation, which exemplifies a well known and more general phenomenon, occurring, for instance, for the NECKER cube and SCHRÖDER staircase illusions.

Our last example concerning continuation across the blind spot is this page itself. When the right eye is focused on the left side of the page, the perceptual impression conveyed is that the rest of the page, including the region covered by the blind spot, contains printed text characters, although it is of course not possible to identify which character or characters the blind spot covers.[1] This example suggests that pattern continuation applies to discrete and complex, as well as continuous and homogeneous, patterns. Of greater importance, it suggests that the cognitive information which corresponds to a pattern *not*

---

[1] In special cases of particular concentration, trained indirect vision, and using the redundancy of natural language, it may be possible to predict the unseen characters, but for randomly selected characters on a page the visual effect persists although character prediction is impossible.

under detailed scrutiny is generic and nonspecific, providing merely a superficial and partial description, possibly statistical, which is insufficient for its precise reconstruction. That is, most of the information constituting the pattern is omitted without our conscious awareness.

Everyday examples come easily to mind. One may readily visualize the appearance of a familiar building, but even gross features leave no more than a general impression; thus the number of columns that front a Greek Revival bank or library or similar edifice must be consciously counted in order to be retained in memory. Similarly, features of even the most familiar face are not normally retained in its mental representation unless they are particularly distinctive for recognition purposes. The countless examples of this type suggest that little of what we think we are seeing has the detail necessary for reconstructing anything but the coarsest approximation to reality.

5.5 Stabilized Vision

for there is no quality in this world that is not what it is merely by contrast. Nothing exists in itself.
—Herman MELVILLE

In the previous section we considered an illustration of the mind's ability to *continue patterns* in order to fill in the gap in the visual manifold caused by the presence of the "blind spot." In this section, we will consider another, more general, class of pattern continuation phenomena that are intimately connected with one important technique for the selective omission of information which reduces the brain's computing load and also extends the range of effectiveness of the sensory system.

It has long been suspected that the human eye is primarily sensitive to changes in light intensity rather than to the magnitude of the intensity itself. LETTVIN, MATURANA, McCULLOCH, and PITTS verified this hypothesis and were able to analyze its consequences for the information characteristics of the frog's vision system at a level of detail that is still beyond our capabilities for investigating human vision.

The extent to which the analysis of images by the human brain relies on changes in light intensity was dramatically revealed in 1953 by experimental studies of images that are entirely or partially stabilized relative to the retina. Even when the human eye focuses on a particular point it indulges in small, irregular, and involuntary, so-called *saccadic*, motions and the scene image consequently oscillates on the retinal screen. The question arises as to what one would see were these images stabilized, so that there would not be relative motion of image and retina. By means of an ingenious system of mirrors and contact lenses, RIGGS, RATLIFF, CORNSWEET, and CORNSWEET showed that within a few seconds after they are stabilized, all images are perceived as a uniform gray field; only *changes* in retinal images are perceived. This fundamental result was verified by YARBUS, using the more accurate but less appealing technique of gluing observing instruments to the eyeball in order to eliminate the slippage of the contact lens which caused occasional momentary reappearance of images in the experiments of RIGGS and his colleagues. More recently, the problem of residual relative motion of stabilized images has largely been eliminated by instruments that rely on low power laser beams reflected from the transparent PURKINJE surface covering the lens of the eye to track the motion of the eye's axis and provide feedback to a motor driven mirror whose operation is directed by a computer. It has also led to new experimental opportunities.

Changes in the images can occur in space and also in time. If an image window is fixed relative to the retina, then each photosensitive receptor will observe the same picture element of the image, and if the latter is not changing with time, as would a motion picture or a view of a dynamically varying scene through the image window, then the higher cognitive centers will not receive image information. For this reason, the vision system incorporates eye movements in order to produce variations in the photoreceptor inputs even when the viewed image does not change with time. This in turn means that the *spatial* input differences in the (static) image are converted into *temporal* input differences, since a given photosensitive receptor will now be exposed to one distribution of light and then to another.

This information about differences among the picture elements of the image which are captured by the time variation in the signal received by a given receptor must be reconverted into positional information. This is accomplished by the parallel (i.e., simultaneous) processing of the data gathered about a number of picture elements by neighboring receptors. While this yields a representation of the image data that is once again in the space domain rather than the time domain (by which we mean that there is a correspondence between the sensory data for all points of the image and a collection of simultaneously generated neural signals), these aggregated neural signals, nevertheless, refer to image states at more than a single instant.

The local analysis of the sensed data which identifies the places in the image where changes occur is accomplished through processes of *lateral inhibition* and *lateral excitation* that combine the signal received by a receptor with those of nearby receptors to eliminate the absolute magnitudes of the stimuli intensity and enhance the response in regions where the image changes.

Saccadic and other eye motions can be decomposed into vertical and horizontal components relative to the usual coordinate axes for the seeing eye. An apparatus conceived by CORNSWEET and refined by CRANE and others at SRI, International, makes it possible to selectively cancel either motion component and thereby provides the opportunity to perform experiments unlike any accessible to the classical methods of experimental observation.

Consider, for instance, the normal perception of a uniform square of one color (say, red) centered on a uniform background of another color (say, green) viewed by one eye. Let neither component of the eye's motion be cancelled. When the eye focusses near one of the vertical sides of the square, the horizontal component of its motion sweeps across the boundary thereby permitting the red-green variation to be perceived. The viewer sees green to the left and red to the right of the left hand vertical side of the square, and green to the right and red to the left of the right hand vertical side. Similarly, when the gaze is directed to a horizontal side of the square, the vertical component of the eye's motion sweeps across the boundary permitting the red-green variation to be perceived. The viewer sees green above and red below the top side of the square, and green below and red above the bottom side.

If the gaze is directed to a point within the square such that the saccadic motion does not cause the foveal gaze to cross any of the sides then, according to the results of RIGGS et al., no information is provided to the higher cortical centers, so a uniform neutral gray field should be seen. In fact, the pattern of green near but exterior to the boundary of the square and red near but interior to the boundary of the square is *continued* to a pattern of red throughout the interior and, in a similar way, to a pattern of green throughout the region exterior to the square.

Pattern continuation of this type minimizes the quantity of information that must be added by the brain to fill out the sensed data in order to provide a stable and nonfluctuating cognitive representation of the image.

Suppose that the vertical component of the eye's motion is cancelled. Near the vertical boundaries of the square the horizontal component of motion will act as before and the vertical edges will be perceived as before. Now consider the horizontal edges of the square. They will not be perceived because the vertical motion of the eye has been cancelled. Hence, near a horizontal edge, there will be two possibilities for pattern continuation. Proceeding from the far left or far right, pattern continuation will extend the background green until an obstruction (i.e., a reason to stop or otherwise modify the pattern continuation, based on sensed information) is encountered. Thus, the entire far background will appear to be green and, because there are no horizontal boundaries, green will be continued into the central region of the square where red will ultimately be encountered as an obstruction. The second possibility for pattern continuation starts from the red central region of the square and continues red through the (unperceived) horizontal edges into the exterior region where, ultimately, green will be encountered as an obstruction.

How is this incompatibility resolved? Which, if either, of these two pattern continuations does an observer actually see? CRANE at SRI, International, reports[2] that observers find it difficult to give a precise description of what they see. The subjective boundary between red and green is, they say, both

---

[2] Personal communication.

red and green but not a reddish green or greenish red. There is a certain lack of definiteness and perhaps a hue alteration, as if one were viewing red and green objects through a highly refracting medium such as a translucent shower room door.

For our purposes, the significant feature of the observers' descriptions is that the boundary is ill-defined and its neighborhood partakes of the image characteristics of the alternatives on both sides of the boundary. They also report that the vertical edges of the square are perceived as curved. Both results are consistent with the idea that in such regions the image as perceived at the cognitive level (as distinguished from, e.g., the physical signal received by the retinal input transducers) is cognitively constructed by a pattern continuation process in accordance with some general principles of information processing which uses as "initial conditions" the unambiguous image properties perceived in regions where the input visual stimuli vary. Nevertheless, as the curved appearance of the vertical edges suggests, the sensed data will not necessarily be preserved unmodified. This is, of course, the general basis for visual illusions.

What principles might govern the perceptions reported by these observers?

First, at the physiological level, the processes of lateral inhibition and excitation act as an edge and vertex (high curvature point) detector, with the latter having precedence in the delivery of information for processing in the visual cortex. At the psychophysical level, the stabilized retinal image experiments are consistent with lateral inhibition and excitation processes. They demonstrate that only changes in the visual field are detectable by the vision system. For simple images, this means that boundaries separating optically homogeneous regions are directly detectable but that homogeneous regions themselves are not. Furthermore, these experiments, as well as ordinary visual experiences, show that a cognitive process of *pattern continuation* completes the mental representation of an image in optically homogeneous regions, including the region which corresponds to the projection of the blind spot on the visual scene. The pattern continuation process acts to fill in and complete the sensed image in a way which minimizes the information needed to supplement the sensed data in order to provide a mental representation of the complete scene. The global cognitive representation thus appears to be derived from local sensed data through a hierarchy of transformations that in general converts the original data to a much more condensed but much more highly structured form.

Let us examine the information processing implications of the stabilized image experiments in greater detail. The retina can be considered as a matrix of photosensitive receptors. Suppose that the eye views a fixed (unmoving) scene that consists of two half-planes of different colors (say, black and white) which meet along a common line, as shown in Figure 5.5.1. The instantaneous image of this scene on the retina will bathe some photosensitive receptors with "white light" reflected from the white half-plane (that is, with light whose

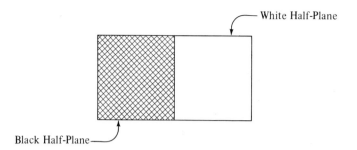

Figure 5.5.1   Black and white half-planes.

spectral characteristics and luminance are similar to those of the light which illuminates the scene), and it will illuminate the other photosensitive receptors with the much smaller amount of light reflected from the black half-plane. If the "black" half-plane were perfectly black, it would be a perfect absorbing medium and not reflect light at all. For the sake of simplicity in the discussion we will assume that this is the case.

If the image of the scene illustrated in Figure 5.5.1 is stabilized on the retina of the eye, then each photosensitive receptor will receive illumination that does not change with the passage of time. Under these circumstances, according to the experimental evidence, the higher visual centers of the brain will fail to report the presence of an image: the scene will not be perceived.

If the image of the scene moves relative to the retina, either as as result of motion of the scene itself, or due to movement of the eye, as in saccadic oscillation, then, depending on the details of the relative motion, some retinal receptors will experience changes in illumination while others will continue to detect light which is unvarying. In general, the photosensitive receptors which will experience changes of illumination are those that lie on and near the image of the line which separates the two planes. For instance, if the eye oscillates along a line perpendicular to the boundary between the two planes, with a small angular excursion, then there will be a vertical band centered on the boundary line, illustrated in Figure 5.5.2 by the horizontally hatched region, where the illumination varies from the "white" state to the "black" state as the eye moves. If the motion of the eye is small, as it is for saccades, then points on the retina which are distant from the line which separates the halfplanes will continue to be illuminated or unilluminated according as they lie in the white or black half-plane, respectively. We know that, in this case, the viewer will observe the scene as it truly is: black and white half-planes which meet along a common line. But the stabilized vision experiments demonstrate that receptors distant from the line which separates the half-planes, i.e., those corresponding to points outside the horizontally hatched region in

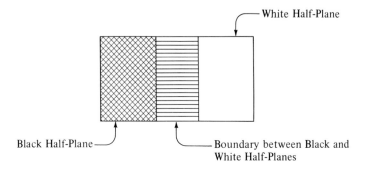

Figure 5.5.2   Boundary between half-planes.

Figure 5.5.2, will not contribute information about the image whereas those corresponding to points which do lie within the horizontally hatched region where the illumination varies with time will contribute information which is in fact the sole sensory basis for perception of the scene. Here we implicitly make the assumption that distant photoreceptors are not connected to one another although nearby ones may be.

Several important conclusions can be drawn from these observations. First, the spatially varying information in the scene is converted into temporally varying information for each point in the retinal image by the human vision system. If the image on the retina remains fixed in time, it is not cognitively perceived. It follows that the time constants for the recovery of the photoreceptors to their original state after signalling the incident illumination must be an important parameter which governs and limits human vision. It is evident from our previous considerations that the spatially varying scene information is encoded into a spatiotemporally varying signal which in effect reports the local *differences* of illumination intensity for each of the photosensitive receptor types; the original illumination magnitudes are lost. Since local differences in scenes which consist of regions each of which is uniform in appearance (such as the scene in Figure 5.5.1) correspond to boundaries of regions, we conclude that the vision system has a built-in "edge" or "contour" detector based on the differencing process. But we should not assume that the edge detection mechanism operates by something akin to the calculation of local tangents to the boundary between regions of different illumination. We will return to this subject in Chapter 6.

It is convenient to think of the differences between the illumination at nearby points as proportional to the spatial derivative of the scene illumination in the direction of the vector joining the points. In the case of Figure 5.5.1, this derivative is zero at every point that does not lie on the line separating the half-planes regardless of the direction considered. For points on the boundary line, the derivative varies from 0 in the direction of the

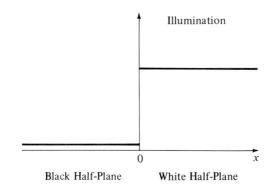

Figure 5.5.3   Ideal illumination on a horizontal line in Figure 5.5.1.

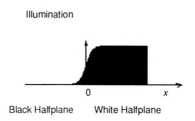

Figure 5.5.4   Realistic illumination on a horizontal line in Figure 5.5.1.

boundary itself, to an extremum for directions perpendicular to the boundary. If we examine a horizontal cross section of image illustrated in Figure 5.5.1, then the intensity of illumination varies as shown in Figure 5.5.3, where the coordinate system has been chosen so that the points on the line which lie in the black half-plane correspond to $x < 0$, those in the white half-plane to $x > 0$, and $x = 0$ corresponds to the boundary. The vertical axis indicates the constant illumination which falls on the photosensitive retinal receptors for each of the two regions. In reality, because of the finite diameter of the receptors, those on which the image of the boundary line falls will receive illumination from both half-planes. This means that the sensed illumination levels will vary continuously and smoothly from 0 (corresponding to the black half-plane) to the positive level corresponding to the white halfplane as suggested by the curve in Figure 5.5.4. The derivative of this function (Fig. 5.5.5) represents some of the information which is captured by the photosensitive receptors and is available for higher level cognitive processing, and the position of the peak of the derivative coincides with the boundary between the black and white regions.

The process of replacing an image by its derivative is one important way of omitting information contained in the original scene while selectively retaining the portion that is most necessary for recognition and evaluation

Scaled Rate of Change
of Illumination

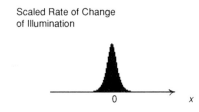

Figure 5.5.5   Derivative of illumination function displayed in Figure 5.5.4.

purposes. In fact, this method of omission is primarily a method of *recoding* the information, because the original input data (or at least for an approximation to them which is sufficiently accurate for cognitive processing needs) can be recovered if illumination intensity information (stimulus magnitude) is retained for points close to the edges and other boundaries. Each connected region of constant illumination is completely specified if the *shape* of the region is known and if the magnitude of the illumination is known for (any) one point of the region, which merely corresponds to the fact that the derivative of a constant is 0. This can be put another way which is more revealing because it provides an example of the general principle of using group invariant functions and "symmetries" of the observational circumstances to reduce information requirements. If $x \mapsto f(x)$ is a function and if $f(x_1) = f(x_2)$ for $a < x_1 < x_2 < b$, (i.e., f is constant on the interval from a to b), then for any number t, $f(x + t) = f(x)$ as long as both x and $x + t$ lie in the interval, so f is an *invariant* of the translations $x \mapsto x + t$ for sufficiently small t. Thus f is invariant for those elements of the group of translations that do not stray too far from their initial point, and hence it is an invariant of the "infinitesimal" Lie algebra of the group. It follows that the derivative (df/dx) not only will be constant, but is actually 0 for those same points; it will not require any storage capacity or channel capacity in an information processing system.

Images are 2-dimensional; the variation of image intensity differences and of derivatives of functions depend on direction. Because they are small except at boundary points of regions and largest in directions perpendicular to the boundary, the most significant information for scene reconstruction and pattern recognition is already provided by the directional derivative evaluated in the direction where its magnitude is most extreme. This further omits information in a selective way without important losses, but it does raise the question of whether the brain can rapidly and concurrently calculate the directional derivatives in a number of different directions at a point in order to ascertain the direction for which the derivative is a maximum. Put another way, since direction is a parameter at each point, it would appear that the directional derivative depends on one variable in addition to the two coordinates of the point itself. This raises the question of whether there is a function like the directional derivative in the sense that it is zero for places where the illumination does not vary and depends only on illumination near

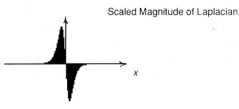

Figure 5.5.6   Laplacian applied to function from Figure 5.5.4.

a point but not on a direction at the point. Functions having these properties are not difficult to find. The simplest of them is called the *Laplacian differential operator*, denoted Δ:

$$\Delta f = \frac{\partial^2 f}{\partial x^2} + \frac{\partial^2 f}{\partial y^2}. \tag{5.1}$$

This generalization of the second derivative combines information about the variation of the function f in both the x- and the y-variable directions. In Chapter 6 we will see that the Laplacian plays an important role in edge detection and pattern recognition. Here, let us merely observe that when it is applied to the illumination exhibited in Figure 5.5.4 (which corresponds to a section of the 2-dimensional image shown in Figure 5.5.1), the Laplacian is 0 far from the boundary (where the illumination is constant) but near the boundary it behaves as shown in Figure 5.5.6. Notice that on this section (where the y-variable is constant) the Laplacian simplifies to the second derivative, $(d^2 f/dx^2)$. It is nonzero only where the illumination and the rate of change of the illumination both vary. If the rate of change does not vary, then the illumination itself can vary linearly with x; hence, a linear variation in the intensity of illumination (a "ramp") could not be observed by a Laplacian detector.

By recoding scene information into derivatives, the human vision image processing system eliminates the sensed magnitudes of illumination except, possibly, at points close to edges. Yet we appear to see scenes in their entirety, with even large uniform regions of illumination "filled in." Thus another important conclusion that can be drawn from stabilized image experiments is that the higher processing centers of human vision must cognitively calculate the "filled in" parts of an image. This is a kind of *pattern continuation* process that uses "boundary conditions" in both the image structure sense and the sense in which this term is used in physics or mathematics to refer to systems whose behavior is determined by general laws that are instantiated by the prescription of specified constraints on particular parts of the total system. The behavior of the remainder of the system is a consequence of the general laws and can be determined by deductive analytical processes. Something like this evidently occurs in human image processing of uniform black and white scenes, where boundary information is used to characterize the complete scene. A note of caution is appropriate at this point, for the determination of "boundaries" in typical complex images is by no means a straightforward matter. Variations of brightness, hue, and saturation (terms which describe the subjective perception of color) interact with variations of texture (about which more will be said in Chapter 6) in ways that are far from being fully understood today.

5.6 Information Content of Contours

The purpose of this section is to study how information is coded in contours. More precisely, we will be interested in how the human vision system organizes the information associated with a contour by eliminating some of it and retaining the rest to be acted upon by the higher mental processing centers.

By a *contour* we mean a simple closed curve appearing in an otherwise uniform field of vision. The field of vision can be modeled by the Euclidean plane of elementary geometry. Contours and collections of contours are idealizations of the boundaries of physical objects which ignore color, shading, and textural properties. A cartoon consisting of black lines drawn on a white sheet of paper is an example of a collection of contours in the sense in which the term is used here.

The perception of contours is an important factor in the perception and interpretation of images. Contours may be explicitly realized as curves appearing against a uniform background as described above but more often they occur together with, and are determined by, variations in texture and color. Thus, a region of one uniform texture seen against a background of a different uniform texture determines a contour which coincides with the curve that separates the two textures; regions of uniform but different colors (here we include gray scale variation as a variation of color) define contours in a similar way.

The study of the information content of contours was begun by the psychologist ATTNEAVE in 1954. He showed that the information associated with a contour or curve is not uniformly distributed along it but "bunches up" at certain points. It is the information associated with these points and their nearby neighbors that is essential for pattern classification and recognition. This is a fundamental result. The elegant experiment upon which he based his work can be summarized as follows. Suppose that a simple closed curve C, such as the one shown in Figure 5.6.1, is given and that an experimental subject is asked to place a fixed number of points, say N, on a blank sheet of paper so as to provide the most accurate approximation to the curve C. Where will the subject place the N points? If too few points are available, it will not be possible to provide an acceptable approximation to the curve. For instance, 2 points are certainly too few to depict curve C in the figure above. If the points are numerous, let us say $N = 1$ million, then after a while their place-

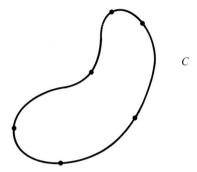

Figure 5.6.1   A simple closed curve.

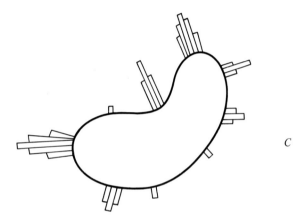

Figure 5.6.2   Histogram for approximations to C.

ment will become tedious and not yield much new information at all. But for an intermediate range of values of N, which depends on the given curve itself, it will be a challenging but possible task to provide a reasonably accurate picture of the form of curve C by properly placing the points. As ATTNEAVE discovered, people tend to place the points in similar relationships to one another. By subdividing the curve C into a sequence of small arcs and accumulating the points intended to correspond to points in each arc along the given curve C placed by numerous experimental subjects, ATTNEAVE constructed a histogram along the curve which showed how many subjects had placed points in each small arc segment. The result might look something like Figure 5.6.2 for the curve C given in Figure 5.6.1. The histogram shows that if only $N = 6$ points were available, a person would most likely place them as shown in Figure 5.6.3 in order to approximate the curve C "best."

ATTNEAVE drew the fundamental conclusion that the approximating points were most likely to be placed at those points of the curve where the absolute

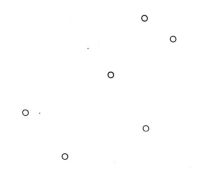

Figure 5.6.3   Approximation to C with $N = 6$.

Figure 5.6.4   "Cheshired" ATTNEAVE's cat.

value of the curvature was a local maximum: where the curve "turns" most. Since this means that the information in a contour must be concentrated at these points, ATTNEAVE took the additional step of realizing that the perception and recognition of objects represented by contours depended primarily upon the information contained in points of the contour near the points of extreme curvature (in many cases that information is in fact sufficient). To illustrate this observation he connected the points of extreme curvature in a picture of a cat by straight line segments to obtain the famous and still clearly recognizable ATTNEAVE cat shown in Figure 1.9.1 of Chapter 1. Figure 5.6.4 shows that the illusion of a cat remains if large portions of the line segments are erased: the information really is concentrated in the neighborhood of the points of extreme curvature.[3]

These results can be confirmed analytically. To do so it is necessary to know how much information is gained from a measurement of a *direction*. This corresponds to the measure of information gain for observations of *position* along a line which was derived in Chapter 2.

---

[3] The presence of the cat is greatly diminished if only the central portions of the line segments are retained. Subjective contours play a crucial role in this context; cf. section 5.7.

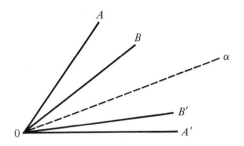

Figure 5.6.5   Two measurements of direction D.

Suppose that we attempt to measure a certain direction D emanating from a point O. Since measuring instruments are inexact, it will not be possible to measure the direction D exactly. The best that can be done is to obtain, as the result of the measurement, a certain angle AOA' such that we can be sure that the direction D lies somewhere within this angle. If a second measurement is made which refines the first, then the unknown direction D will be confined to some smaller angle BOB' contained within AOA'; this situation is illustrated in Figure 5.6.5. How much information is gained from the second measurement compared with what was known at the end of the first measurement? This question can be answered in either of two ways: by directly applying the information measure defined in Chapter 2, or by considering the properties which a measure of directional information should possess and using them to determine the measure. Let us do both.

First we will apply the results of Chapter 2. Think of a circular gauge centered on point O whose periphery contains angular calibrations marked off (like a protractor) in proportion to the length of arc subtended by an angle with its vertex at O.

Let angle AOA' be denoted by α and BOB' by β. Then angle α will correspond to a segment of arc of length $S(\alpha)$ and angle β will correspond to a segment of arc of length $S(\beta)$; the latter segment will be a subset of the former because the angle corresponding to β is contained within the angle corresponding to α. We will measure the directional information gain in terms of the information gained from the measurements of the corresponding arc lengths. From Chapter 2 we have for the latter (see Figure 5.6.6)

$$I = \log_2\left(\frac{S(\alpha)}{S(\beta)}\right).$$

This expression can be simplified to one in which the angles α and β appear explicitly. From elementary geometry we know that the length of arc of a circle of radius r subtended by an angle θ is $S(\theta) = r\theta$ if θ is measured in radians. Hence we may write $S(\alpha) = r\alpha$ and $S(\beta) = r\beta$ where r denotes the radius of the gauge circle, and therefore

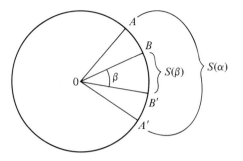

Figure 5.6.6   Pair of direction measurements.

$$I = \log_2\left(\frac{S(\alpha)}{S(\beta)}\right) = \log_2(\alpha/\beta) \text{ bits;} \tag{5.2}$$

the information gained is independent of the radius of the measuring gauge, as of course it should be.

Information gain is always a *relative* quantity. For this reason, the information gained from the measurement of an interval must always be considered relative to some prior measurement of an interval. But for directions and the angles that correspond to them, the situation is different because we know *a priori* that the angular measure of a direction must lie between 0 and 2π. Therefore, a single measurement of direction which limits the unknown angle to lie within a sector of angle α can be considered relative to the *a priori* limitation that it surely lies within a sector of angle 2π. Substitution in (5.2) yields a formula for the information gained by a single measurement of direction which limits it to lie within a sector of angle α. The formula is:

$$I = \log_2\left(\frac{2\pi}{\alpha}\right) \text{ bits.} \tag{5.3}$$

Now suppose that two measurements are made in succession, resulting in angles α and β with $\alpha > \beta$. The absolute information obtained from the first measurement is $\log_2(2\pi/\alpha)$ and the absolute information gained from the second is $\log_2(2\pi/\beta)$. The information gained by the second measurement compared with what was known upon completion of the first measurement is

$$\log_2\left(\frac{2\pi}{\beta}\right) - \log_2\left(\frac{2\pi}{\alpha}\right) = \log_2\left(\frac{\alpha}{\beta}\right)$$

using properties of the logarithm; this is just our original formula (5.2).

The axiomatic approach to determining the measure of directional information is similar to the procedure followed in section 2.3 and brings the role of group invariance to the forefront once again. Suppose that $I(AOA')$ is the information gained by a measurement which limits a direction to lie within an angle AOA', relative to the knowledge that the direction issued forth at some angle from O. The measure of information $I(AOA')$ cannot depend on the unit of distance in the plane, nor can it depend on the orientation of the angle AOA'. Both constraints are most easily represented analytically by introducing complex numbers. Each point in the plane corresponds to a complex number. Let us choose coordinates so that O corresponds to the number 0. We can also assume without loss of generality that A and A' are at the same distance from 0. Then angle AOA' is defined by the complex numbers that correspond to A and A', which we will denote by A and A'. So we may write $I(A, A')$ to show that I is a function of the complex numbers A and A'. The statement that $I(A, A')$ is independent of the orientation of angle AOA' and of the distance in the plane means that, for any complex number z different from 0,

$$I(zA, zA') = I(A, A')$$

because multiplication by a complex number is equivalent to a rotation followed by a magnification. Choose $z = 1/A$ in the equation to find

$$I(1, A'/A) = I(A, A');$$

the information depends only on the ratio A'/A. Since A and A' are equidistant from 0, their ratio is the complex number $e^{i\alpha}$ where α is the magnitude of angle AOA'. Hence the information gained depends only on the angle α. Notice that if α is augmented by 2π, the information remains unchanged since $e^{i(\alpha+2\pi)} = e^{i\alpha}$. So let us write $I(\alpha)$ for information $I(A, A')$, understanding that α is confined to the range $0 \le \alpha < 2\pi$.

If a second, more exact, observation is made, it will result in an information gain $I(\beta)$ corresponding to an angle β which is smaller than α. Recall that an angle is measured by the length of the arc it subtends in a circle of radius 1. Then α and β are arc lengths. Since $\alpha = 2\pi$ implies that no information has been gained, $I(2\pi) = 0$. Now define the relative information gained from the measurement β following the measurement α to be

$$I(\alpha, \beta) = I(\beta) - I(\alpha).$$

This relative information must be independent of the unit of measure (whether angles are measured in radians or degrees cannot affect the angular information content of the measurement), so $I(c\alpha, c\beta) = I(\alpha, \beta)$ for $c \ne 0$; equivalently,

$$I(c\beta) - I(c\alpha) = I(\beta) - I(\alpha).$$

Select $c = 2\pi/\alpha$ to find $I(2\pi\beta/\alpha) = I(\beta) - I(\alpha)$. Replace β by $2\pi\beta$ and α by $2\pi\alpha$ to obtain

$$I(2\pi\beta/\alpha) = I(2\pi\beta) - I(2\pi\alpha).$$

Following the procedure in Section 2.3, we discover that $I(\alpha)$ is proportional to $\log_2(2\pi/\alpha)$, and if we choose to measure information in bits, then the constant of proportionality is 1, which yields formula (5.3).

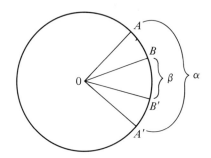

Figure 5.6.7   Angular measurement intervals.

Before turning to apply this formula to study the information content of a contour and ATTNEAVE's results, let us consider several simple examples of how this measure can be used.

EXAMPLE 5.6.1. If $\alpha = \pi$, the angle is a straight line and the information content is $I(\pi) = \log_2(2\pi/\pi) = \log_2 2 = 1$ bit; the measurement has merely specified that the observed direction lies in one of two half-planes.

EXAMPLE 5.6.2. $I(\pi/2) = 2$ bits and more generally, $I(\pi/2^n) = n + 1$ bits. Successive bisection adds one bit at each stage. In particular, smaller angles correspond to greater information. This explains (in part) why arrowheads (e.g., "→") are used as pointers and information markers: they catch the attention because they contain more information than smooth contours or large angles.

EXAMPLE 5.6.3. Excluding considerations of size, the triangle which provides least information is equilateral. That is, if A, B, C are the angles of a triangle, then $A + B + C = \pi$ and the corresponding information is

$$I = \log_2(2\pi/A) + \log_2(2\pi/B) + \log_2(2\pi/C);$$

I is minimal if $A = B = C = \pi/3$. This can be proved using the calculus of functions of several variables.

More generally, the N-sided polygon providing least information is equi-angular: rectangles provide less information than parallelograms or trapezoids, etc. This result has applications to understanding visual illusions.

Let us apply this train of ideas to the problem of locating those parts of a contour where the information is concentrated. Suppose a portion of a contour is given and that it is subdivided into short segments of equal length from some initial point P to some final point Q as shown in Figure 5.6.8. Each segmenting point (excluding P and Q themselves) can be thought of as the vertex of an angle formed with the two neighboring points. In the figure, AOA'

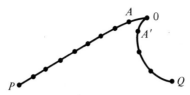

Figure 5.6.8   Linear approximation of a contour.

Figure 5.6.9   A right angle.

is such an angle. Associated with each angle is its measure of information given by formula (5.3) and, for each successive pair of angles along the path from P to Q, (5.2) prescribes the gain (or loss) of information upon passing from one angle to the next. Suppose, for instance, that the portion of the contour consists of two adjacent sides of a rectangle and the included right angle POQ (Figure 5.6.9). As we move by small steps from P to O along PO, we see that each angle is a straight angle whose corresponding information is $\log_2(2\pi/\pi) = \log_2 2 = 1$ bit but that the information gain in passing from one straight angle to the next is $\log_2(\pi/\pi) = \log_2 1 = 0$ bits since the angle remains unchanged. When the right angle with vertex at O is finally reached, there is a positive gain in information, equal to $\log_2(\pi/(\pi/2)) = \log_2 2 = 1$ bit. At the next step, passing from the right angle to the straight angle, there is an information *loss*, equal to $\log_2((\pi/2)/\pi) = \log_2(1/2) = -1$ bit, which is interpreted as meaning that information is gained passing from the straight sides *to* the right angle, and lost upon passing *from* the right angle to the straight sides. The right angle is the only place where the contour is "curved" (that is, changes its direction), so we conclude that only there, at the point of extreme curvature, is information concentrated.

   This example is special because it consists of straight line segments and angles, but it does not have nontrivial smoothly curved portions. The general case can be treated the same way. One finds that any curved part of a contour yields a greater information gain than any straight segment (which yields comparatively none at all), that the more strongly curved portions yield more information, and that "corners," i.e., angles, yield the greatest amount with the more acute angles providing more information than the less acute ones.

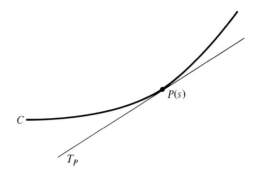

Figure 5.6.10   Curve with tangent line.

All this can be combined into the single statement that the information content of a contour is concentrated in the neighborhood of points where the absolute value of the curvature is a local maximum.

The qualitative formulation given previously can be supplemented by a quantitative one based on an explicit definition of the term *curvature* which has been used so freely thus far. Suppose that a curve in the plane is given together with some fixed point O on it, and further suppose that the points of the curve are put into correspondence with their distance from O along the curve, where distance is measured as positive for points of the curve lying on one side of O and negative for points on the curve lying on the other side. Let $P(s)$ be a point on the curve at distance s from O. We will suppose that the curve has a tangent line at each if its points. This means that it is "smooth" and does not have corners. Figure 5.6.10 exhibits such a curve C along with the tangent line $T_{P(s)}$ to the curve at point $P(s)$. If Δs is a small increment in s, there will be a point on the curve at distance $s + \Delta s$ from O. Denote this point by $P(s + \Delta s)$; it will lie close to $P(s)$ along the curve, the closer to $P(s)$ the smaller Δs is. The curve will have a tangent line $T_{P(s+\Delta s)}$ at $P(s + \Delta s)$, and the tangent lines $T_{P(s)}$ and $T_{P(s+\Delta s)}$ meet at a small angle $\Delta \alpha$ as shown in Figure 5.6.11. For smooth curves (curves without corners), the angle $\Delta \alpha$ will tend to 0 as Δs tends to 0, that is, the closer $P(s)$ and $P(s + \Delta s)$ are to each other. In the language of the calculus, the ratio $\Delta \alpha / \Delta s$ is an approximation to the curvature of the curve at $P(s)$; the curvature $k_{P(s)}$ is expressed precisely by the defining equation

$$k_{P(s)} = \lim_{\Delta s \to 0} \frac{\Delta \alpha}{\Delta s} = \frac{d\alpha}{ds}, \tag{5.4}$$

where $(d\alpha/ds)$ denotes the *derivative* of α with respect to the distance s measured along the curve.

Neither the human eye nor any other measuring instrument can distinguish infinitesimal quantities so the best that can be done is to approximate curvature by the ratio $\Delta \alpha / \Delta s$ for Δs near the limits of resolution of the instrument:

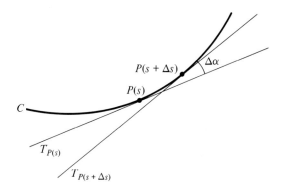

Figure 5.6.11   Angle between nearby tangent lines.

say 1 jnd of distance for the human eye. Then for real situations, (5.4) will replaced by the approximation

$$\frac{\Delta\alpha}{\Delta s} = k_{P(s)}. \tag{5.5}$$

Now suppose that two points are selected on the curve, corresponding to distances s_1 and s_2 from O, and let the corresponding curvatures be estimated using the same small value of Δs. For the point $P(s_1)$ we will have

$$k_1 = k_{P(s_1)} = \Delta\alpha_1/\Delta s \tag{5.6}$$

where we have abbreviated the curvature at $P(s_1)$ by k_1 and $\Delta\alpha_1$ is the angle between the tangent lines at $P(s_1)$ and $P(s_1 + \Delta s)$. Similarly, at $P(s_2)$ we find the curvature estimate

$$k_2 = k_{P(s_2)} = \Delta\alpha_2/\Delta s$$

with corresponding notational conventions. Dividing these two equations term by term yields the fundamental relationship

$$\frac{\Delta\alpha_1}{\Delta\alpha_2} = \frac{k_1}{k_2}. \tag{5.7}$$

We already know how to measure the information gain which results from measurements of two angles: by (5.2) we obtain for the information gain corresponding to measurement of angle $|\Delta\alpha_2|$ compared with measurement of $|\Delta\alpha_1|$ (where we use absolute values to ensure that the angles lie between 0 and 2π),

$$I = \log_2\left(\frac{|\Delta\alpha_1|}{|\Delta\alpha_2|}\right)$$

and by (5.7) this can be expressed in terms of the curvatures at the corresponding points of the curve by the relation

$$I = \log_2\left(\frac{|\Delta\alpha_1|}{|\Delta\alpha_2|}\right) = \log_2(|k_1/k_2|). \tag{5.8}$$

If, for example, we fix the point $P(s_2)$ so that the curvature $|k_2|$ is very small, then a point $P(s_1)$ where the curvature $|k_1|$ is larger will yield an information gain, and equation (5.8) provides the means for measuring that gain quantitatively.

It is important to realize that since real instruments cannot distinguish points which are infinitesimally close together, it follows that they also cannot distinguish between curves that have "corners," and curves that are "smooth" and have tangent lines at each of their points. For this reason curves with "corners," such as polygons, can be considered as idealizations of what the human eye or other instruments are actually capable of observing. This remark is necessary for the proper interpretation of (5.8) because the curvature of a "corner" is, of course, infinite although an infinite gain in information cannot be realized by any information processing system.

The expression (5.8) confirms the results ATTNEAVE deduced experimentally. Information contained in a contour is concentrated in the neighborhood of points of extreme curvature, for if S_2 is kept fixed at a point where k_2 is small, then in a neighborhood of that point I will be greatest where the curvature is most extreme. Thus, if the problem of describing the shape of a contour is decomposed into a collection of "local" problems of determining the best place to put points in various short intervals along the curve, then it will be optimal to place the points at the local extrema of curvature.

Thusfar we have studied the distribution of information along a contour from a phenomenological standpoint, without any concern for whether or how the necessary computations could be performed by the neural net. Now we will turn to consider this question briefly.

The vision system of the horseshoe crab *Limulus* has been intensively investigated. It appears to share many features with the human vision system although *Limulus's* eye is compound, consisting of approximately 1,000 ommatidia. Let us number the ommatidia for ease of reference.

When the ith ommatidium is illuminated by light of intensity I_i and all other ommatidia remain unilluminated, its output fiber fires with a frequency e_i. The firing frequency increases and decreases along with the intensity of illumination. When several ommatidia are illuminated simultaneously, the firing frequency of the ith ommatidium will depend on illumination of the other ommatidia as well and will generally be less than e_i. This effect is called *lateral inhibition*: "inhibition" because the firing frequency is generally decreased and "lateral" because it acts orthogonally to the direction of propagation of the sensed signal which travels from the photosensitive receptor to the brain.

Let the firing frequency of the ith ommatidium in the presence of illumination be denoted by x_i. The results of empirical studies are summarized in the HARTLINE-RATLIFF equations:

$$x_i = e_i - \sum_{j=1, j=i}^{n} k_{ij} \max(0, x_j - t_{ij}), \qquad i = 1, \ldots, n. \qquad (5.9)$$

Here n denotes the total number of ommatidia, $\max(u, v)$ denotes the larger ("maximum") of the numbers u and v, k_{ij} is the coefficient of inhibition of the ith ommatidium by the jth, and t_{ij} is a threshold value which depends on the ith and j^{th} ommatidia. The constants k_{ij} and t_{ij} are nonnegative.

The inhibition coefficients k_{ij} decrease, and the threshold values t_{ij} increase, as the distance between the ith and the jth ommatidia in the eye increases. In general, $k_{ij} = k_{ji}$ and $t_{ij} = t_{ji}$; that is, the inhibitory effect of the jth ommatidium on the ith and of the ith on the jth may be different, and the threshold value for inhibition of the ith ommatidium by the jth may differ from the threshold value for inhibition of the jth by the ith.

The system of equations (5.9) is nonlocal. Its implications are not directly apparent because the output—and unknown—firing frequencies $x_1, \ldots, x_n$ appear on both sides of the equations. In order to gain an understanding of what consequences they have for vision we will make some simplifying assumptions.

Let us assume that the threshold value t_{ij} for inhibition of the ith ommatidium by the jth actually depends only on the properties of the inhibiting j^{th} ommatidium but not on the ith. Then we can write t_j in place of t_{ij}. Let us also assume that the firing frequency x_i is greater than the threshold value t_i for $i = 1, \ldots, n$, i.e., for every ommatidium. Then $\max(0, x_j - t_j) = x_j - t_j$ and equation (5.9) can be written in the simpler form

$$x_i = e_i - \sum k_{ij}(x_j - t_j). \qquad (5.10)$$

Subtraction of t_i from both sides and replacement of the difference $x_i - t_i$, $i = 1, \ldots, n$ by the symbol u_i yields the system of equations

$$u_i = (e_i - t_i) - \sum k_{ij} u_j, \qquad i = 1, \ldots, n. \qquad (5.11)$$

The use of matrix algebra greatly simplifies the rest of the analysis. Let $u = (u_i)$, $c = (e_i - t_i)$ denote vectors whose components are as indicated, let $K = (k_{ij})$ denote the matrix whose elements are k_{ij}, and denote the unit matrix (i.e., the matrix whose diagonal elements are 1 and all others are 0) by $\mathbf{1}$. Then the system of equations (5.11) is equivalent to the vector equation

$$(\mathbf{1} + K)u = c. \qquad (5.12)$$

It follows that the solution of the equation is

$$u = (\mathbf{1} + K)^{-1} c = c - Kc + K^2 c - \cdots \qquad (5.13)$$

if the inverse matrix $(\mathbf{1} + K)^{-1}$ exists.

If the coefficients of inhibition are small, then the adjusted firing frequencies u can be approximated by

$$u_i = c_i - \sum K_{ij} c_j, \qquad (5.14)$$

from which we see that each component adjusted firing frequency will, in general, be *less* than the uninhibited adjusted firing frequency. Let us therefore accept the conclusion that the effect of illumination is to reduce the firing frequency—the output—of the latter, and that the effects of the neighboring ommatidia are independent and combine by simple summation.

Suppose that the image of a uniformly illuminated convex region falls on the compound eye of *Limulus*, and that the intensity of illumination is greater than any threshold values. For an ommatidium all of whose nearby neighbors are exposed to interior points of the convex region, the effect of lateral inhibition will be greatest and the response of the ommatidium will be least amongst those exposed to the illumination.

Further, suppose that an ommatidium is illuminated by a point on the boundary of the convex region. Then only those of its neighbors which lie inside the region will be illuminated and the inhibitory effect will be proportionally *reduced*. Consider, for instance an ommatidium situated at the vertex O of an angle whose interior is illuminated. The neighboring ommatidia which lie within a circle of some fixed radius centered at O will inhibit the ommatidium at O if they are illuminated. If the ommatidia are isotropically distributed with respect to O, then the fraction of ommatidia within the circle with are illuminated will be equal to the ratio of the angle to 2π. Hence inhibition will be *less* for more *acute* angles.

A refinement of this argument shows that the portions of the contour where the curvature is extreme will produce the least inhibitory effect. The scale on which variations in the contour occur must of course be sufficiently large so that significant changes in curvature can be detected by the vision system.

The process of lateral inhibition in *Limulus* detects just those configurations along a contour where, according to our previous investigations, the information is concentrated. Similar processes are no doubt at work in higher animal and in human vision. Thus, a universal principle of information processing is realized in the biological realm by this particular implementation.

5.7 Subjective Contours

Sensory illusions are numerous and varied, especially for the modality of vision. We have already discussed the two most remarkable although rarely noticed ones: the illusion of continuity of the visual manifold despite the presence of the "blind spot," which demonstrates that the manifold of vision is a mental construction; and the illusion of direct perception of light intensity belied by the stabilized vision experiments which demonstrate that only stimuli differences are perceived. The illusion of *subjective contours*, which is the subject of this section, is well known and readily demonstrated. It provides special insight into the way information is organized in the cognitive centers, and how little of it is actually needed to provide the impression of perception.

It also demonstrates that the "creative filling in" of omitted information in images is routinely performed as image information is processed. The study of subjective contours also provides another application of the measure of directional information which was developed in the previous section.

Subjective contours have been described in numerous books and articles about visual illusions, and they have been subjected to intensive study in many psychophysical investigations dating back to at least 1904, and including relatively recent ones by CoREN, GREGORY, KANISZA, and ULLMAN. The important discovery by VON DER HEYDT, PETERHANS, and BAUMGARTNER in 1984 that subjective contours have physical correlates in the visual cortex further blurs the boundary between reality and illusion, for they observed cortical cells in Area 18 of the rhesus monkey that respond to subjective contour stimuli, i.e., to stimuli that induce the subjective impression of contours for human observers in regions where the stimulus luminance is constant. They also showed that cells in Area 17 do not respond to subjective contour stimuli. Thus one must conclude that subjective contours have a "reality" that is antecedent to knowledge-based, learned pattern recognition although they are in part independent of physical sensation: they lie between physical reality and cognitive abstraction, modifying the former and mediating the latter. Under certain conditions, the visual system will create contours which bridge the gaps in incomplete patterns. Figure 5.7.1 exhibits several examples of these *subjective contours*.

Studies have shown that the surfaces which are bounded by subjective contours can appear as much as 11% brighter than the background. Indeed, this artifact of brightness differential is the mechanism which the visual system uses to create the impression of contours where there are none.

Subjective contours appear to arise when, for example, an ordered set of points is given and two directions are specified at each point by excising a

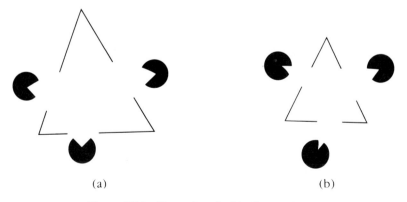

(a)   (b)

Figure 5.7.1   Examples of subjective contours.

P

Q

Figure 5.7.2   Concentration of information at endpoints of contours.

sector from a small black disk centered at each point. If the points are appropriately positioned and the sectors appropriately oriented, then consecutive points will appear to be joined by a contour tangent to a direction defined by a side of the sector at each of them. Consider such a pair of points and directions, as shown in Figure 5.7.2. These artificial contours are evidence that much of the information contained in real contours is concentrated in the neighborhood of certain points—in this case, the endpoints.

There are infinitely many curves that connect P to Q and are tangent to the arms of the sectors shown in the figure. How does the brain make its selection from these riches when it creates a subjective contour? This question has been treated by ULLMAN; we will obtain his results in another way by applying general ideas about information.

We will assume that the higher cognitive centers select a curve that joins P to Q and has a given direction, i.e., tangent vector, at each of these points by minimizing the additional information that is required to define it. For if some other curve were selected, only part of the information expressed by the subjective contour would represent sensed external information and other, in principle unnecessary, information would arise from purely internal processing and thereby provide the organism with confusing artifacts in place of information which accurately describes its environmental situation. It is clearly advantageous to avoid polluting the sensed data with internally generated data that bears no relation to reality.

Suppose that a curve C joining P and Q and having the direction T_P at P and T_Q at Q is a candidate subjective contour; this situation is illustrated in Figure 5.7.3.

We have already seen that the information contained in the curve is concen-

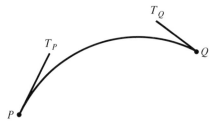

Figure 5.7.3   Initial conditions for subjective contour.

trated in the neighborhood of its points of extreme curvature. More precisely, the information gain comparing curvatures at points P_1 and P_2 is measured by $I = \log_2(|k_1/k_2|)$ where k_1 and k_2 denote the corresponding curvatures. The information gain will be *greatest* for that pair of points where $|k_1|$ is greatest and $|k_2|$ is least. Denote these curvatures by $|k|_{max}$ and $|k|_{min}$, respectively, where the subscripts refer to the maximum and minimum values of the absolute value of the curvature $|k|$.

$$I = \log_2(|k|_{max}/|k|_{min}) \tag{5.15}$$

is the greatest information gain that can be obtained by comparing the neighborhoods of two points on the curve. If the subjective contour is selected so as to make this quantity as small as possible, then it will add the least possible amount of spurious information to the information provided by the initial sensed data.

So we seek that curve C (in Figure 5.7.3) which minimizes (5.15). A version of this problem, arising in an entirely unrelated context, was solved by the mathematician RADEMACHER in his last paper, published posthumously in 1967. Application of RADEMACHER's method to the present circumstances yields the solution to our problem: the curve C consists of two circular arcs C_P and C_Q; C_P passes through P tangent to the direction T_P, C_Q passes through Q tangent to the direction T_Q, and the circular arcs C_P and C_Q meet at a point and have a common tangent line there. The same result was obtained by ULLMAN, using other methods and related but different assumptions. ULLMAN has suggested a computational algorithm by means of which the neural net could construct this curve.

We will not present the details of RADEMACHER's solution to the contour determination problem, which are likely to be of greatest interest to the mathematician. But some discussion of the construction itself and its implications will be helpful in clarifying the role it plays in the mental construction of the illusion of reality.

First consider three disks, each with a sector of angle $\pi/3$ removed, arranged on the plane so as to form an equilateral subjective triangle. In this case each side is a subjective contour which consists of a straight line, as is indicated in Figure 5.7.4. The straight line is a degenerate special case of RADEMACHER's construction in which both circles C_P and C_Q coincide and have infinite radius.

If the three sectoral angles in Figure 5.7.4 are each increased or decreased by a fixed amount, the equilateral concave and convex curvilinear triangles displayed in Figure 5.7.5 result. In both cases each side consists of a circular arc. For these situations the two circles in RADEMACHER's construction coincide.

The general case of a subjective triangle arises when the three sectoral angles are all unequal and are oriented so that no two of them can be joined by a single circular arc or straight line. Figure 5.7.6 exhibits this case and identifies the circular arc segments that make up the subjective contour. p, q, and r are, respectively, the points on the subjective contour opposite angle P,

Figure 5.7.4   A subjective equilateral triangle.

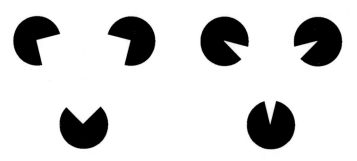

Figure 5.7.5   Concave and convex subjective equilateral triangles.

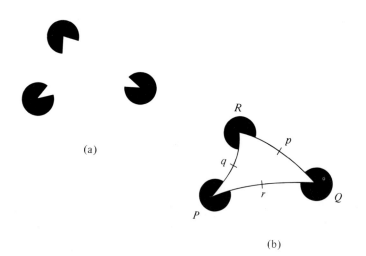

(a)

(b)

Figure 5.7.6   Subjective triangle (a) showing circular arc construction (b).

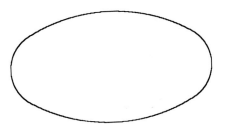

Figure 5.7.7   RADEMACHER'S oval.

Q, R respectively where the two circular arcs of RADEMACHER's construction are tangent to each other. In this case the information gained upon passing along the contour from, say, P to Q is $\log_2(r_{max}/r_{min})$ where r_{max} is the radius of the *smaller* circle (which has the *greater* curvature) and r_{min} is the radius of the *greater* circle (which has the *smaller* curvature). This is the least information that can result from joining P to Q by a smooth curve having the prescribed directions at P and at Q. In the special cases discussed previously $r_{max} = r_{min}$ so the information increment which results from the subjective contours is 0: the optimal situation.

Figure 5.7.7 shows an oval constructed according to RADEMACHER's procedure which is tangent to the sides of a rectangle, whose ratio of width to height is 15/8, at the midpoints P, P', Q, Q' of the sides of the rectangle. According to the theory, this oval coincides with the subjective oval that results from presentation of points P and P' with vertical directions, and Q and Q' with horizontal directions. Notice the (four) points where pairs of circular arcs of different radii meet smoothly. It is at these points that the curvature "jumps" from one value to another; according to ATTNEAVE's result, these are the concentration points for information about the oval.

Examination of Figure 5.7.7 makes it obvious that the human vision system can detect the discontinuous curvature change. It would be of interest to repeat ATTNEAVE's experiment with RADEMACHER ovals of varying eccentricity to determine the extent to which points placed by experimental subjects to approximate the oval are located at the discontinuities of curvature. Along each of the four circular arcs one might expect to find a uniform distribution of points with the exception of the four points where the circular arcs are joined and the curvature changes, where a much denser concentration is to be expected.

If the information content of a contour is concentrated at points where the magnitude of the curvature is maximal, then it should not be possible to visually detect a point on a curve where the curvature is a continuous function but its derivative is discontinuous. Confirmation of this prediction would support the hypothesis that the human vision system is not able to estimate derivatives of order greater than two of the intensity of illumination.

5.8 Models of Human Color Perception

The purpose of vision is to see things, not to judge light.

—Jerome LETTVIN

The perception of color has remained an enigma from the earliest recorded musings of philosophers even unto the present day. PLATO, ARISTOTLE, DESCARTES, HOOKE, NEWTON, YOUNG, GOETHE, MAXWELL, HELMHOLTZ, ROOD, SCHRÖDINGER, and LAND are but a few of the many who have attempted, with varying degrees of success, to penetrate the secret of this wonderful sense, whose subjective properties differ so markedly from objective reality.

A prism of clear glass will decompose a beam of sunlight into its constituent spectral hues which, as NEWTON demonstrated, cannot be further decomposed by passage through another prism: no new colors emerge although the hues that were present in the incident colored beam are spread further apart after their passage through the second prism. Hues vary continuously throughout the spectrum; nevertheless, it has become traditional in recent Western culture to partition the spectrum into six hue categories: *red*, *orange*, *yellow*, *green*, *blue*, and *violet* in order of decreasing wavelength. The categories are not separated by intrinsically defined boundaries but are determined by convention in an approximate way.

When spectral hues of various intensities are combined by superposition the result is a light whose hue matches some spectral hue or that of white light of some intensity. This demonstrates that hues are subjective percepts rather than physical observables. More precisely, every spectral hue, i.e., wavelength of visible light, corresponds to some hue but a hue (even if it is not white) need not correspond to a single indecomposable spectral light. Thus the hue of a light is not a measure of its wavelength or any other single physical characteristic.

Color names have been used by different cultures in widely differing ways. The reader should be wary when reading, in translation, descriptions that employ color terms. For instance, although we may think we understand the poet's meaning when HOMER speaks of "rosy-fingered dawn," what are we to make of a sky which is "bronze" or "iron" by day; of a sea variously "black," "white," "gray," "purple," or "wine-dark" but never blue or green; of the "violet-dark" fleece of the Cyclops' sheep; of PINDAR's "blue" earth, and AESCHYLUS's "green" blood? Faced with uses of color terminology which disagreed so completely with their own perceptions and usage, GOETHE and later scholars generally adopted one of two theses: that the Greeks were color blind, or that Greek color terminology was defective. Some, testing the boundaries of reason and of the then new DARWINian theory of evolution, went so far as to conclude that the ability to see the full range of colors gradually evolved during the historical period, with the Greeks presumably to be found at an intermediate stage of development. Color terminology in Greek poetry is treated at length and with scholarly devotion by IRWIN in her dissertation, where she concludes that the Greeks ordered colors in a way very different from that which is common now. Whereas we tend to distinguish colors according to their hue, the (subjective) variables of brightness and saturation being secondary, they tended to create a 1-dimensional classification corresponding more nearly but not exactly to variation of brightness. For them, the most compelling feature of blue as it is normally seen was its darkness. For certain hues, amongst which green stands out, other non-color features associated with "green" objects dominated their identification with the hue name and lent that name a broader and non-color-specific primary meaning. Green, for the Greeks, was basically "moist" and associated with life and youth as "dry" was associated with death and age, a use we preserve in much diminished status when we speak of "green" wood or a "greenhorn."

One reason why color terminology has varied in so inconsistent a way from one time and culture to another must surely be that the space of perceived colors is multidimensional and consequently does not have an intrinsically defined linear order. Since there is no natural or unique way to arrange colors sequentially, it should not be surprising that different peoples have singled out different characteristics and developed a primary ordering of colors in terms of that characteristic. For our purposes, the importance of this remark lies in the possibility it suggests that our own color terminology merely reflects yet another subjective selection which places greater emphasis on one aspect of color variation than either objective physical properties or the full structure of the subjective space of variation can justify. With this cautionary remark in mind, we will turn to consider various theories of color and models of how it is perceived.

Philosophers and scientists were not always as confident of their ability to explain the phenomena of color as they are today. PLATO is no doubt the most eminent representative of the earliest opinion about the subject: that the mystery posed by the mutability of color transcends human comprehension. In *Timaeus* he wrote:

> The law of proportion according to which the several colors are formed, even if a man knew he would be foolish in telling, for he could not give any necessary reason, nor indeed any tolerable or probable explanation of them.

and, concerning the mixtures of colors, he had the following to say:

> He, however, who should attempt to verify all this by experiment would forget the difference of the human and the divine nature. For God only has the knowledge and also the power which are able to combine many things into one and again resolve the one into many. But no man either is or will be able to accomplish either the one or the other operation.

This passage contains a concise and accurate description of the celebrated experiments performed by NEWTON more than 2,000 years later which ushered in the modern and fundamentally mathematical view of color perception.

PLATO's view obviously could not lead anywhere, and it played no further role in the development of the understanding of color. ARISTOTLE's opinions, however, continued to influence thinkers until the middle of the nineteenth century, and were indirectly responsible for some of the most disgraceful and counterproductive episodes in the annals of scholarship.

In the eighteenth century, one finds the influential French Jesuit Louis Bertrand CASTEL accusing NEWTON of stupidity and bad faith, as well as faulty observation. In 1810, the great poet GOETHE who, as HELMHOLTZ observed, was usually courtierlike and even-tempered, referred to NEWTON as "incredibly impudent" and a "Cossack Hetman" and called his work on color "mere twaddle," "ludicrous explanation," and "admirable for children in a go-cart." GOETHE ultimately attacked NEWTON's veracity with the words "but I see nothing will do but lying, and plenty of it." Evidently the study of color was once pursued with greater passion than is customary today.

The Aristotelian theory which called forth such vigorous defense was presented to the world in *On the Soul, Sense and the Sensible*, and *Meteorologica*. In the first work, ARISTOTLE concludes the necessity of a medium between the eye and the object it views; in the second he asserts that

> as vision would be impossible without light (between the object and the eye), so also would it be impossible if there were no light inside (the eye)

and proceeds to propose the theory, adopted much later by GOETHE, that color is a mixture of black and white. The third work concludes, insofar as color is concerned, that the rainbow exhibits precisely the three colors red, green, and violet, in that order. The

> appearance of yellow is due to contrast, for the red is whitened by its juxtaposition with green.

This passage from *Meteorologica* foreshadows in a curious way the three color theory of Thomas YOUNG and focuses attention, perhaps for the first time, on the mutability of colors as a function of their surroundings.

It is important to recognize that not only was ARISTOTLE's theory (that color is a mixture of white and black) wrong, but it did not lead and could not have led to a quantitative description of the results of color mixing experiments.

The first theory which associated distinct colors with distinct and quantifiable physical states appears to have been the creation of DESCARTES. In assessing his work, one should bear in mind that he lived in an age of great intellectual ferment due to events which altered the prevalent conception of the nature of physical reality. The discovery of the Americas and their pagan civilizations, the circumnavigation of the globe, the invention of the telescope and consequent discovery of the miniature solar system consisting of Jupiter and its moons all conspired to destroy scholasticism and to demand the establishment of a new intellectual order. DESCARTES undertook this task. With regard to the particular problem of accounting for action at a distance, the mutual interactions of bodies not in immediate contact with each other, he was compelled to postulate the existence of an invisible pervasive medium—the *aether*—whose mechanical properties transmitted the contact forces of pressure and impact which were then the only ones well understood and supported by experience and experiment. The constituent particles of the aether were assumed to be continually in flux but, due to the pervasiveness of the aether, the motion of one particle entailed the motion of a chain of others. When applied to the rainbow with its "diversities of color and light," this theory suggested to DESCARTES that, in the words of E.T. WHITTAKER,

> the various colors are connected with different rotatory velocities of the globules, the particles which rotate most rapidly giving the sensation of red, the slower ones of yellow and the slowest of green and blue—the order of the colors being taken from the rainbow. The dependence of color on periodic time is a curious foreshadowing of a great discovery which was not fully established until much later.

According to this theory, each perceived color must correspond to a state of rotation of aether globules and distinct hues correspond to distinct angular velocities. Thus the essential distinction between the subjective perception of colors and correlated physical states had not yet been made.

In elaboration of his theory, DESCARTES supposed that the rotatory motion perceived as color was acquired through successive refraction of the light impulse as it was transmitted through the aether by static pressure. This notion, and the entire conception of color on which it is based, was attacked by Robert HOOKE in his *Micrographia*. As a result of his studies of the colors of thin plates (the colors seen when light falls on a thin layer of air bounded by two parallel transparent plates), HOOKE demolished DESCARTES's theory and proposed one of his own. HOOKE's theory is transitional between DESCARTES's views and the fully developed wave theory of light. He concludes that light "must be a *Vibrative* motion," propagated in waves, and that the origin of color lies in the deflection of the wave front from the perpendicular to the direction of motion of the light pulse by characteristic reflections and refractions. In this instance the light pulses may be supposed

> *oblique* to their progression, and consequently each Ray to have potentially *superinduc'd* two properties, or colours, viz. a *Red* on the one side and a *Blue* on the other, which notwithstanding are never actually manifest, but when this or that Ray has one or the other side of it bordering on a dark or unmov'd *medium*, therefore as soon as those Rays are entered into the eye, and so have one side of them each bordering on a dark part of the humours of the eye, they will each of them actually exhibit some colour.

we may collect these short definitions of Colours: *That Blue is an impression on the Retina of an oblique and confus'd pulse of light, whose weakest part precedes, and whose strongest follows.*

A similar definition of red follows, in which the roles of "strong" and "weak" are interchanged. HOOKE recognizes that according to this theory there are "but two Colours" and argues

that the *Phantasm* of Colour is caus'd by the sensation of the *oblique* or uneven pulse of light which is capable of no more varieties than two that arise from the two sides of the *oblique* pulse, thus each of those be capable of infinite gradations of degrees (each of them beginning from *white*, and ending the one in the deepest *Scarlet* or *Yellow*, the other in the deepest *Blue*).

The colors are hereby identified with the real line as a geometrical continuum, and may be parametized, for instance by the tangent of the angle of obliquity of the incident wave front, with white corresponding to normal (perpendicular) incidence.

HOOKE's theory was short-lived, for in 1671 NEWTON published the results of his elegant and decisive experiments which showed that, by means of a transparent prism, a beam of light could be decomposed into primitive "spectral" constituents which resisted further decomposition, that superposition of the constituents of a light resulted in reconstitution of the original light, and that there are perceived colors—specifically, white—which do not occur as spectral constituents.

NEWTON's color circle, which has been used by countless artists and is known to everyone, makes its first appearance in the solution to Proposition VI Problem II of his *Opticks*, which asks

In a mixture of primary colors, the quantity and quality of each being given, to know the color of the compound.

By *compound* is meant "superposition" and here it must be borne in mind that we are speaking of mixtures of colored *lights*, not of pigments. The problem asks for that primary, i.e., spectral, hue which *perceptually* matches the superposition, and also for a quantitative measure of

its distance from whiteness,

that is, of the *perceptual* saturation of the color. NEWTON's solution arranges the spectral hues along the circumference of a circle in their natural order of occurrence in the spectrum with the red and indigo extremes conceived as gradually passing into a common violet. White he placed at the center of the circle. Given various spectral hues, NEWTON constructs circles whose centers are the given hues and whose areas are proportional to the number of rays of each color in the given mixture, i.e., to the intensity of illumination of each.

The centroid of the resulting configuration specifies a point in the color circle whose distance from the origin (white) is a measure of the saturation of the superposed resultant; its spectral hue is determined by the intersection with the circle of spectral hues of the radius which passes through the centroid. The quantity of the resultant light is the sum of the areas of the circles corresponding to its constituents, so that a *perceived light* depends on three variables: *quantity*, *hue*, and *saturation*. NEWTON's assertions mean that the set of perceived lights can be conceived of as a cone-shaped volume in 3-dimensional Euclidean space.

One immediate consequence of this theory is that infinitely many distinct mixtures of light must be perceived as identical. Therefore, a theory of color perception cannot be simply a theory of the physical properties of light; the nature of the visual sensory system must play an essential role.

NEWTON was aware that his geometrical rule is but an approximation; it was, he said

> accurate enough for practice, although not mathematically accurate.

In NEWTON's work there is still confusion between physics and psychophysics; he appears to have assumed that the eye, considered as an instrument, measures physical differences perceived as colors by means of a one-to-one correspondence between the set of physical light stimuli and the set of sensory responses, although his algorithm for determining the color of a mixture of lights already implies that the correspondence is at best many-to-one. A similar confusion persists in the work of EULER, who in 1752, identified the variation of physical color stimuli with frequency variation but also failed to recognize that the frequency of a light wave is not determined by its subjective color.

The formulation of a theory that explicitly recognized the distinction between the physical properties of a light stimulus and the correlated subjective sensations it produces was one of the many achievements of the multifaceted Thomas YOUNG. He proposed a physiological mechanism for converting light into the sensation of color which has since been confirmed in its essentials and is the foundation upon which current accepted theory is supported. In 1801 he wrote:

> As it is almost impossible to conceive each sensitive point of the retina to contain an infinite number of particles, each capable of vibrating in perfect unison with every possible undulation, it becomes necessary to suppose the number limited, for instance, to three principal colors.... Each sensitive filament of the nerve may consist of three portions, one for each principal color.

YOUNG's theory lay dormant for 50 years until HELMHOLTZ in Germany and MAXWELL in Great Britain rescued it from oblivion. They confirmed and strengthened its principal features in a series of ingenious experiments and thereafter the development of the theory of color perception proceeded rapidly.

In this section, we will discuss aspects of color which are concerned with the invariance of perception under the action of some group of symmetries of all or part of the space of perceived colors and, indirectly, with processes which selectively omit various types of information present in light stimuli.

The human eye is sensitive of light whose wavelength lies between about 0.4 and 0.7 micrometers (1 micrometer $= 1 \ \mu m = 10^{-6}$ meter). From the relation (wavelength) $\times$ (frequency) $=$ velocity and the value $\sim 3 \times 10^8$ meters/second for the velocity of light, we find that this range of variation of wavelength corresponds to a variation of frequency between about 4.25×10^{14} hertz (Hz) and 7.5×10^{14} Hz, from which it follows that a single photon of visible light carries, according to PLANCK's formula, an energy $E = h\nu$ which ranges from $(6.62 \times 10^{-27}) \times (4.25 \times 10^{14}) \cong 2.8 \times 10^{-12}$ erg to $(6.62 \times 10^{-27}) \times (7.5 \times 10^{14}) \cong 5 \times 10^{-12}$ erg, that is, between about 2.8 and 5 electron-volts. Thus a single photon in the visible light range carries enough energy to interact at the chemical level with the material constituents of the eye, so we should expect that the effect of a photon incident on a retinal receptor would be to change the state of some chemical substance in it.

In fact, the light sensitive photopigment in retinal receptors is based on the chemical substance *retinal* which, stimulated by an incident photon of visible light, changes from the 11-*cis* form to the all-*trans* form illustrated in Figure 5.8.1.

The 11-*cis* form of retinal is capable of combining with a compound called

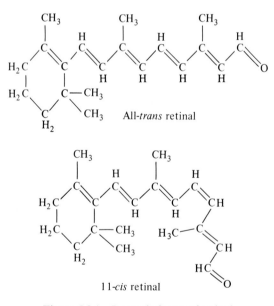

Figure 5.8.1   Isomeric forms of retinal.

opsin to form the photosensitive pigment; the all-trans form of retinal is converted into *retinol*, which is responsible for initiating the neural impulse that signals reception of the light stimulus. The sequence of biochemical events has been described by PEPPERBERG, BROWN, LURIE, and DOWLING.

Radio waves, which are a less energetic form of electromagnetic radiation, correspond to photons whose energy lies in the kilohertz range for "AM" broadcasts and the range near 100 megahertz for "FM" broadcasts. The more energetic FM photons each carry only about 6.6×10^{-7} electron-volts of energy, which is much too little to cause chemical changes of state or even to affect the thermal background (which averages about $kT \cong 0.04$ eV at ordinary temperatures) unless T is but a few thousandths of a degree above absolute zero. Thus we should not expect to ever find a chemistry-based life form that is directly sensitive to radio broadcasts unless it has a built-in conducting antenna to concentrate the incident electromagnetic energy.

The spectral composition of daylight varies greatly depending upon the time of day, cloud cover, and season. Figure 5.8.2 shows how light energy is distributed amongst the spectral hues for some typical circumstances. It is remarkable that the appearance of color does not seem to depend very much on the conditions of illumination.

Already in 1876 HELMHOLTZ had brought attention to the paradox of the painter's ability to represent greatly different states of illumination intensity with pigments. The painter of a picture in the classical or realist styles does

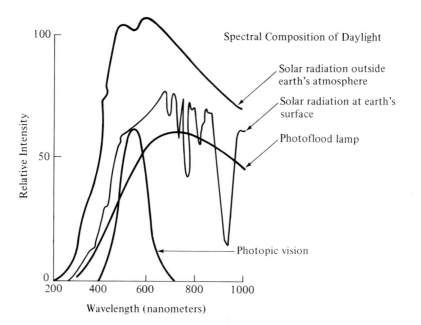

Figure 5.8.2   Spectral composition of daylight for various conditions.

not attempt to produce the same distribution of light and color that would be incident upon the eye if the original scene were viewed. That would not be possible. Indeed, were two pictures hung adjacent to each other in a gallery, one of a noonday desert scene exhibiting various degress of whiteness and dark shadow and the other of a moonlit night or other dark interior scene with whitish highlights, then the actual quantities of light reflected to the eye of a beholder by bright (*respectively*, dark) parts of either picture will be approximately the same per unit area since in both cases the same white pigment will have been used. Moreover, the brightest white on a painting in a gallery as it would be ordinarily lit is perhaps but $1/40$ the brightness of that white when lit by direct sunlight: if viewed in the desert, the painting of the desert as lit in the gallery would appear dark grey. More remarkable still, the brightest white pigment reflects only about 100 times as much light as the darkest black, whereas the sun's disk is about 80,000 times as luminous as the disk of the full moon although both subtend about the same angle at the eye. The artist consequently has no hope of reproducing the true light distribution with his palette of limited hues and ranges of brightness; therein lies his art. For according to FECHNER's work, the perceived brightness $b(x)$ of a light of physical intensity x is proportional to $\log x$ and therefore the difference in brightness $b(x_1) - b(x_2)$ of lights of intensity x_1 and x_2 will be proportional to

$$\log x_1 - \log x_2 = \log(x_1/x_2).$$

This implies that brightness differences are invariant under the simultaneous modification of light intensity

$$x_1 \mapsto ax_1, \qquad x_2 \mapsto ax_2, \qquad a > 0.$$

"Hence," HELMHOLTZ wrote,

> on the whole, the painter can produce what appears an equal difference for the spectator of his picture, notwithstanding the varying strength of light in the gallery, provided he gives to his colours the same *ratio* of (intensity) as that which actually exists.

Moreover, it is this *ratio* which characterizes the state of illumination of pairs of familiar objects and thereby enables us (exclusive of extreme states of illumination for which FECHNER's formula fails) to recognize both the real conditions of illumination and their artistic representations.

Thus the painter can so arrange the conditions of his stimuli as to achieve much the same illusion of reality as the eye obtains from the light stimuli from the original scene.

Let us imagine a fictitious "space of perceived colors" or "color space" whose individual points correspond to the subjectively distinct colored lights. Lights which are *perceived* as being different in any way correspond to different points of color space, but lights which may have very different physical constitution (i.e., be different combinations of spectral hues) correspond to the same point in color space if they are perceived as being the same. The properties of color space correspond directly to the properties of the human color vision system, and only indirectly to the physical properties of light stimuli. Our objective in this section is to study the properties of color space, not of physical light stimuli.

Colors can be arranged according to their degree of closeness to each other. For instance, if some colors differ solely in their *brightness*, then it is possible to say which differ least and we can even conceive of colors whose brightness varies continuously between two fixed levels. It is the same for hues which, as we have already observed, can be ranged along a continuum, and of saturation. In general, we can suppose that color space is equipped with a notion of closeness of the colors represented by its points, and that points in color space that are close to each other correspond to lights which are perceived as similar.

Suppose that a scene be viewed. Each point (more precisely, each small pixel) of the scene will reflect some light to the eye and that light corresponds to some point in color space. The light from different points in this scene may be perceived to be the same in which case they will correspond to the same point in color space. So the arrangement of points in color space does not correspond to the arrangement of objects in the scene; it merely represents the perceived colors that are present in the scene.

Suppose that the spectator views a second copy of the scene which differs from the first only in that the intensity of illumination is different; it is some multiple of the intensity of illumination for the first scene. Then every point in the scene will, in effect, appear brighter so the corresponding point in color space will move from its original position to another which represents a light which differs from the first only in brightness. Thus, the change in the intensity of the background illumination will result in a motion of the points in color space. The change in background illuminant may be perceived, relative to the original illuminant, to involve simultaneous changes in its hue, brightness, and saturation, not merely changes in one of these perceptual variables.

Common experience provides considerable evidence that any point of color space can be displaced to any other nearby point by a suitable (small) change in the background illumination. (A space which possesses this property is said to be *locally homogeneous*.) For instance, since not only the intensity but also the spectral composition of the illumination in a daylight-lit gallery varies depending on the time of day, the cloud cover, and the season (cf. Figure 5.8.3) without significantly affecting the relative hues, saturation, or brightness of components of the exhibited paintings, we must conclude that their relative properties are independent of the state of illumination throughout a large region of color space which includes all normal conditions of illumination; indeed, artists never tell us the lighting conditions under which a painting was created nor do they insist that it be displayed in the same background illumination in which it was painted. That it is unnecessary to relearn color relationships when one dons tinted sunglasses or drives an automobile with a tinted windshield provides additional corroboration, and the essentially inter-

changeable use of incandescent and fluorescent illumination supports this
claim in yet another way. Local homogeneity and its fundamental role in the
visual perception process is compatible with the stabilized vision experiments
reported in section 5.5. They demonstrated that if the small involuntary
saccadic eye motions were cancelled so that the image of a scene is motionless on the retina, then the stabilized image at first

> looks very sharp and clear, but then it rapidly fades out and disappears and
> the field looks uniformly grey. Stabilized patterns disappear within seconds,
> or even fractions of a second after being presented. After the image has dis-
> appeared if it is moved across the retina, ... it will reappear and then quickly
> disappear again. (CORNSWEET 1970)

This is just the result that would be anticipated were color space *locally
homogeneous*, for in that case single colors would not be discriminable, and
the stationary neural activity of each photosensitive light receptor in the retina
would be interpreted as a fixed color independent of the actual composition
of the physical illumination it receives.

In other words, there does not exist a color which can be absolutely dis-
criminated by the human visual system; all colors are equivalent; and only
certain relationships which depend on *pairs* of colors presented to the retinal
light receptors within sufficiently brief intervals of time can have any perceptual
significance.

The distinction between models of visual information processing which
posit an absolute color space whose elements can be perceived and identified
without regard to the effects of others, and a processing model wherein all
perceived observables are relative quantities which cannot be identified in
isolation is analogous to the difference between NEWTON's concept of absolute
space and time and EINSTEIN's relativistic space-time. Just as relativity theory
assures us that there is no privileged observer in physics, so does the description of color perception in relative terms imply that there can be no privileged
observer of colors, that is, no one who can specify the color of a perceived
light in an absolute sense.

Our knowledge of the properties of light is derived from the outputs of one
or another type of transduction system. If the system consists of the human
eye and brain, we speak of color perception, whereas if the transducer is conceived as consisting of the collection of all apparatus currently used in physical optics, we speak of the "physical" or "optical" properties of the light. The
task of a theory of color perception is to relate the output signals of the ideal
perceptual transducer to the physical properties (i.e., to the output of the
optical transduction system) for a given light input.

For our purposes, a physical light signal can be characterized by describing
how the intensity of the light varies as a function of its frequency. For visible
light there is a minimum frequency v_{min} which corresponds to the reddest
spectral hue and a maximum frequency v_{max} which corresponds to the most

violet spectral hue. Each visible spectral hue corresponds to a frequency v intermediate between those two. A particular distribution of light will be specified by assigning to each frequency v a nonnegative number $I(v)$ which measures the intensity, i.e., quantity of light, in an infinitesimal frequency interval dv at v. Figure 5.8.2 exhibits several curves which display the variation of light intensity with frequency for various common circumstances.

When two physical lights are combined by superposition (say, by using mirrors to cause two light beams to be joined into one), the intensity distribution function which describes the resultant physical light is the sum of the functions which describe its constituents. This simply means that the quantity of light in the superposition is the sum, for each spectral hue, of the quantities which were combined. When physical lights are thought of in this way, it follows that an arbitrary physical light can be imagined to consist of the superposition of suitable quantities of monochromatic light of all possible frequencies between v_{min} and v_{max}. The monochromatic spectral hues are themselves indecomposable, as NEWTON's experiments suggested. Thus, a complete description of a physical light requires the specification of an intensity—represented by a nonnegative number—corresponding to each frequency. Since such a specification would require knowledge of infinitely many numerical quantities (one for each of the infinitely many monochromatic spectral hues), it would be inaccessible to any finite observing system, including the human eye. It should not come as a surprise that the first and most basic task of the vision system is to deal with this problem by selectively omitting information to bring the result into compass with the capabilities of the visual information processing system. Phrased in a more mathematical way, the space of physical lights is infinite dimensional, but the space of perceived colors can be only finite dimensional. In fact, the space of perceived colors is 3-dimensional for normal daylight vision.

Let us denote physical lights by boldface lowercase letters $\mathbf{x}, \mathbf{y}, \ldots$ and write $\mathbf{x}(v)$ for the intensity of light at frequency v in the infinitesimal range dv. The perceived light that corresponds to the stimulus $\mathbf{x}$ will be denoted by the same letter in the ordinary lowercase font, thus x.

It is both convenient and natural to denote the physical light that results from the superposition of $\mathbf{x}$ by $\mathbf{x} + \mathbf{y}$, and the perceived light which corresponds to the physical light $\mathbf{x} + \mathbf{y}$ will be denoted by $x + y$. It is important to remember that the "$+$" signs are used in different ways in the expressions $\mathbf{x} + \mathbf{y}$ and $x + y$, and that the properties of this "sum" for perceived lights must be determined from experiments. Our choice of this notation is meant to suggest what experiment confirms: that perceived lights combine the same way physical lights do when the latter are combined by superposition.

Suppose that two copies of the physical light $\mathbf{x}$ are superposed (e.g., by mirrors); their superposition is represented by the expression $\mathbf{x} + \mathbf{x}$ which can be more conveniently denoted $2\mathbf{x}$. The corresponding perceived light is $x + x$, which is conveniently denoted $2x$. In general, if $a > 0$ is any number, $a\mathbf{x}$ stands for that physical light whose intensity at any frequency is a times the intensity of $\mathbf{x}$ at that frequency: in symbols, $(a\mathbf{x})(v) = a(\mathbf{x}(v))$ (the actual superposition can be arranged by suitably configured mirrors). The corresponding perceived light will be symbolically denoted by $a\mathbf{x}$.

The principal results obtained by the pre–twentieth century investigators of color perception can be expressed in terms of the notions of addition of perceived lights and the multiplication of a perceived light by a positive number. We can formulate them as axioms from which the properties of the space of perceived colors are to be deduced.

Axiom 1. Every positive multiple of a perceived light is a perceived light.

Expressed symbolically, Axiom 1 asserts that $a\mathbf{x}$ is a perceived light whenever x is a perceived light and a is a positive number. Since the relative intensities of pairs of spectral hues in $a\mathbf{x}$ is the same as the relative intensities of the same pairs in $\mathbf{x}$, we expect that x and $a\mathbf{x}$ will differ only in brightness. The axiom is an idealization: for very large values of a, the viewing eye will be destroyed; for very small values, "background noise" will destroy the relationship which persists between $\mathbf{x}$ and x for normal daylight vision. Axiom 1 asserts that the space of perceived colors is a (generalized) *cone*.

Axiom 2. No superposition of perceived lights produces the absence of perceived light, i.e., blackness.

The perceived light which corresponds to the zero physical light is denoted 0, and has the properties one expects zero to have. Symbolically, Axiom 2 states that if x is a perceived light, there does not exist a perceived light y such that $x + y = 0$. Axiom 2 implies that the space of perceived colors is a *convex cone*.

Axiom 3 (GRASSMANN 1853, HELMHOLTZ 1866). Each point on the line segment which joins two perceived lights in color space corresponds to a perceived light.

Symbolically, if x and y are distinct perceived lights and if $0 < t < 1$, then each point $tx + (1 - t)y$ denotes a perceived light. This axiom implies that the convex cone which represents perceived lights is *connected*, i.e., consists of one piece, and is *simply connected*, i.e., has no "holes" in it.

Axiom 4 (GRASSMANN 1853). If three perceived lights x, y, z be given, then there is a fourth perceived light w such that $x + y = z + w$, i.e., $x + y$ and $z + w$ are the same perceived light.

This axiom implies that the cone which represents perceived colors has at most three dimensions. It is worthwhile noting that although some superpositions of three colors can be matched by a fourth, there are colors for which this is not true; but the matching statement expressed by Axiom 4 appears to be generally valid: the superposition of any pair of colors can be matched by the superposition of a third color with some fourth color which is determined by the three given ones.

The local homogeneity of color space under changes of background illumination has already been discussed. This property is restated as

Axiom 5. The space of perceived colors is locally homogeneous with respect to changes of background illumination.

In order to give formal meaning to this statement something must be said about the mathematical interpretation of a change of background illumination.

If an illuminant is reflected from an object in a scene, the reflected light will depend on the characteristics of the incident illumination and on the reflective properties of the object. If the intensity of the illuminant at frequency v is $I(v)$ and the reflectance of the object corresponding to that frequency is $R(v)$, then the reflected intensity which would be observed by a spectator would be $R(v)I(v)$. From this we see that a change of background illumination which modifies $I(v)$ to, say, $I'(v)$, will result in a reflected light characterized by $R(v)I'(v)$ in place of $R(v)I(v)$. Hence the observed light varies linearly with changes in background illumination.

If the reflected light $\mathbf{x} = RI$ (which has intensity $R(v)I(v)$ at frequency v) corresponds to the perceived light x in color space and $\mathbf{x}' = RI'$ corresponds to x', then we will suppose that the linear change from I to I' corresponds to a linear change from x to x' in color space. The meaning of Axiom 5 can be expressed as follows: color space is a convex cone in a (3-dimensional) vector space, and changes of background illumination correspond to linear transformations of the 3-dimensional space that carry the color space subset onto itself.

With this interpretation of changes of background illumination, it can be proved that there are only two geometrical structures for color space that are compatible with the axioms. One of these was first proposed by HELMHOLTZ in 1891; the other is somewhat more complicated and appears to be a less likely candidate for representing color space for reasons which will be set forth later.

We have already remarked that people do not find it necessary to relearn the relative relationships of colors when a change of background illumination occurs. This implies that the perceptual invariants are independent of changes of background illumination. More precisely, if x, y, ... are perceived colors and if $f(x, y,...)$ is a perceptual invariant, that is, a function of the perceived colors which does not change when the background illumination is changed, then f must be a function which satisfies the equation

$$f(gx, gy,...) = f(x, y,...),$$

where g denotes any change of background illumination and $gx, gy,...$ denote the perceived colors into which x, y, ... are transformed by the illumination change g. The collection of all g's forms a *group G*—the group of background illumination changes. The "product" $g_3 = g_2 g_1$ of two elements g_1 and g_2 of the group is that change of background illumination which is obtained by first introducing the change g_1 and then following it by the illumination change g_2; what this means in a concrete sense will be made clear momentarily.

When the problem of identifying the perceptual invariants is formulated this way, it emphasizes the role of group invariance and assumes a form analogous to the discussion of the quantitative measure of information presented in Chapter 2.

Our first objective is to show that perceptually invariant functions cannot depend on only one variable point of color space; a perceptual invariant, like information itself, is a relative quantity which must depend on at least two variable points—that is, on a pair of observed colors. The fundamental point-pair perceptual invariant can be thought of as providing a measure of distance between the points of color space, such that the distance between a pair of points remains unchanged as the points move about in color space under the influence of changes in background illumination. This perceptual distance will reduce to FECHNER's psychophysical function in the case that the pair of colors differ only in brightness.

Let us begin by idealizing the problem so that it can be studied without requiring an elaborate mathematical apparatus. Suppose that color space for normal daylight vision were 2-dimensional rather than 3-dimensional. For this case Axioms 1 to 3 would imply that color space is a 2-dimensional cone, that is, the part of a plane bounded by some angle. If we introduce a Cartesian coordinate system and take one side of the angle to coincide with the positive x-axis, color space might be modeled by the hatched region in Figure 5.8.3.

A change of background illumination is a linear transformation of the (x, y) space which carries the space P of perceived colors into itself. Thus, a change of background illumination is characterized by the constants a, b, c, d in the

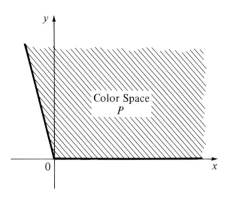

Figure 5.8.3   2-Dimensional space of perceived colors.

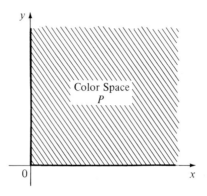

Figure 5.8.4   2-Dimensional space of perceived color in normalized coordinates.

system of equations

$$x_1' = ax_1 + bx_2,$$
$$x_2' = cx_1 + dx_2,$$
(5.16)

where the point with coordinates (x_1', x_2') belongs to P whenever the point with coordinates (x_1, x_2) does. This simply means that perceived colors become (other) perceived colors under the transformation (5.16). If $x_2 = 0$, then either $x_2' = 0$ or (x', x_2') lies on the other boundary line. In the former case, we find $0 = x_2' = cx_1 + d \cdot 0 = cx_1$ whatever the value of x_1, so $c = 0$. This restricts the kinds of linear transformations that can correspond to changes of background illumination.

Let us stretch and rotate the coordinate system on the plane to force the region P to coincide with the first quadrant. This can be accomplished without loss of generality because it merely amounts to selecting a convenient set of units in which to measure the coordinates of points. But if it is done, it follows that color space P will be identified with the hatched region in Figure 5.8.4 and a change of background illumination represented by (5.16) must have one of the forms

$$x_1' = ax_1, \qquad x_2' = dx_2 \quad a > 0, d > 0,$$

or

$$x_1' = bx_2, \qquad x_2' = cx_1 \quad b > 0, c > 0,$$

in order to satisfy the condition that a change of background illumination take P onto itself. This completely determines the possible kinds of changes of background illumination, and so it also determines the group G of background illumination changes. Indeed, if we express a change of background illumination by the coordinate equation for its effect on points of color space, then

$$(x_1', x_2') = g(x_1, x_2) = (ax_1, dx_2) \quad \text{or} \quad = (bx_2, cx_1)$$

according to (5.16).

Suppose that $f((x_1, x_2))$ is a perceptually invariant function of the single point in color space whose coordinates are (x_1, x_2). Then

$$f((x_1, x_2)) = f((ax_1, dx_2))$$

for any positive numbers a, b, c, d. Following the method of Chapter 2, we see that $f((x, y)) = f((1, 1))$ by selecting $a = 1/x_1, d = 1/x_2$, so f does not depend at all on the point of color space with coordinates (x_1, x_2); thus, a function of one point is not a suitable perceptual invariant.

If a hypothetical perceptual invariant f depends on a pair of points x and y in color space which are given in terms of their coordinates by $x = (x_1, x_2)$, $y = (y_1, y_2)$, then

$$f(x, y) = f((x_1, x_2), (y_1, y_2)) = f((ax_1, dx_2), (ay_1, dy_2)).$$

Select $a = 1/x_1$ and $d = 1/x_2$ to find

$$f(x, y) = f\left((1, 1), \left(\frac{y_1}{x_1}, \frac{y_2}{x_2}\right)\right):$$

the perceptual invariant depends only on the *ratios* of corresponding coordinates of the pair of points in color space.

One kind of perceptual invariant is an *invariant metric* or *measure of distance* between points in color space. Just as the ordinary Euclidean distance between a pair of points in the plane remains unchanged or invariant when the points are simultaneously moved in the same way by any combination of rotations and translations, so the invariant distance between points in color space is left unchanged by a change of background illumination which simultaneously "moves" the points.

The distance between x and y in color space will be denoted $d(x, y)$. To say that it is *invariant* simply means that $d(x, y) = d(gx, gy)$ where g denotes the transformation induced by a change of background illumination. In terms of coordinates, $x = (x_1, x_2)$ and gx is expressed by the formula $gx = (x_1', x_2')$ where $x_1' = ax_1, x_2' = dx_2$, and $a > 0, d > 0$. The distance between x and y is given in terms of the coordinates of the points by means of the formula

$$d^2(x, y) = \alpha^2 \log^2(x_1/y_1) + \beta^2 \log^2(x_2/y_2), \tag{5.17}$$

where α^2 and β^2 are positive constants which are related to the properties of the viewer's retinal receptors; we shall have more to say about this later. The analogy with the familiar Euclidean formula for distance in the plane, (distance between x and $y)^2 = (x_1 - y_1)^2 + (x_2 - y_2)^2$, is evident.

It is easy to check that $d(x, y)$ is in fact a perceptual invariant under changes of background illumination, for if $(x_1, x_2), (y_1, y_2)$ are replaced by (ax_1, dx_2), (ay_1, dy_2) in (5.17) the expression remains unchanged because the distance is a function only of the ratios of corresponding coordinates of the points in color space.

Suppose that $y = kx$, where $k > 0$ is some positive number. If a physical light $\mathbf{x}$ corresponds to x in color space, then the physical light $k\mathbf{x}$ will correspond to kx. Since $\mathbf{x}$ and $k\mathbf{x}$ have proportional distributions of intensity as a function of spectral frequency, they should differ only in brightness at the perceptual level. Let us calculate the perceptually invariant distance $d(kx, x)$. From the defining formula (5.17) one finds

$$d^2(kx, x) = \alpha^2 \log^2(kx_1/x_1) + \beta^2 \log^2(kx_2/x_2) = (\alpha^2 + \beta^2) \log^2 k$$

whence

$$d(kx, x) = \sqrt{\alpha^2 + \beta^2}|\log k|,$$

where the vertical bars denote the absolute value. The perceptually invariant distance between x and kx is proportional to the logarithm of k; but this is just FECHNER's psychophysical function for brightness, which asserts that the perceived distance between illuminants that differ only in intensity is proportional to the logarithm of the intensity ratio.

This interpretation shows that rays in color space that emanate from the origin of coordinates correspond to colors that differ only in brightness. Differences of saturation (and, in the full 3-dimensional model of color space, of hue) correspond to variations in directions that are, relative to the perceptually invariant metric, perpendicular to the rays of brightness variation. It can be shown that the points of color space that correspond to a fixed brightness (relative to a given light) lie on the curve whose equation is $x_1^\alpha x_2^\beta = k$. In case $\alpha = \beta = 1$, it is an hyperbola; the general case can be reduced to this one by a suitable change of coordinates.

In this 2-dimensional version of color space, the points on each of the x_1- and x_2-axes correspond to a single ideal hue whose brightness increases with distance from the origin. The two hues are complementary, so variations along the curves of constant brightness correspond to variation in *saturation* of one hue and its complement. The perceptual distance between points x and y that both have the same brightness can be calculated from the distance formula (5.17) in a similar way. Suppose that $x_1^\alpha x_2^\beta = k$ and $y_1^\alpha y_2^\beta = k$; the common value k characterizes the common magnitude of brightness of the two colors. Then $x_2 = kx_1^{-\alpha/\beta}$ with a similar expression for y_2, and the distance formula yields

$$d^2(x, y) = \alpha^2 \log^2(x_1/y_1) + \beta^2 \log^2(kx_1^{-\alpha/\beta}/ky_1^{-\alpha/\beta})$$

$$= \alpha^2 \log^2(x_1/y_1) + \beta^2(-\alpha/\beta)^2 \log^2(x_1/y_1)$$

$$= 2\alpha^2 \log^2(x_1/y_1),$$

so

$$d(x, y) = \alpha\sqrt{2}|\log(x_1/y_1)|. \tag{5.18}$$

If y is fixed, then the saturation of x relative to y increases indefinitely as x_1 increases indefinitely or decreases to zero. Notice that the relative saturation is independent of the brightness since k does not appear in equation (5.18).

Having considered an ideal 2-dimensional color space, let us now turn to the full 3-dimensional space of perceived colors which corresponds to normal daylight vision. A point x belonging to this space can be characterized by its three coordinates relative to some fixed but arbitrary Cartesian coordinate system: $x = (x_1, x_2, x_3)$. The color space consists of those points x whose coordinates are all positive numbers: that is, the first octant of the coordinate system. The linear equations which correspond to a change of background illumination and generalize (5.16) from 2 to 3 dimensions are simply

$$x_1' = a_1 x_1, \qquad x_2' = a_2 x_2, \qquad x_3' = a_3 x_3,$$

where a_1, a_2, a_3 are positive numbers and $x = (x_1, x_2, x_3)$ is moved to $x' = (x_1', x_2', x_3')$ under the influence of the change of background illumination.

If x and y are two points of color space, the invariant distance between them, which will continue to be denoted by $d(x, y)$, is determined by the formula

$$d^2(x, y) = \alpha^2 \log^2(x_1/y_1) + \beta^2 \log^2(x_2/y_2) + \gamma^2 \log^2(x_3/y_3), \quad (5.19)$$

which generalizes the formula (5.17) for the invariant distance for the 2-dimensional example. Here there are *three* constants α, β, γ that depend upon the observer's eye. This model of the space of color perception was introduced by HELMHOLTZ in 1891. He identified the constants α, β, γ that determine the perceptual metric with the parameters derived from the absorption spectra of the red, green, and blue sensitive photopigments in the retinal receptors. As a result, the three dimensions of the space of perceived colors are identified with the three channels of retinal color perception in a concrete way.

Following our previous line of reasoning, we can determine the distance between colors x and kx which correspond to physical lights $\mathbf{x}$ and $k\mathbf{x}$. Since the latter differ only in intensity, the corresponding colors (ideally) differ only in brightness and $d(x, kx)$ should be a measure of that perceived difference. We find from (5.19) that

$$d^2(kx, x) = \alpha^2 \log^2(kx_1/x_1) + \beta^2 \log^2(kx_2/x_2) + \gamma^2 \log^2(kx_3/x_3)$$

$$= (\alpha^2 + \beta^2 + \gamma^2) \log^2 k$$

so

$$d(kx, x) = \sqrt{\alpha^2 + \beta^2 + \gamma^2} |\log k|;$$

this is FECHNER's formula in the 3-dimensional color space.

Colors that are perceived as equally bright lie on a 2-dimensional surface in color space. The equation of such a surface is

$$x_1^{\alpha} x_2^{\beta} x_3^{\gamma} = k, \qquad (5.20)$$

where the parameter k characterizes the brightness. By judiciously modifying the coordinatization of the underlying 3-dimensional space, we can select units of measurement such that the constants α, β, γ can all be set equal to 1; for this case, the surfaces of constant brightness are hyperboloidal, satisfying equations of the form $x_1 x_2 x_3 = k$. For the sake of simplicity of presentation, let us suppose that $\alpha = \beta = \gamma = 1$.

Each surface of constant brightness can be likened to NEWTON's color circle corresponding to some relative brightness level. But, since color space is locally homogeneous, we cannot speak of the hue or the saturation of a color in an absolute sense. Only *relative* saturations or hues have observable meaning.

How can relative hues and saturations be drawn forth from the geometry of color space? The easiest way to understand the geometry and its relation to traditional and intuitive notions is to change the coordinate variables to convert the surfaces of constant brightness into planes. This can be easily accomplished: set $u_1 = \log x_1$, $u_2 = \log x_2$, $u_3 = \log x_3$. Then the surface of constant brightness whose equation in the original coordinates is $x_1 x_2 x_3 = k$ becomes, after taking logarithms of both sides and substituting the new variables,

$$u_1 + u_2 + u_3 = \log k.$$

This is the equation of a plane that intercepts the three coordinates axes in the points $u_1 = \log k$, $u_2 = \log k$, $u_3 = \log k$. The effect of a change of background illumination is easily expressed in terms of the u-coordinates. If the change of background illumination moves the point $x = (x_1, x_2, x_3)$ to $x' = (x_1', x_2', x_3')$ according to the equations

$$x_1' = a_1 x_1, \qquad x_2' = a_2 x_2, \qquad x_3' = a_3 x_3,$$

then

$$\log x_1' = \log x_1 + \log a_1,$$

$$\log x_2' = \log x_2 + \log a_2,$$

$$\log x_3' = \log x_3 + \log a_3.$$

Writing $u_1' = \log x_1'$, etc., and $b_1 = \log a_1$, $b_2 = \log a_2$, $b_3 = \log a_3$, these relations become

$$u_1' = u_1 + b_1, \qquad u_2' = u_2 + b_2, \qquad u_3' = u_3 + b_3;$$

the effect of a change of background illumination is to move each point in the space of the u-coordinates in the direction of the vector whose coordinates are (b_1, b_2, b_3) a distance equal to the length $\sqrt{b_1^2 + b_2^2 + b_3^2}$ of that vector; see Figure 5.8.5. This motion will leave the brightness of points unchanged if and only if $b_1 + b_2 + b_3 = 0$.

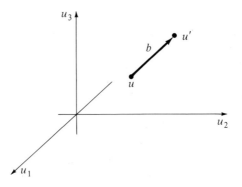

Figure 5.8.5   Effect of change of background illumination in u-coordinates.

With this description of changes of background illumination in hand, consider the plane of points of common brightness corresponding to $u_1 + u_2 + u_3 = \log k$ and changes of background illumination for which $b_1 + b_2 + b_3 = 0$. This plane can be thought of as the usual plane, and the brightness-preserving changes of background illumination correspond to motions of its points a given distance in a given direction. In terms of the coordinates on this plane, the perceptually invariant distance is the same as the usual distance familiar from euclidean geometry, since the Euclidean distance between points is not affected if the points are simultaneously moved in the same direction by the same distance.

Now suppose that a point O in this plane is selected. Relative to O, the spectral hues correspond to the various *directions* from O and the saturation of a color corresponding to a point P relative to O is measured by the Euclidean distance from P to O (in terms of the u-variables). Complementary colors relative to O are distinct colors which lie at equal distance from O on a line which passes through O.

Let us consider an example of how these geometrical relationships can be used to solve problems of color perception. Figure 5.8.6 shows an experimental

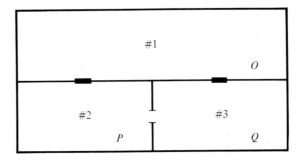

Figure 5.8.6   Experimental situation.

arrangement of three rooms. The experimenter occupies room *No.* 1 whose walls are painted with color *O*. Rooms *No.* 2 and *No.* 3 have walls whose colors are close to but distinct from *O* and from each other. The common wall of rooms *No.* 2 and *No.* 3 is pierced by a small (circular) hole which appears to an occupant of either of these rooms to be a spot of color on the common wall. The common wall between rooms *No.* 1 and *No.* 2 and between rooms *No.* 1 and *No.* 3 each have a small one-way aperture which permits the experimenter to observe the colors of rooms *No.* 2 and *No.* 3. The problem is to specify the colors that occupants of rooms *No.* 2 and *No.* 3 perceive when viewing their common wall.

If we assume (for covenience) that all colors are equally bright, then the situation can be represented in the plane of colors of constant brightness described above. The directions *OP* and *OQ* are the hues of room *No.* 2 and room *No.* 3 respectively, relative to *O*, i.e., the hue of room *No.* 1. This is the standpoint adopted by the experimenter. Thus the difference in the hues of rooms *No.* 2 and *No.* 3 will be measured by the experimenter as given by angle *POQ*. The corresponding color space diagram is shown in Figure 5.8.7.

The occupants of rooms *No.* 2 and *No.* 3 adopt quite different standpoints, however. The occupant of room *No.* 2, for example, will draw the color space diagram as in Figure 5.8.8. The hue of *Q* will be measured by the direction *PQ*. Similarly, the occupant of room *No.* 3 will use Figure 5.8.9. If all three diagrams are combined, we obtain Figure 5.8.10.

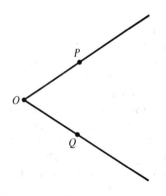

Figure 5.8.7   Color space diagram of experimental situation.

Figure 5.8.8   Experimental situation from *No.* 2's standpoint.

Figure 5.8.9   Experimental situation from *No.* 3's standpoint.

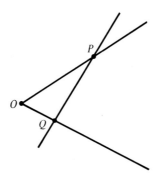

Figure 5.8.10   Geometry of combined viewpoints.

Since the direction of P as seen from Q and the direction of Q as seen from P lie along the same line, it follows that the hue spots seen by the occupants of rooms *No.* 2 and *No.* 3 on the wall separating them are complementary. Moreover, since angle POQ is less than π radians, it follows that the experimenter sees the hues represented by P and Q as closer to each other than they appear to the occupants of rooms *No.* 2 and *No.* 3.

Finally, suppose that the experimenter's peepholes become 2-way holes. Then all of the occupants are on equal footing; each can consider himself to be "the" experimenter observing two subjects. The circumstances being completely symmetrical amongst the three observers, we can conclude that any two observers perceive the spot of color on their shared wall as having complementary hues, and each sees the spots corresponding to the other two rooms as differing less than the occupants of those rooms perceive them to differ.

There is another type of transformation of color space which preserves the color cone and the perceptual metric, although it is not a linear transformation. Consider

$$x'_1 = \frac{1}{x_1}, \qquad x'_2 = \frac{1}{x_2}, \qquad x'_3 = \frac{1}{x_3}.$$

This carries the color space onto itself. If y' denotes the same function of the y-coordinates, then

$$d^2(x', y') = \alpha^2 \log^2\left(\frac{1/x_1}{1/y_1}\right) + \beta^2 \log^2\left(\frac{1/x_2}{1/y_2}\right) + \gamma^2 \log^2\left(\frac{1/x_3}{1/y_3}\right)$$
$$= \alpha^2 \log^2(y_1/x_1) + \beta^2 \log^2(y_2/x_2) + \gamma^2 \log^2(y_3/x_3)$$
$$= \alpha^2 \log^2(x_1/y_1) + \beta^2 \log^2(x_2/y_2) + \gamma^2 \log^2(x_3/y_3)$$
$$= d^2(x, y);$$

the perceptual distance is not changed when a pair of points are moved by this transformation. What is the interpretation of such a motion of points in color space? It certainly does not correspond to a change of background illumination, for colors corresponding to large values of the product $x_1 x_2 x_3$ (and hence to large relative brightness) are transformed into colors corresponding to small values (and hence of small relative brightness) and, relative to the point with coordinates (1, 1, 1), hues are transformed into their complements. This geometrical relationship is qualitatively the same as the relationship between a scene and a photographic *color negative* of it.

Earlier we remarked that the axioms for color space are satisfied by two different geometrical structures. The one described previously, credited to HELMHOLTZ, is the simpler but also the more natural of the two and appears to provide a geometrical description which is generally consistent with observed phenomena. The other structure is more interesting from a purely geometrical point of view but, although it shares many features with the simpler geometry, it does not provide a satisfactory model for color perception. Nevertheless, because of its intrinsic interest, we will describe it briefly.

In this model, the points of the 3-dimensional color space can most conveniently be represented by a symmetric 2×2 matrix:

$$x = \begin{pmatrix} x_1 & x_3 \\ x_3 & x_2 \end{pmatrix}.$$

Such an x belongs to the color cone if and only if the matrix is *positive*; this means $x_1 x_2 - x_3^2 > 0$ and $x_1 > 0$. The transformations that correspond to changes of background illumination can be most conveniently expressed in terms of matrix multiplication in the form

$$x = A^t x A$$

where

$$A = \begin{pmatrix} a_1 & a_2 \\ a_3 & a_4 \end{pmatrix}$$

is a matrix of constants and

$$A^t = \begin{pmatrix} a_1 & a_2 \\ a_3 & a_4 \end{pmatrix}$$

is the *transposed matrix*. It turns out that the surfaces of constant brightness are equivalent to the non-Euclidean plane with the distance function which turns it into a space of constant negative curvature. We will not enter into a discussion of the properties of this space except to remark that the treatment of the experimental situation illustrated in Figure 5.8.6 leads to the same qualitative conclusions as were found for the simpler model, although the quantitative details are quite different.

5.9 The Gaze as a Flying-Spot Scanner

Previous sections have illustrated some ways in which the principles of group invariance and selective omission of information reveal themselves in biological information processing systems. In this and the following section the emphasis is on examples of the employment of hierarchical structures for the processing of sensed data by diverse biological systems.

In this section the human foveal gaze will be considered as an instance of a 3-level hierarchical structure whose purpose is to partition the limited "bandwidth," i.e., channel capacity, of the vision system so that it can serve the conflicting needs for high spatial resolution and wide aperture in sensing the visual environment. This will reduce the information load on the vision processing system while still providing a mechanism for high resolution image analysis.

In the context of constrained channel capacity, there is a duality between high spatial resolution and wide field of view (we will examine another aspect of this duality in section 5.12). Many biological organisms employ a hierarchical structure to split the available channel capacity between sensors of various types and to couple separate subsystems that are most sensitive in one or another of the competing alternatives. Thus one subsystem may provide low resolution but a wide field of view, and another may provide the complementary high resolution but narrow field of view capability. These subsystems are combined by means of a control system that utilizes feedback, especially from the low resolution system, to direct the activity of the high resolution subsystem efficiently in accord with the properties of the sensed image signal.

Although it is by no means universal throughout the realm of animal vision systems, the method of *scanning* does occur in widely varying circumstances as a solution to the problem of using a low channel capacity image analysis system to monitor a high channel capacity input sensor. We will consider three examples: the human foveal gaze, the acoustic fovea of the horsehoe bat *Rhinolophus ferrumequinum*, and the copepod *Copilia*. We will begin with a brief discussion of the latter.

The curious anatomy of the eye of *Copilia*, which lives in the Bay of Naples, was described by EXNER in 1891. GREGORY and his colleagues studied the creature more carefully three-quarters of a century later; we follow his account.

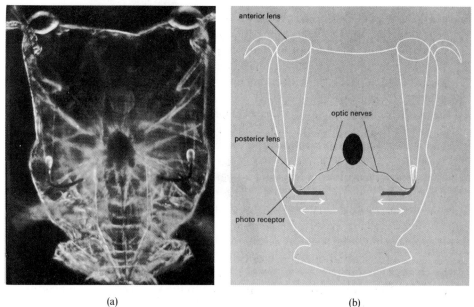

(a)   (b)

Figure 5.9.1   (a)–(b) *Copilia quadrata*. (From R.L. Gregory, *Concepts and Mechanisms of Perception*. Copyright © 1974 R.L. Gregory. Reprinted with permission of Charles Scribner's Sons, a division of Macmillan, Inc. and permission of Duckworth and Company, London.)

The female of the species *Copilia quadrata* is about 3 millimeters in length, including the tail, and about 1 millimeter in each of the dimensions of width and maximum thickness of body. Figure 5.9.1 illustrates the body, excluding the tail, from above. The bulk of its body is transparent and contains two pairs of lenses, symmetrically situated with respect to the central axis of the body, each pair consisting of a fixed anterior lens of diameter about 0.15 millimeter and a movable posterior lens located about 0.65 millimeter behind the anterior lens. The posterior lenses lie deep within the animal's body and are each joined to a pigmented bow-shaped symmetrical structure that contains the photosensitive receptors and is in turn attached to the optic nerve. GREGORY remarks that the whole structure "is essentially the same as a single ommatidium of a conventional compound eye, except that the distance between the corneal (anterior) lens and the crystalline cone is greatly increased."

The anterior lens is rigidly fixed in the carapace of the animal but the posterior lens is suspended within the body in a system of muscles and ligaments that move it in a plane perpendicular to the optical axis of the lens system and which is assumed to coincide with the focal plane of the fixed anterior lens. The posterior lenses move in a symmetrical and synchronized manner, moving toward the centerline of the animal and each other more rapidly than they move apart. At their farthest separation, which is the resting state, the optical axes of the pair of lens systems are parallel. GREGORY notes that the optical axes never converge and concludes therefrom that the mobility of the posterior lenses is not intended to provide range-finding information. He concludes that it is a *scanning* mechanism for serially sampling the image plane of the anterior lenses. The maximum observed frequency of scanning is about 15 scans per second but evidence concerning the variability of the scanning frequency is not conclusive. This unusual arrangement provides a means for increasing the resolution of *Copilia's* eye by trading time for space and is an interesting illustration of the uncertainty principle for measurement. If the scenes that are of interest to *Copilia* vary slowly, then by scanning the image plane of an anterior lens with a single "flying-spot scanner," a complete image can be constructed within the limits of resolution, image plane location, and memory storage capabilities of the posterior lens and the vision apparatus which it stimulates. In this way *sequential signal processing* substitutes for the *parallel signal processing* more commonly found in more complex organisms.

Let the area of the light sensitive portion of a posterior receptor be denoted by P; it can be thought of as corresponding to an elementary picture element ($=$pixel) whose image content is transmitted to the vision processing neural apparatus as a signal which is undifferentiated in space. If the effective image plane area of the corresponding anterior receptor is A, then A can be partitioned into $N = A/P$ nonoverlapping pixels. The organization of this vision system can be expressed as a 2-level hierarchy, whose root node corresponds to the image produced by the anterior lens considered as an entity, while the N pixels are nodes belonging to the next level (see Figure 5.9.2). The neural apparatus attached to the posterior lens may have an internal structure which consists of a number of photosensitive receptors whose position is fixed relative to the position of the posterior lens. If it does, and if there were M receptors in the image plane of the posterior lens, then each pixel of area P would in effect be subdivided into "retinal" pixels of area P/M, and would correspond to a third level in the hierarchy which corresponds to the organization of the vision system, as illustrated in Figure 5.9.3. In this representation it is implicitly assumed that the scanning process can be tolerably approximated by a procedure which captures the image information from successive pixels discretely rather than as a continuous analogue signal. Although the latter form of signal acquisition is more likely on physiological grounds, the higher processing stages may well perform the equivalent of a discretization—

$N = A/P$ Posterior Lens Pixels

Figure 5.9.2   Hierarchical representation of *Copilia's* vision system.

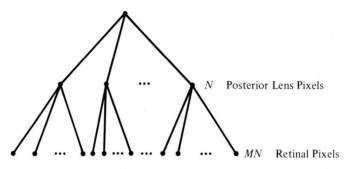

N Posterior Lens Pixels

MN Retinal Pixels

Figure 5.9.3   Hierarchical representation of *Copilia*'s vision system with hypothetical retinal pixels.

analogous to analog to digital conversion. But the precise details of the form in which the image signal is captured is irrelevant to the purposes of the present section. We are also implicitly assuming that *Copilia*'s scanning mechanism is uniform in the sense that it periodically visits each pixel; clearly, this is also but an approximation.

If the preceding assumptions are accepted for the sake of argument, then the vision mechanism of *Copilia* in effect executes a sequential search of the image produced by the anterior lens. We have previously calculated that a sequential search for a 3-level hierarchy is optimal (in the sense that it minimizes average search time) if the number of nodes $N(1)$ and $N(2)$ are related to the search rates v_1 and v_2 (where the integers denote the level number; recall that the root belongs to level 0) by (cf. (4.4)):

$$N(2) = v_1 v_2 \left(\frac{N(1)}{v_1}\right)^2 .$$

In the present case $N(1) = N$, and $N(2) = MN$. Substitution yields the relationship

$$M = \left(\frac{v_2}{v_1}\right) N$$

for an optimal system. In this equation N can be estimated from the ratio of areas A/P, and v_1 can be estimated from the scanning frequency of 15 Hz observed by GREGORY. From these data, the ratio M/v_2 can be estimated. M can in principle be determined from a physiological investigation. v_2 is essentially the same as the average processing time required by the "retinal pixels" regardless of whether they are "wired" sequentially or in parallel; if this number can be determined by experiment, it would be possible to verify whether *Copilia*'s vision system is optimal in the sense of its hierarchical structure.

Our discussion of *Copilia* has been laced with suppositions. The situation is less hypothetical for the human vision system. Although the physical structures are quite different, from a sufficiently abstract standpoint one can conclude that the systems function along quite comparable lines. In the case of human vision, the retina is covered by an array of some 125 million photosensitive rods which are primarily employed in "scotopic" vision in low levels of illumination; they are sensitive to luminance variation but do not discriminate color. Superimposed on this array is the collection of color sensitive cones that function in normal daylight vision; their density diminishes as the distance from the fovea increases; Figure 5.3.3 illustrates the density variations for the rods and cones as a function of distance from the fovea for the human eye. The high resolution component of the human vision system is the fovea itself, whose central portion consists, as we have previously remarked, only of color sensitive cone receptors packed in an approximation of a hexagonal lattice which achieves the densest possible packing.

Let us ignore the rod subsystem in our analysis. The cones are most densely packed in the foveola, a disk of radius 200 micrometers which contains about 20,000 cones. Outside the foveola the density of cones rapidly decreases to an average value of about 6×10^{-3} cones/square micrometer. The array of extrafoveal cones can be thought of as a low spatial resolution subsystem relative to the subsystem consisting of the foveola. .

If a physical scene is stationary, that is, unvarying as time passes, then by sequentially directing the foveal gaze to various points of the image it will be possible to increase the amoumt of information compared with what nonfoveal portions of the retina of the same area supply. Thus we are led to conceive the human vision detection system as a hierarchy whose levels correspond to low resolution nonfoveal detection refined by a level of foveal scanning. Let the area of the foveola be denoted F and the area of the retina by R. If the retina can be conceived of as composed of $N = R/F$ disjoint regions of foveola size, then the low resolution signal detection system consists of these N patches each of which contains an array of about $6 \times 10^6/N$ rods. It was stated in section 5.3 that $R = 10^9$ square micrometers and the foveola has an area of about 1.25×10^5 square micrometers whence $N = R/F = 7,950$ regions. Since there are about 6 million extrafoveal cones all told, each patch of foveal size contains about 750 cones. Thus the ratio of cone photoreceptors in the foveola to extrafoveal cone photoreceptors in a region of the same area is about $27:1$; this ratio does not reflect the projection of the photoreceptor cells to common neural transmitting cells which is $1:1$ for the foveal receptors but no less than $8:1$ for the extrafoveal cones. Hence, at the higher cortical levels of the vision system the ratio of foveal to nonfoveal channel capacity for regions of the same area may be greater than $200:1$.

5.10 Biological Echolocation Systems

It is well known that evolution has enabled organisms to accomplish similar ends by employing widely varying means. The echolocating bats offer an interesting example of this kind of adaptation. During daylight hours, flying insects can be identified by their predators by means of vision systems that are sensitive to visible light but at night other means must be used. The echolocating bats have developed a fully adequate alternative based on the use of sonar—reflected acoustic waves. Light and sound could hardly be more different from the point of view of the physicist. Light travels through empty space, and is impeded by air and more ponderous physical bodies; sound cannot be transmitted in the absence of matter. Light is an electromagnetic oscillation whose vibrations are transverse to its direction of propagation; sound is an oscillatory variation of pressure in the transmitting medium whose vibrations are along the direction of propagation. The wavelength of light, which is a characteristic measure of the size of the objects it can directly affect, depends on its frequency but varies between 0.4 micrometers and 0.7 micrometers for light that is visible to humans, whereas the wavelength of sound varies between 0.5 centimeters and 15 meters for sounds audible to humans, and between 2 centimeters and 2 millimeters for echolocating bats. Hence, the smallest objects that a bat can detect are much larger than the smallest objects which are in principle detectable by light sensing systems.

Despite the dissimilarity of the physical means used to transmit information, human vision based on light and bat echolocation based on sound serve similar purposes. One of the most remarkable features of these dissimilar systems is that they employ strikingly similar information processing principles to organize the sensed data and extract positional and other information about their surroundings from it. The purpose of this section is to highlight the common information processing principles which underlie the different physical embodiments that are necessitated by the physical differences between light and sound.

In the previous section we discussed the role of the gaze in human vision as a component of a hierarchical system of information processing. One of its purposes is to reduce the computational demands on the vision system by limiting its resolution while still providing a means for acquiring image data in a high resolution mode when more refined information is needed in order to evaluate or classify the scene. Here we will describe an "acoustic fovea" utilized by the echolocating horseshoe bat *Rhinolophus ferrumequinum*, whose information processing skills have been intensively studied by NEUWEILER and his school.

Echolocating bats use their larynx for the production of ultrasonic sounds and their ears for its subsequent detection. Both function in a manner generally similar to that for other mammals, including man. The horseshoe bats emit a pure tone which is narrowly tuned to a pitch in the range of 81 to 86 kHz controlled to within an accuracy of 50 to 200 Hz. When these sounds are reflected by objects in the environment, the pitch of the echo will be shifted if the reflecting object is in motion relative to the bat. The bat uses this DOPPLER shift to identify and track flying insects which are its prey.

The auditory systems of bats and humans are physiologically similar but there are several important differences in the basilar membrane. The bat's ear is sensitive to frequencies varying from a few kHz up to about 86 kHz with

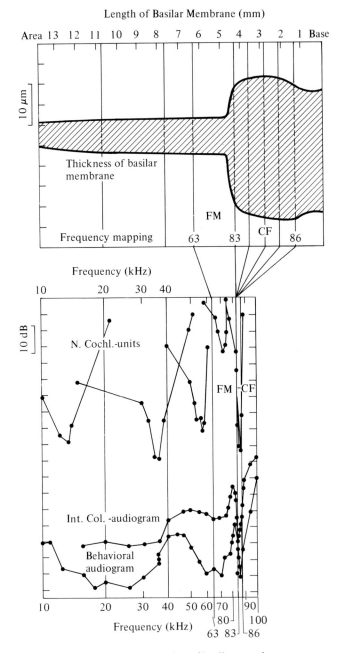

Figure 5.10.1   Diagrammatic representation of basilar membrane response for *Rhinolophus ferrumequinum*. (From Neuweiler, Bruns, & Schuller, *Acoustical Society of America 68*, pp. 741–753. Copyright 1980 American Institute of Physics. Reproduced with permission.)

an insensitive region at about 81 kHz. Figure 5.10.1 is a diagrammatic representation of the basilar membrane response for the horseshoe bat. The graph shows the thickness of the basilar membrane as a function of distance from its base to its apex (indicated on the upper abscissa). Positions along the membrane respond to different frequencies (indicated on the lower abscissa). Of its total length of about 13.5 millimeters, about two-thirds is devoted to frequencies between about 10 kHz and 80 kHz, but the 4.3 millimeters of the basilar membrane closest to the oval window of the ear is specialized to form a mechanical filter with exceptional sensitivity in the range 82 kHz–86 kHz. In other parts of the basilar membrane a distance of 4.3 millimeters corresponds to a frequency range of more than 1 octave; e.g., the range from 20 kHz to 40 kHz corresponding to locations between about 7.5 and 10.75 millimeters (see Figure 5.10.1). But the narrow interval of 4,000 Hz allocated to the thickened initial 4.3 millimeters of the basilar membrane is less than a semitone at these frequencies. Thus the horseshoe bat is able to achieve great relative precision in discriminating frequencies within the range 82 kHz–86 kHz.

The high resolution of information acquired by the specialized portion of the basilar membrane is preserved by the neural transmission network to the auditory cortex. Of the 16,000 afferent fibers in the cochlear nerve of *Rhinolophus*, about 20% come from that part of the organ of Corti served by the thickened basilar membrane. For these reasons and others which will be discussed below, the thickened, more sensitive, portion of the basilar membrane is called an *acoustic fovea*.

The natural shape of the basilar membrane is a spiral. Figure 5.10.2 shows how the frequency response corresponds to location, with the great expansion for the range 82 kHz–86 kHz.

Figure 5.10.3 displays the relationship between the number of neurons per octave and the best frequency to which the neurons are sensitive for neurons belonging to the spiral ganglion and to the inferior colliculus. In both cases

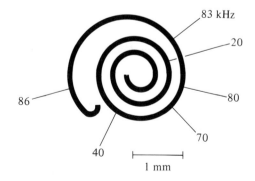

Figure 5.10.2   Projection onto transmodiolar plane of basilar membrane with frequency response. (From Neuweiler, Bruns, & Schuller, *Acoustical Society of America* *68*, pp. 741–753. Copyright 1980 American Institute of Physics. Reproduced with permission.)

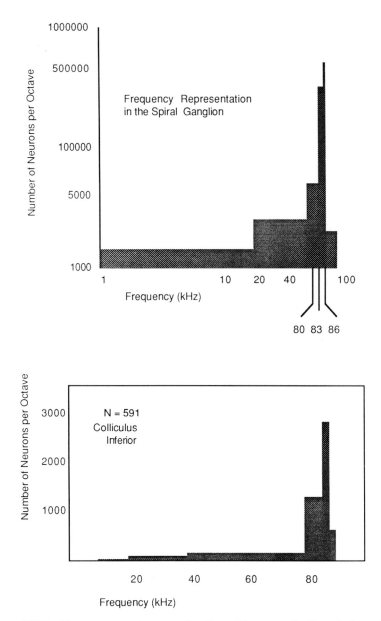

Figure 5.10.3   Neurons per octave as a function of frequency in the spiral ganglion and the inferior colliculus. (From Neuweiler, Bruns, & Schuller, *Acoustical Society of America 68*, pp. 741–753. Copyright 1980 American Institute of Physics. Reproduced with permission.)

the overrepresentation of the foveal frequencies is striking. This shows that the high resolution of the mechanical filter which senses the acoustic signal is preserved along the higher auditory pathways so that some form of this refined information is available to the cognitive centers.

Investigations suggest that the frequency analysis of the acoustic signal is performed entirely by the mechanical filtering mechanism of the cochlea, and that the neural network does not perform this function.

The true justification for the term *acoustic fovea* lies not in its morphological similarity to the fovea of the human vision system but in its functional similarity. Just as the general motion sensitive receptors of the human retina function as a low resolution detector which directs the foveal gaze to regions of interest in the image, so does the general broad-band but low-resolution part of the bat's basilar membrane detect regions of interest in a scene for more detailed exploration by the acoustic fovea. But whereas human vision relies on one light source for input to both the low resolution and the foveal high resolution detection apparatus, a common acoustic signal cannot serve both functions for the bat. There are several reasons. Whereas the wavelength of light is so small that the finest features of macroscopic objects are detectible, the wavelength of sound in the audible range of the bat is so variable that with the lower frequencies, only coarse measures of position and conformation are possible. Since high resolution acoustic measurements depend upon high frequency echos, the acoustic fovea should be concentrated at the high frequency end of the auditory spectrum, as it is.

The refinement in human foveal vision is the result of the denser packing of photoreceptors which in effect increase the spatial response frequency of the foveal region. This high resolution scanner is trained on the region of interest in the image by moving the eye so as to gaze at the region. The acoustic fovea of the bat cannot be trained on an arbitrary portion of the auditory manifold as the gaze can be trained on the visual manifold. Nevertheless, the method employed by the bat, as we shall describe, is based on similar principles adapted to different circumstances.

The bat normally emits a pure tone of approximately 83 kHz, which lasts for about 60 milliseconds. Echoes from an object which is stationary relative to the bat will have the same frequency. But as the bat approaches an object the DOPPLER shift of the reflected echo will increase its pitch. The echoes start to return before the bat has finished vocalizing. Thus the auditory information processing system must distinguish between the emitted sound and the returning echo which are mixed in the sensed signal. First tones normally suppress a response to an overlapping second tone, but the horseshoe bat has adapted to its peculiar needs by enhancing rather than inhibiting response to a second tone when the first lies in the range 82 kHz–86 kHz, although the usual inhibition occurs outside the specialized part of the auditory spectrum.

The time overlap of the emitted signal and the echo return permits the bat to perform a frequency comparison. This is an important factor in the effectiveness of the acoustic fovea because as a hunting bat approaches its prey, the frequency of the DOPPLER-shifted echo increases. The bat can fly as fast as 12 meters per second, which would produce a DOPPLER-shifted echo of between 86 kHz and 87 kHz, outside the range of the acoustic fovea. The bat has solved this problem by an ingenious feedback system which serves the same purpose as the muscular feedback loop which couples the human fovea to moving object in a scene. Whenever the frequency of the return echo is greater than a frequency which is nearly identical to the center frequency of the acoustic fovea, the bat decreases the frequency of the next emitted sound by the same amount. The following echo will consequently lie closer to the center frequency of the fovea. By this means the bat is able to maintain the echo frequency within several hundred hertz of the central foveal frequency. As NEUWEILER has observed, "This feedback system effectively uncouples the bat's echolocation system from its own motion. Just as the oculmotor muscles produce tracking motions to keep an image on the fovea of the retina, the laryngeal muscles adjust their tension to maintain the echo frequency in the acoustic fovea."

The information gained by an acoustic fovea observation compared with an observation by the nonfoveal part of the auditory sensing system can be estimated by comparing the number of afferent nerve fibers connected to the two parts of the organ of Corti. It has already been remarked that of the 16,000 fibers about 20% are connected to the acoustic fovea. Thus the information gain will be $\log_2(12,800/3,200) = \log_2 4 = 2$ bits.

An estimate in terms of the relative sensitivity of the basilar membrane to frequency variations can be made as follows. Since the foveal portion of the basilar membrane is about half as long as the nonfoveal portion, it can be considered as half as sensitive for a given relative frequency range. If there are $2N$ just noticeable differences possible for the nonfoveal range, there will be N jnd's for the foveal range. Then the relative frequency descrimination in the nonfoveal range will be $(1/2N)(80\,kHz/10\,kHz) = 4/N$ and in the foveal range, $(1/N)(86\,kHz/82\,kHz) = (23/21N)$. The corresponding information gain for a foveal observation compared with a nonfoveal one is $\log_2((4/N)/(23/21)) = \log_2(84/23) = 1.87$ bits, approximately the same as the estimate derived from a count of afferent nerve fibers. Since the bat will emit sounds about 16 times per second, the acoustic fovea provides an information gain of about 30 bits per second. In this regard it is interesting to note that human speech phonemes are transmitted at a rate of about 50 bits per second, which, when the internal redundancy of natural language is eliminated, amounts to an information tansfer rate of only about 10–15 bits per second.

5.11 A Catalog of Visual Illusions

The study of what are called illusions of the senses is, however, a very prominent part of the psychology of the senses; for it is just those cases which are not in accordance with reality which are particularly instructive for discussing the laws of those processes by which normal perception originates.

—Hermann VON HELMHOLTZ

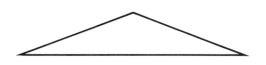

Figure 5.11.1   Base of isoceles triangle appears to sag.

That there is often a discrepancy between appearance and reality was certainly known to the ancient Greeks, for their architects compensated for the peculiar distortions created by the human vision system. Because the horizontal base of an isoceles triangle seems to sag (Figure 5.11.1), the architrave of the east face of the Parthenon incorporated a compensating upward bow; because a bright column seen against a dark background appears broader than it is, and a dark column seen against a light background appears narrower, they built the angle columns, which are seen against the background of the bright sky, broader than those seen against the building's dark interior.

Despite their early recognition in special cases, the systematic study of visual illusions had to wait until the nineteenth century for the birth of psychology as a scientific endeavour. The first scholarly paper devoted to the subject appears to have been published in 1854 by J.J. Oppel, who introduced and studied the illusion of divided space (see Illusion 4.2, in the catalog appended to the end of this section) in which a divided interval appears to be of greater length than the same interval undivided. According to Coren and Girgus, by 1900 more than 200 papers on visual illusions had been published. By 1950 the total had grown to more than 1,000. But to place this very considerable growth in proper perspective, during the period from 1915 to 1950, whereas there were some 4,500 papers in psychology published annually, on the average only about 4 papers concerned with visual illusions appeared.

Most of the basic visual illusions were discovered by the end of the nineteenth century, but even today few of them have been explained. As more has been learned about the neurophysiology of the vision system, it has become possible to explain some illusions satisfactorily and to separate the constituents of others into optical, retinal, and cortical categories. But, if anything, it has become increasingly certain that information processing strategies rather than limitations and deficiencies of the optical and retinal components of the vision system are primarily responsible for illusions.

Evidence that animals experience similar visual illusions has also been steadily accumulating. Fish, chicks, white rats, and various kinds of monkeys appear to encounter the same discrepancies between appearance and reality that people do. This may support the hypothesis that there are certain universal information-processing principles which govern all organisms in their relation to sensory stimuli.

Not all apparent discrepancies between the appearance of a stimulus and its physical reality are normally considered illusory. For example, the effect of perspective, which makes the same object appear large or small as its distance from the viewer changes is understood to be the natural consequence

Figure 5.11.2   Fɪᴄᴋ's illusion.

of the laws of optics and generally is not considered an illusion. One may say that a visual stimulus is illusory if the discrepancy cannot be explained. In this sense, the goal of a theory of illusions is their elimination. But we will not insist on this narrow definition.

In 1851 Fɪᴄᴋ had called attention to the fact that vertical lines appear longer than horizontal lines of the same length (Figure 5.11.2) as an example of the asymmetry of the visual field; it was Oᴘᴘᴇʟ who recognized this "asymmetry" as an illusion in his path-breaking publication 3 years later. In fact, measurements show that a vertical line appears as much as 7% to 14% longer than a horizontal line of the same length. The remarkable anomaly does not depend on alignment with the vertical determined by the gravitational force but on the vertical defined on the retinal image by a line perpendicular to the plane determined by the optical axes of the two eyes focusing on a given point.

Among the various illusions several stand out because they are particularly simple, or striking, or because they appear to be constituents of other more complicated illusions. In 1860 Zöʟʟɴᴇʀ submitted an article to the *Annalen der Physik* discussing an illusion of direction (Illusions 4.12, 4.13). Referring to the illustrations, one sees that the long parallel lines appear to tilt in a direction determined by the angle of the short parallel transversal fins which cut them. As the angle between a line and the fins become smaller, the tilt of the line appears to increases, but when the angle becomes less than about 10 degrees, the nature of the illusion changes: the tilt reverses direction and the lines appear to consist of segments that do not abut each other precisely. This effect is greatest when the angle is about 2 degrees (Figure 5.11.3).

The illusion of Zöʟʟɴᴇʀ actually consists of two parts: the tilted lines, which he had observed, and an illusion of discontinuity of the short cross-line fins which was first recognized by Pᴏɢɢᴇɴᴅᴏʀꜰ, editor of the *Annalen der Physik* to which Zöʟʟɴᴇʀ had submitted his paper. Pᴏɢɢᴇɴᴅᴏʀꜰ's illusion (Illusion 4.6 and Figure 5.11.4) is encountered in many contexts.

In 1889 F.C. Müʟʟᴇʀ-Lʏᴇʀ introduced a striking illusion (Illusion 4.1 and Figure 5.11.5) which has formed the basis for many investigations because of its simplicity, the strength of the effect, and its role as a constituent in other illusions. Within a decade of its publication, at least 6 different explanations for it were propounded, all of them unsatisfactory. Müʟʟᴇʀ-Lʏᴇʀ argued for

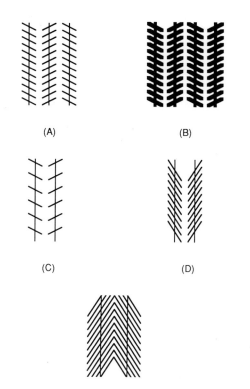

(A)   (B)

(C)   (D)

(E)

Figure 5.11.3   Variants of ZÖLLNER's illusion.

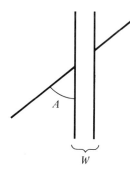

Figure 5.11.4   POGGENDORF's illusion.

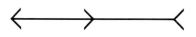

Figure 5.11.5   BRENTANO's version of the MÜLLER-LYER illusion.

Figure 5.11.6   Another variant of the MÜLLER-LYER illusion.

a theory of "total impression." In 1890 LASKA thought it could be explained by the "tendency to form closed figures." In 1896 THIERY proposed a theory based on perspective cues which GREGORY has espoused and enriched in recent years. LIPPS argued, in 1897, that aesthetic feelings of freedom and constraint held the key. A more sophisticated but no more successful theory of actual or implied eye movements was proposed by HEYMANS in 1896 and by WUNDT in 1898; and in the latter year, EINTHOVEN attributed the illusion to optical limitations of the eye.

More recently, as knowledge about lateral inhibition and other aspects of early visual processing by the neural network have increased, theories have been suggested which rely on these information processing functions of the retina. Although the evidence suggests that retinal processing can produce the effects of the MÜLLER-LYER illusion, ingenious experiments using subjective contours and random dot stereograms have convincingly demonstrated that higher cortical processes independently interpret the MÜLLER-LYER stimuli in a way which produces the same illusory effect. These developments will be discussed below.

The horizontal line segment with the "outward" pointing fins in the MÜLLER-LYER illusion appears as much as 25% to 30% longer than the line segment of the same length with "inward" pointing fins. The effect depends on both the length of the fins and their angle with the horizontal, it being greatest when the fins are about 20% to 35% the length of the line segment and decreasing thereafter.

The "perspective" theory of this illusion seems to be eliminated by the following version of it (Figure 5.11.6), published by BRENTANO in 1892. The illusion can be further simplified by eliminating the horizontal line segments: the apparent distances between adjacent vertices (the "extent of space" as some writers call it) appear unequal. The illusion evidently has something to do with angles!

A remarkable family of illusions of quite another type was invented by FRASER in 1908. These so-called *twisted cord* illusions, of which the parts of Illusion 4.18 are examples, are formed by playing out a "twisted cord" consisting of black and white strands over a patterned background. Figure 5.11.7 exhibits a different variation on this theme: the letters are formed from vertical and horizontal twisted cords, yet appear to be tilted. These illusions have not received as much attention as they deserve.

These illusions all involve angles in one way or another. Beginning in 1891 with JASTROW, investigators have studied the relationship between the apparent and actual magnitude of an angle. The relationship can be described in qualitative terms as follows:

Figure 5.11.7 Fraser's twisted cord illusion. (From Luckiesch, *Visual Illusions, Their Causes, Characteristics and Applications* p. 88. Copyright 1965 Dover Publications, Inc. Reproduced with permission.)

1. Small angles tend to be overestimated; maximum overestimation occuring for angles between 10 and 20 degrees;
2. Large angles tend to be underestimated; maximum underestimation occuring for angles between 125 and 160 degrees; i.e., the supplementary angle is overestimated;
3. 90 degree angles are estimated accurately;
4. As an angle increases to 45 degrees, the overestimation effect "weakens and then shifts to an underestimation until it reaches 90 degrees." The cycle repeats between 90 and 180 degrees.

In 1948 Berliner and Berliner proposed a formula to account for all cases of angular distortion. If A' denotes the apparent magnitude, A, the true angle, and C, and appropriate constant which depends on the circumstances, then

$$A' = A + C \sin 4A. \tag{5.21}$$

Observe that as a function of A, this expression is periodic with period $\pi/2$ (90 degrees) and that the distortion $A' - A = D$ increases from $D = 0$ at $A = 0$ to its maximum value at $A = \pi/8$ (22.5 degrees). The formula is consistent with the rough qualitative description given previously but does not provide an accurate account of the relationship.

We see that small angles tend to be overestimated, implying that the information gained from observation of the apparent angle is less than would be gained from observation of the true angle, since the gain in information is measured by $I = \log_2(2\pi/\theta)$ for an angle θ. Thus observation of small angles appears to incorporate a mechanism which systematically and selectively omits information.

The discrepancy between the apparent and true magnitudes of angles can result from a combination of independent information processing characteristics of different components of the vision system. Three levels of processing structure are of particular interest: (1) the eye as an optical system, (2) retinal information processing, and (3) higher level cortical information processing.

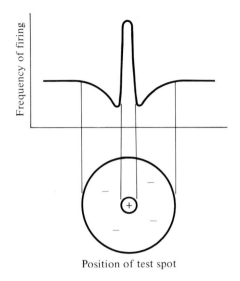

Figure 5.11.8   Effect of lateral inhibition on neural firing frequency.

The optical level is the least interesting. Although it is true that certain illusions result from the optical properties of the eye, especially the nearly spherical shape of the retinal matrix upon which an image falls, these are all well understood, and their effects can be eliminated by suitable compensatory means.

The other two sources of illusion are more interesting, and it is also more difficult to distinguish their consequences.

Lateral interaction in retinal information processing can produce discrepancies between the true and apparent magnitudes of angles. Recall the discussion of lateral inhibition in section 5.6, whose results can be summarized and supplemented for our present needs in the following way. When a small test spot of light is moved across an otherwise uniformly illuminated retina, the frequency of firing varies approximately as shown in Figure 5.11.8.

This pattern can be thought of as resulting from a combination of the excitation of nearby receptors by the stimulated receptor and their inhibition according to the profiles indicated in Figure 5.11.9. This type of receptor is called "on-center." In mammalian eyes, "off-center" receptors, for which inhibition outweighs excitation at the stimulated point, are about equinumerous with on-center receptors. Figure 5.11.10 illustrates how such a receptive field might result from the combination of excitation and inhibition.

The characteristic shapes of these curves should be compared with the graph displayed in Figure 5.5.6, which is the second derivative of the light intensity curve shown in Figure 5.5.5; we shall have occasion to return to explore this connection in greater detail in Chapter 6.

Now suppose that an angle is viewed against a uniform background and that its image on the retina produces a distribution of on-center neural firing

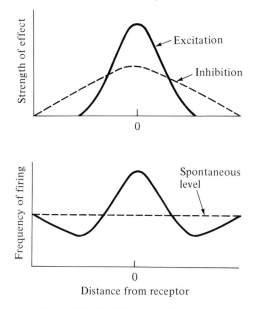

Figure 5.11.9   "On-center" receptor.

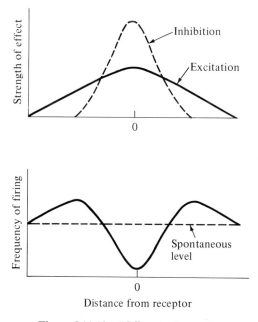

Figure 5.11.10   "Off-center" receptor.

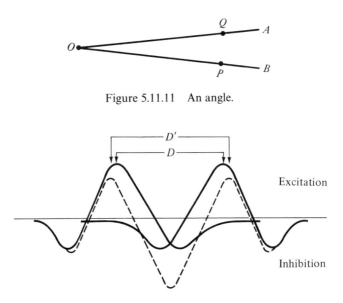

Figure 5.11.11   An angle.

Figure 5.11.12   Displacement of maxima of overlapping neural response functions.

response functions. The peak of such a function, which corresponds to the maximum firing rate, lies over a point on the angle. Thus, for a point such as *P* in Figure 5.11.11, which lies far from ray *OA*, the on-center response in a direction perpendicular to *OB* will look like the curve shown in Figure 5.11.9. The same is true for the response at *Q* considered in a direction perpendicular to *OA*. If *P* and *Q* are moved toward the vertex *O*, they will approach each other and their response functions will overlap. This effect will initially (i.e., when *P* and *Q* are close enough to the vertex) increase the distance between the maxima of the two response functions, thus creating an apparent increase in distance between *P* and *Q*. This situation is illustrated in Figure 5.11.12, where *D* denotes the distance between the maxima of the independent response function, and *D'* the distance between the maxima of their sum.

As the points *P* and *Q* continue to approach *O* and each other, their response functions overlap to an increasing degree, and ultimately, close to *O*, begin to reinforce each other as shown in Figure 5.11.13. When this occurs, the two distinct response maxima can no longer be distinguished so the rays *OA* and *OB* can no longer be distinguished: thus the vertex *O* appears to have been displaced into the interior of the angle. The combined effect of these lateral interactions is to distort the shape of the angle so that it appears somewhat like the dotted curve shown in Figure 5.11.14.

It is evident that the magnitude of angle *AOB* will appear to be greater than it actually is. Moreover, the position of the vertex *O* will appear to have been displaced into the interior.

The latter result implies that the apparent length of the bars in the MÜLLER-LYER illusion and the distance between adjacent vertices in the BRENTANO

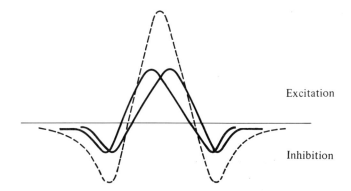

Excitation

Inhibition

Figure 5.11.13 Reinforcement of maxima of overlapping neural response functions.

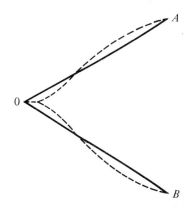

Figure 5.11.14 Apparant shape of angle as a result of lateral interactions.

version of it should differ from their actual lengths and distance as they are observed to do. But this argument does not prove that lateral interactions in the retina can account for the entire discrepancy between appearance and reality.

The apparent increase in magnitude of the angle due to lateral interactions can be used to explain the ZÖLLNER and POGGENDORF illusions. But again, other forces may be at work so that lateral interactions may account for only part of the observed effects.

In order to determine whether higher cognitive processing participates in the illusions of MÜLLER-LYER and POGGENDORF based on mis-estimation of the magnitudes of angles, some means for eliminating the effects of retinal lateral interactions must be found. Several ingenious methods have been proposed.

● One can attempt to create the illusions by means of subjective contours; since the contours do not correspond to retinal excitation, an illusion must, it is argued, be the result of higher level processes.

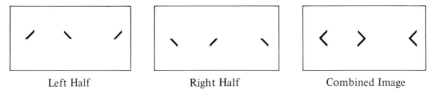

Left Half   Right Half   Combined Image

Figure 5.11.15   Stereo pair stimuli.

- One can observe the stimuli using binocular stereopsis in such a way that neither eye alone is exposed to an angle as a stimulus. Since retinal excitations include only straight line segments but not angles (e.g., as in Figure 5.11.15), lateral interactions cannot play a role and the illusion must, it is argued, be the result of higher level processes. Binocular rivalry makes this procedure an uncertain and difficult experimental technique.
- One can embed the left- and right- angle-free components of the stimuli described above and illustrated in Figure 5.11.15 in random dot sterograms. This technique, effectively exploited by JULESZ, has everything to recommend it.

All three methods support the conclusion that angle-dependent effects such as the MÜLLER-LYER and POGGENDORF illusions occur even if retinal lateral interactions are eliminated. Hence we must conclude that higher level information processing strategies are able to produce the illusions.

The work of HUBEL and WIESEL on cortical feature detectors in the cat has provided the neurophysiological foundations upon which an explanation of the cortical provenance of these illusions can be based. Area 17 of the cerebral cortex of the cat contains cells that respond to elongated receptive fields of both the on-center and off-center type (illustrated in Figure 5.11.16) which are tuned to detect directions in the visual field. The regions marked with a " + " sign excite the neural firing frequency and the regions marked with a " − " sign inhibit it. Only a finite number of different directions are represented by receptive fields with the consequence that not all directions will be perceived with the same intensity of response: the detectors are relatively narrowly tuned to particular directions. This fact provides the basis for an assessment of the cortical response to an angle imaged on the retina.

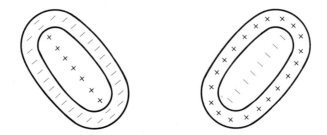

Figure 5.11.16   Directional feature detectors in the cerebral cortex.

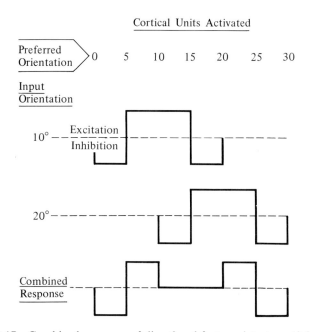

Figure 5.11.17   Combined response of directional feature detectors. (Adapted from Coren and Girgus, *Seeing Its Deceiving: The Psychology of Visual Illusions.* Copyright Lawrence Earlbaum Associates, 1978.)

We will follow the discussion given by CoREN and GIRGUS. Suppose for the sake of argument that a directional feature detector is tuned to respond to an input orientation of 10 ± 5 degree (relative to some standard direction) and inhibits response for angles that lie between 0 and 5 degrees, and between 15 and 20 degrees. Suppose that another excites a response for angles in the range 20 ± 5 degrees and inhibits response to angles between 10 and 15 degrees, and between 25 and 30 degrees. These conditions are illustrated in the upper two graphs in Figure 5.11.17. For a stimulus consisting of a pair of directions, oriented to 10 and 20 degrees, the response will be the sum of the two directional feature detector responses, which is shown as the third graph in the figure. The summation of excitation from one stimulus detector and inhibition from the other reduces the excitation level for directions between 10 and 20 degrees. As a result, the average peaks of excitation correspond to directions of approximately 7.5 and 22.5 degrees; the angle formed by the two directions appears to be 15 degrees rather than its correct magnitude of 10 degrees.

In 1977 BURNS and PRITCHARD measured the cortical response to directional stimuli in the cat. They isolated 36 cells in the visual cortex which have a maximum response to a specific direction and the little response to directions that differ from the specific one by more than 15 degrees. They measured the response of these detectors when a second line which intersected the first at an angle of 30 degrees was added to the retinal image. Their results are dis-

Illusion 1.2   Multivalued pattern continuation (Figure 5.4.3).

2. *Stabilized images*: These cannot be illustrated; see the discussion in section 5.5.
3. *Subjective contours.*

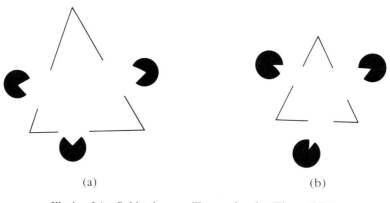

(a)   (b)

Illusion 3.1   Subjective curvilinear triangles (Figure 5.7.1).

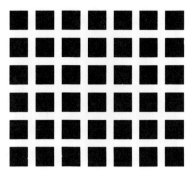

Illusion 3.2   The intersections of the white bars contain gray subjective figures.

Illusion 3.3   A subjective equilateral triangle (Figure 5.7.4).

Illusion 3.4   Concave and convex subjective equilateral triangles (Figure 5.7.5).

Illusion 3.5   Subjective stereo image.

4. Metrical, areal, and angular illusions in 2 dimensions.

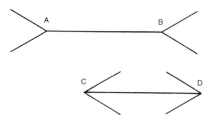

Illusion 4.1   The Müller-Lyer illusion. The intervals *AB* and *CD* are of equal length although *AB* appears to be longer.

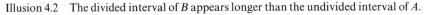

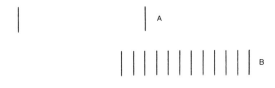

Illusion 4.2   The divided interval of *B* appears longer than the undivided interval of *A*.

Illusion 4.3   The distance from *A* to *B* appears smaller than the distance from *B* to *C* although they are equal.

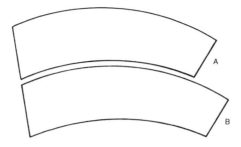

Illusion 4.4   The two figures are congruent.

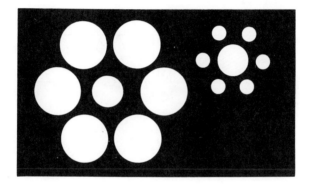

Illusion 4.5   The center circles are congruent. (From Thurston and Carraher, *Optical Illusions and the Visual Arts*, p. 111. Copyright 1966 Van Nostrand Reinhold Company Inc. Reprinted with permission.)

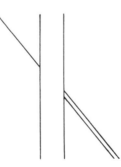

Illusion 4.6   POGGENDORF's illusion: Which of the two lines on the right extends the line on the left?

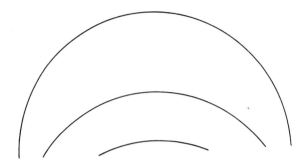

Illusion 4.7   Changes in the subtended angle make circular arcs from one circle appear to have different curvature.

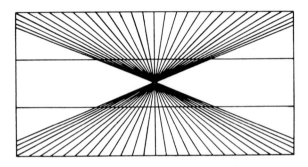

Illusion 4.8   HERRING's illusion of direction. (From Luckiesh, *Visual Illusions, Their Causes, Characteristics and Applications*, p. 80. Copyright 1965 Dover Publications, Inc. Reproduced with permission.)

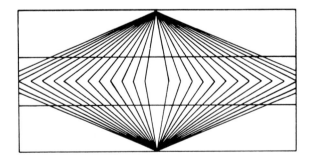

Illusion 4.9   WUNDT's illusion of direction. (From Luckiesh, *Visual Illusions, Their Causes, Characteristics and Applications*, p. 79. Copyright 1965 Dover Publications, Inc. Reproduced with permission.)

Illusion 4.10   Parallel lines which appear curved. (From Luckiesh, *Visual Illusions, Their Causes, Characteristics and Applications*, p. 79. Copyright 1965 Dover Publications, Inc. Reproduced with permission.)

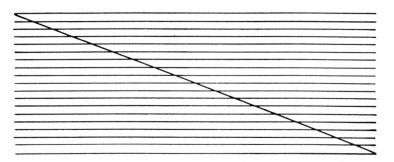

Illusion 4.11   The diagonal straight line appears to consist of discontinuous line segments.

Illusion 4.12   ZOLLNER's illusion of direction. (From Thurston and Carraher, *Optical Illusions and the Visual Arts*, p. 115. Copyright 1966 Van Nostrand Reinhold. Reproduced with permission.)

Illusion 4.13   ZOLLNER's illusion of direction exhibited an illusion of discontinuous short line segments crossing the vertical long lines. (From Thurston and Carraher, *Optical Illusions and the Visual Arts*, p. 115. Copyright 1966 Van Nostrand Reinhold. Reproduced with permission.)

Illusion 4.14   The square is distorted. (From Luckiesh, *Visual Illusions, Their Causes, Characteristics and Applications*, p. 61. Copyright 1965 Dover Publications, Inc. Reproduced with permission.)

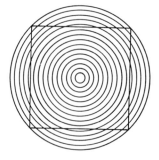

Illusion 4.15   The square is distorted. (From Thurston and Carraher, *Optical Illusions and the Visual Arts*, p. 117. Copyright 1966 Van Nostrand Reinhold. Reproduced with permission.)

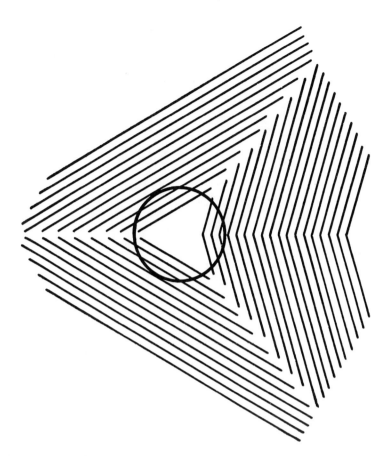

Illusion 4.16   The circle appears to be an oval. (From Luckiesh, *Visual Illusions, Their Causes, Characteristics and Applications*, p. 62. Copyright 1965 Dover Publications, Inc. Reproduced with permission.)

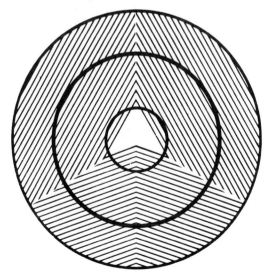

Illusion 4.17   The innermost circle appears to be an oval. (From Thurston and Carraher, *Optical Illusions and the Visual Arts*, p. 117. Copyright 1966 Van Nostrand Reinhold. Reproduced with permission.)

(b)

(a)

Illusion 4.18 "Twisted cord" illusions: (a) straight cords, (b) concentric circles, (c, d) Variations on the *"twisted cords"* theme. (Parts b–d from Lanners, *Illusionen*, pp. 62–63. Copyright 1973 Verlag C.J. Bucher. Reproduced with permission.)

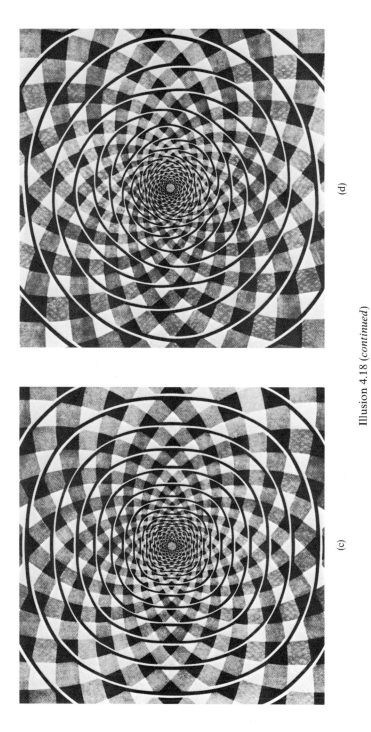

(d)

(c)

Illusion 4.18 (*continued*)

5. *Perspective and other 3 dimensional illusions.*

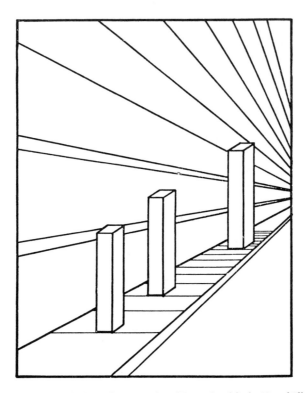

Illusion 5.1   A striking illusion of perspective. (From Luckiesh, *Visual Illusions, Their Causes, Characteristics and Applications*, p. 60. Copyright 1965 Dover Publications, Inc. Reproduced with permission.)

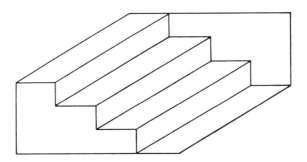

Illusion 5.2   SCHRÖDER's reversible staircase. (From Thurston and Carraher, *Optical Illusions and the Visual Arts*, p. 115. Copyright 1966 Van Nostrand Reinhold. Reproduced with permission.)

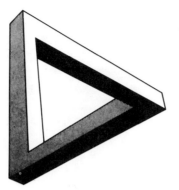

Illusion 5.3   The PENROSE's impossible object. (From Lanners, *Illusionen*, p. 35. Copyright 1973 Verlag C.J. Bucher. Reproduced with permission.)

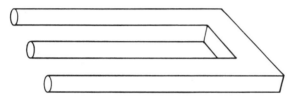

Illusion 5.4   An impossible object. (From Thurston and Carraher, *Optical Illusions and the Visual Arts*, p. 126. Copyright 1966 Van Nostrand Reinhold. Reproduced with permission.)

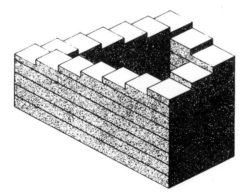

Illusion 5.5   ESCHER's impossible stairs. (From Lanners, *Illusionen*, p. 34. Copyright 1973 Verlag C.J. Bucher. Reproduced with permission.)

Illusion 5.6 CHENEY's unpatentable machine. (From R.L. Gregory, *Concepts and Mechanisms of Perception.* Copyright © 1974 R.L. Gregory. Reprinted with permission of Charles Scribner's Sons, a division Macmillan, Inc. and permission of Duckworth and Company, London.)

Illusion 5.7   Dalmatian Dog. (From Thurston and Carraher, *Optical Illusions and the Visual Arts*, p. 18. Copyright 1966 Van Nostrand Reinhold. Reproduced with permission.)

5.12 Hierarchical Sampling Systems and the Duality between Noise and Aliasing

the visual system avoids the aliasing of high frequencies inherent in any regular arrangement of image sampling elements and simultaneously minimizes sampling noise for low frequencies that fall within its potential NYQUIST bandwidths. These advantages stem from a ... spatial sampling scheme that apparently has not been used in man-made image-recording devices.

—John I. YELLOTT, JR.

In this section we continue our study of the selective omission of information in the context of identifying and interpreting the information in images, or, as it is generally expressed, *understanding* images. In section 4.5 we saw that

SHANNON's sampling theorem makes it possible to reconstruct signals from their periodic samples if the signals are band-limited, i.e., if their FOURIER transform has bounded support. If a signal is not band-limited, or if the sampling process is not regular, i.e., periodic, then the sampling theorem is inapplicable and the question of what would be an effective strategy for attempting to reconstruct the signal from samples of it arises.

Discrete sampling of a continuous signal can introduce two kinds of distortion: *noise* and *aliasing*. Noise due to sampling scatters energy broadly throughout a range of frequencies of the signal. Aliasing converts energy at frequencies greater than the *NYQUIST limit* (half the sampling rate) into energy at lower frequencies and may produce spurious periodic features in the post-sampled signal.

These defects of sampling can be noticed everywhere. In the domain of vision, the raster-scan sampling used to create video images produces the well-known "jaggy" representations of straight lines that are inclined at a small angle to the raster, as well as other more complex moiré patterns. The illusion of motion produced by video and motion pictures will sometimes create bizarre effects due to the limited number of frames per second which may be much less than the frequencies of periodic motions in the represented scene. If the temporal sampling frequency is too low, then the "snapshot" images it yields may provide the impression of impossible sequences of events, such as the often observed apparent counter-rotation of spoked wheels and the stroboscopic illusion of stationary helicopter propellors or fan blades.

Many biological information processing systems acquire their data by discrete sampling of a continuous signal. This is particularly evident for vision systems of insects, whose numerous ommatidia form an array of sample detectors, and for the vertebrates, which rely on hierarchically organized vision systems whose sampling apparatus may consist of many millions of discrete retinal photoreceptors arranged in more than one level of resolution.

Noise and aliasing are dual artifacts of the sampling process. In our study of SHANNON's sampling theorem (section 4.5) we saw that a signal that is *band-limited*, i.e., whose FOURIER transform is zero outside some bounded interval, can be reproduced without loss of information if it is sampled at *regular* intervals at a rate that is equal to at least twice the greatest frequency in its FOURIER transform. In the preceding sentence we have emphasized that the samples must be taken at regular intervals in order for this theorem to hold.

If the Sampling Theorem is applied to a signal that contains energy at higher frequencies than the NYQUIST frequency, then energy corresponding to the higher frequencies will appear to be concentrated at certain lower frequencies in the reconstruction of the sampled signal, giving rise to spurious *aliases* at these lower frequencies in the FOURIER transform of the signal, and to periodic artifacts in the sampled signal. These are the moiré patterns that have often been used with interesting results by artists.

Figure 5.12.1 illustrates how a signal that is regularly sampled at too low a frequency will produce aliases of a still lower frequency.

Figure 5.12.1   Sine waves of different frequencies with the same set of equally spaced sample values.

Figure 5.12.2 is one example of the artist's use of aliasing for aesthetic purposes.

Let us consider in greater detail the illusion of motion that is provided by displaying successive "frames" of a continually varying scene with sufficient frequency.

In the American standard NTSC video format, a television frame consists of 525 scan lines (of which fewer than 500 are visible on the screen) alternately refreshed 30 times per second. This interlacing of refreshed and unrefreshed scan lines is a compromise between increasing the frequency of renewal of the image to 60 Hz and suffering a greater degree of unsatisfactory jerkiness and discontinuity in the apparent motion.

Motion pictures display 24 frames per second but they adopt a special method to increase the perceived degree of apparent continuity which is adequate to sustain the illusion of motion in normal circumstances. Computer terminal displays are normally refreshed at 60 Hz without interlacing. We will use this sampling rate for our examples.

According to the Sampling Theorem, regular sampling at 60 equal intervals will suffice to perfectly regenerate a scene whose temporal variations are due to frequencies not greater than one-half the sampling rate, i.e. 30 Hz. There are special cases for which 60 frames per second undersamples the changing scene, with curious and occasionally comical consequences.

Consider a wheel which has a single spoke, rotating about its center. The spoke is a radius of a circle, and the rotation of the wheel can be gauged by measuring the angle through which the radius has turned as time passes. Let "snapshots" of the wheel be taken at times $t_1, t_2, \ldots, t_n, \ldots$ and let the rate of rotation of the wheel be u Hz, i.e. u rotations per second. If $u = 60k$ Hz with k an integer, then each snapshot frame will find the spoke in the same position: the wheel will appear to be motionless. Hence, rotation at $60k$ Hz is observed as rotation at the lower alias frequency 0 Hz. Thus sampling at too low a frequency may produce the artifact of *absence of motion* although the reality is its presence. If u is expressed in the form

$$u = (60k + v)\text{Hz}, \qquad -30 < v < 30, \qquad (5.22)$$

then successive frames will display the spoke as having rotated the $(v/60)^{th}$ part of one full rotation. If, for example, $v = 1$ Hz, then successive frames will show the spoke apparently displaced $1/60$ of a complete rotation in the positive

Figure 5.12.2   Ludwig WILDING, *Kinetic Structure 5-63*, 1963 (Painted Wood and Glass). (Reproduced with permission of the artist.)

sense of rotation, say counterclockwise. If, however, $u = 59$ Hz, then one can write $59 = (60 \times 1) - 1$ whence $v = -1$ so that successive frames will appear to exhibit a displacement by $1/60$ of a complete rotation in the contrary (clockwise) direction. More generally, if v is negative when the frequency of rotation u is expressed in the form (5.22), then the wheel will appear to rotate in the false, clockwise, sense since the true frequency u will be replaced by the aliased frequency v as a result of the sampling process. According to the sampling theorem, $u = 30$ Hz is the greatest frequency of rotation for which 60 samples per second can faithfully reproduce the true motion. At $u = 30$ Hz the spoke will travel halfway around the circle between successive frames so it will appear as a diameter, from which we can conclude that the rotational frequency must be $(60k + 30)$ Hz for some integer k. For slightly smaller frequencies, say $u = 29$ Hz, the true motion can be reconstructed from the samples, but for slightly greater frequencies, say $u = 31$ Hz, successive frames will show the spoke seeming to advance by $1/60$ of a rotation more than one half a rotation. But this can be interpreted as a rotation in the contrary direction of 29 cycles per second, i.e. $u = -29$ Hz. Thus $u = 31$ Hz has the alias $u = -29$ Hz.

If the sampling frequency is increased beyond 60 frames per second, then it will of course be possible to correctly deduce the motion of the wheel for correspondingly greater rates of rotation without experiencing aliasing. But there is another variable under the control of the observer that can be manipulated without increasing the average sampling rate: the spacing of the intervals between successive snapshot frames.

Suppose that we agree to sample a continuously varying scene 60 times per second but that nonuniform sampling intervals are permitted. Let the sampling times during one second be $t_1, t_2, \ldots, t_{60}$, with $0 < t_1$ and $t_{60} = 1$ second. Let the successive intervals between samples be

$$\Delta t_i = t_i - t_{i-1}, \qquad i = 1, \ldots, 60, \tag{5.23}$$

where we put $t_0 = 0$. There are 60 sampling intervals. If $\Delta t_i = 1/60$ for all i, then we are back to the uniform sampling case considered above.

A pair of frames separated by the interval Δt_i acts as if it were part of a uniform sampling schedule that samples the scene $1/\Delta t_i$ times per second. Observe that

$$\Delta t_1 + \cdots + \Delta t_{60} = 1. \tag{5.24}$$

Hence $\Delta t_i < 1$ for every i, from which it follows that the least rate of a uniform sampling schedule corresponding to some Δt_i is 1 sample per second. But there is no lower bound to the duration of the intervals other than that they all be positive, implying that one of them can be selected so small that the uniform sampling schedule corresponding to it can have as great a rate as may be desired. Thus we should expect that a suitable choice of the intervals Δt_i will enable us to obtain some information about the contribution made to a signal by an arbitrarily high preassigned frequency, although that information will be available for only a part of the sampling cycle, a part that will be the more brief the greater is the corresponding frequency.

In return for (albeit incomplete) information about frequencies above the Nyquist limit (which is equal to 30 Hz in the present example), the *principle of conservation of information* suggests that some information which would otherwise be available must be given up. Let us investigate how this comes about and what characteristics the lost information will have.

Suppose that the following non-uniform sampling schedule is adopted: $\Delta t_1 = 1/2 = \Delta t'$ and the remaining 59 intervals are of equal length. Their common length will be $\Delta t'' = (1/2) \times (1/59) = 1/118$. Half the time this sampling schedule has a maximum sensitivity of 59 Hz. Recall that 60 uniformly spaced samples per second corresponds to a NYQUIST frequency of 30 Hz. During the high frequency sampling phase, relatively great discrimination of temporal variations in the scene will be possible without aliasing. For the rotating wheel, rotational speeds up to 59 Hz can be observed without encountering the possibility of mistaking the direction or the rate of rotation. During the equally long low frequency sampling phase, it will be impossible to accurately assess the rotational rate of the wheel if it is rotating faster than once per second. If, for instance, the wheel is rotating at 50 Hz, then during a $1/2$ second long sampling interval between successive frames the wheel will have completed 25 cycles, but it will appear to be stationary. During the other 59 intervals, each of which lasts but $1/118$ second, the wheel will turn through the $50/118 = 0.42$ part of a complete revolution. Thus the perceived effect will combine the impression of uniform rotation at a constant rate of 50 Hz produced by half the frames, and the impression of a complete absence of motion produced by the samples that correspond to the single $1/2$ second interval. The "part-time" alias of the absence of motion can be thought of as a kind of *noise* produced by the sampling process the obscures the accurate information and makes the observer less confident that the extrapolation of the information provided by the high frequency samples correctly describes the source. Since the absence of motion can never produce the appearance of motion for any sampling schedule, it is evident that the irregular sampling procedure described above does provide more information about the nature of the spectrum above the NYQUIST frequency than regular sampling can provide.

The interpretation of the part-time aliases as noise can be further justified by consideration of more complicated examples. Suppose, for instance, that

$$\Delta t_1 = 1/2 \qquad\qquad = \Delta t',$$

$$\Delta t_2 = 1/3 \qquad\qquad = \Delta t'', \qquad\qquad (5.25)$$

$$\Delta t_i = (1/6) \times (1/58) = 1/348 = \Delta t.$$

This sampling schedule attains even greater maximum sensitivity for a correspondingly smaller part of the sampling cycle. The NYQUIST frequency corresponding to $\Delta t'$ is 1 Hz; to $\Delta t''$ it is 1.5 Hz; and to Δt it is 174 Hz. During the $1/6$ second when sampling at 348 Hz occurs, the 50 Hz continuous rotation of the wheel will be very accurately sampled, successive pairs of frames exhibiting a rotation of the spoke by 0.14 of a complete cycle. During the remaining $5/6$ second, the two sampling schedules will seem to indicate: one, that the wheel is not rotating at all; the other, that the wheel is rotating at $1/3$ Hz.

In addition to selecting the duration of the sampling intervals Δt_i subject only to the constraints that they be positive and sum to 1 second, the observer is also free to permute the order in which they occur in the sampling schedule, and even to change that order from one sampling cycle to the next without changing the average sampling frequency. Increasing the irregularity of re-currence of the low frequency sampling intervals and the aliases they create will further contribute to the impression that they constitute noise rather than real features of the signal (we do not admit signal noise as a "feature"). Thus the introduction of irregularity into the sampling schedule provides a means for converting aliasing consequences of the limited frequency of discrete sampling into noise consequences. In circumstances where structured features are the most important input to later stages of the information-processing strategy, it will be preferable to cope with background noise in the sampled signal rather than with false features produced by aliasing whose structure can not be differentiated from the real features of the original signal.

With these remarks as background, let us examine the effects of sampling an image signal using a spatial sampling rate which is less than twice the greatest spatial frequency that is present in the image. Figure 5.12.3 displays an 8 bit gray scale image of concentric circles whose spatial frequency increases with increasing distance from their common center. The image is a 512×512 square array of picture elements ("pixels") which we will refer to as *the high resolution image* of the concentric circles. Figure 5.12.4 exhibits the result of undersampling the high resolution image by selecting 4 percent of

Figure 5.12.3   Concentric circles. 512×512 high resolution image.

Figure 5.12.4   Concentric circles. 4% Periodic sample and linear interpolation.

the pixels according to the following prescription. Let the pixels correspond to the coordinates (m, n) where m and n run through the integers from 1 to 512 and let a and b be arbitrary but fixed integers in the range 1 to 5 inclusive. Sample the high resolution image by selecting the pixels whose coordinates are of the form $(5i + a, 5j + b)$ where i and j run through nonnegative integers. Thus, 1 pixel is selected from each 5×5 array (with the possible exception of pixels near the boundary of the image). There are 25 possible sampling schedules of this type, corresponding to the 25 possible choices of the pairs (a, b). In order to reconstruct an image that does not have gaps in the places of pixels that were not sampled, gray scale values for the unsampled pixels can be calculated by means of linear interpolation based on a triangulation whose vertices are the sampled pixels. The choice of a triangulation is not unique; hence, the smoothed image that results from linear interpolation will depend on the triangulation. But for triangulations that avoid triangles having very small angles, the differences will in general be imperceptible. Figure 5.12.4 shows numerous circular and oval alias artifacts of varying sizes. It is impossible to distinguish between the undersampled version of the high resolution image displayed in Figure 5.12.3 and the undersampled version of a high resolution image that coincides with Figure 5.12.4. Thus the possibility of confusing the artifacts of aliasing for the image signal arises. This problem can be particularly vexing in a hierarchical signal sampling system that combines sampling arrays of various degrees of resolution. The primate and human eye, for example, combine foveal sampling sensitive to relatively high spatial frequencies with extrafoveal sampling which has substantially lower spatial resolution but samples a much larger solid angle in normal daylight vision. The output of these two subsystems is combined to provide the organism with an efficient means for detecting "interesting" visual events throughout a large solid angle while providing means for more detailed analysis by directing the gaze upon the source of the event. The combination of these subsystems makes more efficient use of the available bandwidth for processing image signal data than would otherwise be possible, but it opens itself to the possibility that the outputs of the two subsystems might be inconsistent: the low resolution extra-foveal subsystem in effect undersamples the image relative to the high resolu-tion foveal system, and may produce alias artifacts that will not be present in the foveal output. In this event, additional calculations may be required in order to rationalize the inconsistent results.

It has long been supposed that the photosensitive cone receptors in the primate and human fovea are arranged in the densest possible packing—a hexagonal lattice—to the extent that the biological manufacturing processes are able to achieve this goal. That is, apart from minor unavoidable irregu-larities that accompany the growth of the retina, the foveal photoreceptors were assumed to form a regular lattice for which a 2-dimensional version of the SHANNON sampling theorem appropriate to hexagonal lattice sampling is applicable (cf. MERSEREAU for a derivation of the latter). It was further assumed that the centers of the extrafoveal cones are also approximately arranged on a hexagonal lattice but that the variations are greater because of the greater difficulty of determining relative positions outside the fovea since the cones do not touch one another. This view was challenged in 1983 by YELLOTT, who argued that departures of the extrafoveal photoreceptor array from the hexa-gonal arrangement has the advantage of reducing aliases at the expense of introducing noise. He argued that this "enabled the visual system to escape aliasing distortion despite a large mismatch between retinal image bandwidth and the NYQUIST limits implied by extrafoveal cone densities."

Figure 5.12.5 displays photomicrographs ($\times 830$) of 3.75×4 micrometer sections of the retina of the *Rhesus* monkey. These sections subtend a visual angle of 15×18 minutes. The sequence of images shows (A_1) cone outer segments in the central fovea; (B_1) cone outer segments in the parafovea at an eccentricity of approximately 6 degrees; (C_1) the same in the periphery at an eccentricity of approximately 35 degrees; and (D_1) the same in the outer periphery. Outside the fovea the sections pass through the tips of the cones, which appear surrounded by empty space; the cones are surrounded by rods which are more or less densely packed in the area unoccupied by the cones. One can see that even in the central fovea, the cones are not strictly arranged

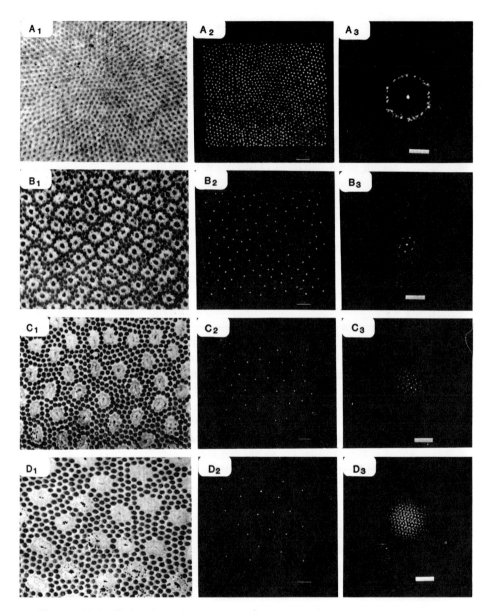

Figure 5.12.5   (Column 1) Photomicrographs ($\times$ 830) of 3.75 by 4.5 μm sections of rhesus retina (7). The sections subtended a visual angle of 15′ by 18′. (Column 2) Pinhole sampling arrays constructed from the cones in column 1. (Column 3) Power spectra (optical transforms) of column 2. Bars represent 120 cycle/deg. (A_1) Cone outer segments in the central fovea. (A_2) Sampling array, 1121 points: bar, 2′ on the retina. (B_1) Cone outer segments in the parafovea (eccentricity, approximately 6°). Cone tips are surrounded by empty space created by their inner segments. Each cone is circled by a ring of rods. (B_2) Sampling array, 123 points: bar, 2′. (C_1) Cone outer segments

on a hexagonal lattice. The question arises as to the consequences of this departure from periodicity in the receptor sampling array. By means of an elegant and simple experiment YELLOTT demonstrated that the FOURIER transform of the 2-dimensional sampling array of photoreceptors is not a periodic lattice but consists instead of a large central peak surrounded by an island wherein the magnitude of the FOURIER transform is relatively small. Outside the island the transform is such as to scatter power above the NYQUIST limit into broadband noise. These results are exhibited in the rightmost column of images in Figure 5.12.5.

Other researchers have challenged YELLOTT's interpretation: HIRSCH and HYLTON have argued that the foveal photoreceptor array does contain strong elements of hexagonal symmetry, and BOSSOMAIER, SNYDER, and HUGHES claim that although irregularities in the sampling array may reduce aliasing, they also make interpolation more sensitive to noise.

The question ultimately turns on the properties of the FOURIER transform of the sampling array. It will be convenient to discuss 1-dimensional sampling arrays first. Let the points of a sampling array be $\{\lambda_n | n \in \mathbf{Z}\}$, where $\mathbf{Z}$ denotes the set of integers. The sampling array is represented by the distribution

$$s(x) = \sum_{n=-\infty}^{\infty} \delta(x - \lambda_n)$$

(where δ denotes the DIRAC impulse distribution) and its FOURIER transform is

$$S(y) = \sum_{n=-\infty}^{\infty} e^{2\pi i \lambda_n y}.$$

If the sampling points $\{\lambda_n\}$ form a periodic sequence, then the FOURIER transform is a periodic sequence of δ-functions also, as shown in Figure 5.12.6. Suppose that a signal, represented by the function $x \mapsto f(x)$, is sampled by the array represented by $s(x)$. The FOURIER transform of the sampled signal will be the convolution of the FOURIER transform F of the signal f with the transform S of the sampling array function s. Thus the spikes at points other than $y = 0$ in Figure 5.12.6 will produce aliases if the width of the graph of F is greater than the distance between successive spikes in the graph of S.

The ideal sampling distribution would correspond to a sampling function s whose FOURIER transform S consists of a single spike at $y = 0$ and is otherwise zero, as illustrated in Figure 5.12.7. Such a sampling array would avoid both

---

◁   in the periphery (eccentricity, approximately 35°). The section passed near the tips of the cones, which are centered in the empty space created by their inner segments. Each cone is surrounded by two or three rows of rods. (C_2) Sampling array, 40 points: bar, 2.5′. (D_1) Cone outer segments from the far periphery (near the ora serrata). (D_2) Sampling array, 23 points: var, 2.5′. (Reproduced with permission of the author and the American Association for the Advancement of Science from "Spectral Consequences of Photoreceptor Sampling in the Rhesus Retina," Yellott, J., *Science* Vol. 221, p. 383, July 22, 1983.)

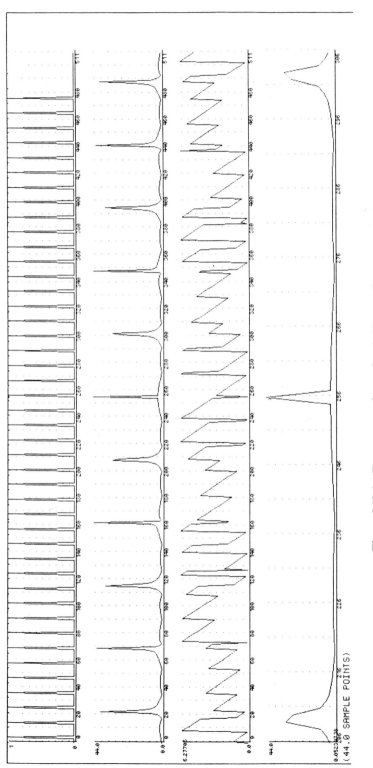

Figure 5.12.6   FOURIER transform of periodic sampling array.

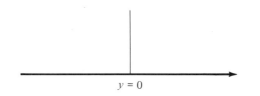

Figure 5.12.7   FOURIER transform of ideal discrete sampling array.

aliases and noise; unfortunately, it is known that a *discrete* set of sampling points cannot lead to a function S of this type. Thus the problem is to arrange the points λ_n of the sampling distribution so that the function S approximates the ideal distribution as closely as possible, recognizing that either aliasing or noise or some combination of the two must occur unless the class of signals is restricted in some way. This is, ultimately, a problem in the theory of *almost periodic functions*.

The periodic sampling distribution eliminates noise but results in aliasing if the bandwidth of the sampled function is greater than the NYQUIST frequency. In the other direction, SHAPIRO and SILVERMAN have shown that POISSON sampling processes, wherein the sampling points λ_n are selected randomly from the uniform probability distribution, are alias-free in a statistical sense (that is, averaged over a collection of signal functions) no matter what the bandwidth of the sampled function is, but this process scatters noise throughout all frequencies. They also show that "jittering" the points of a periodic sampling array by adding a small random magnitude to the coordinate of each sample point does not elminate aliases; it is as if the jittered sample point "remembers" the lattice point whence it came with the consequence that the signal is merely "blurred" by this modification of the array. We see that some modifications of the sampling array may trade aliases for noise, but the question is subtle and mathematically complex.

Figure 5.12.8 illustrates the magnitude and phase of $S(y)$ for a POISSON distribution of sampling points. The large central peak at $y = 0$ captures the signal while the randomly varying phase and statistically constant magnitude of $S(y)$ for $y \neq 0$ produce noise.

Returning to image signals, the positive diameter of the photosensitive cones forces the distance between their centers to be greater than the diameter of a cone. This leads one to consider a 2-dimensional sampling array consisting of points whose coordinates are selected at random subject to the condition that the points are at a distance greater than some preassigned constant, say d, from their nearest neighbor. This distribution is called by YELLOTT the *Poisson disk* sampling array; $d = 0$ corresponds to the POISSON array itself. Figure 5.12.9 displays the FOURIER transform of the 1-dimensional version of this sampling process. We see that the noise level close to the central peak is less than the corresponding level for the POISSON arrays.

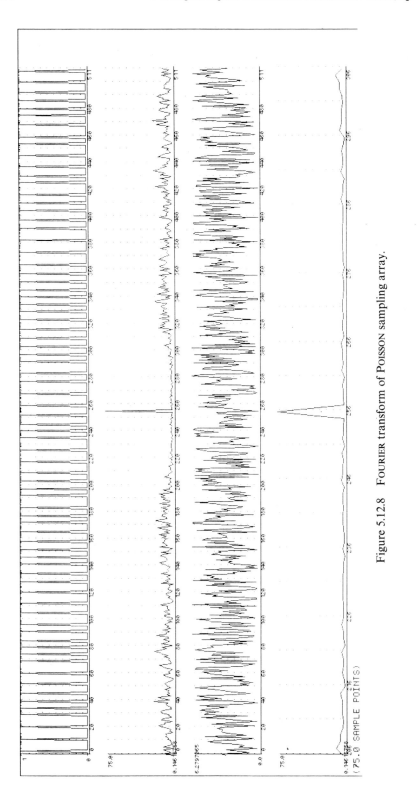

Figure 5.12.8   Fourier transform of Poisson sampling array.

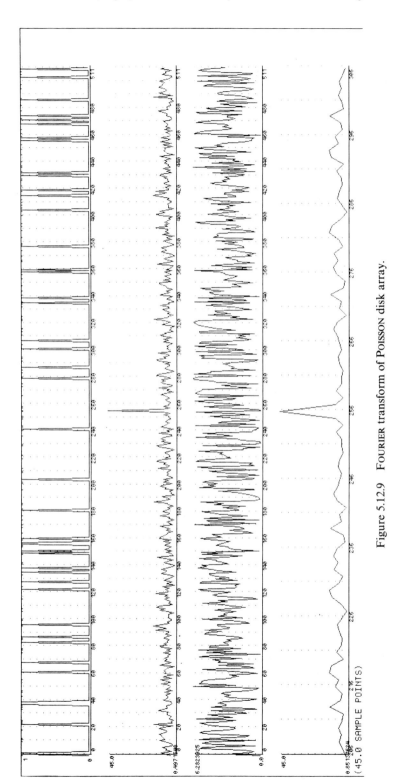

Figure 5.12.9   FOURIER transform of POISSON disk array.

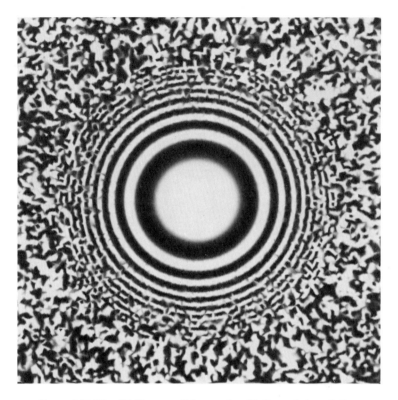

Figure 5.12.10   4% POISSON disk sample with linear interpolation.

Let us apply these ideas to the high resolution image of concentric circles. Figure 5.12.10 shows a linear interpolation reconstruction of a POISSON disk sample consisting of approximately 4% of the pixels of the 512 × 512 image displayed in Figure 5.12.3. The low spatial frequency circles close to the center of the image are reproduced with relatively high fidelity; the high spatial frequencies that lie above the NYQUIST frequency are converted into noise rather than into aliases. The reader should compare this figure with the periodic sampling exhibited in Figure 5.12.4. Since noise does not correspond to structured objects in the physical world, there can be no possibility of confusing the consequences of a limited sampling rate with the presence of objects in the visual field. These remarks apply with equal force to sampling signals of types other than images.

The concentric circles example suggests that certain randomized sampling procedures can successfully eliminate aliases at the expense of introducing noise which obscures features above the NYQUIST frequency but does not introduce artifacts that can be confused with structured portions of the signal. One may compare the relative effectiveness of periodic and randomized sampling (at the same sampling rate) for images that correspond to natural scenes which tend to consist of highly irregular distributions of light intensity. In general, if either the image or the sampling procedure (or both) is "irregular," then samples of the image will not have alias artifacts: it is the *relative* randomness of the sampling process in relation to the image signal that governs the avoidance of aliasing. Thus it follows that appropriately randomized sampling of any image signal will be alias-free.

The use of irregular sampling arrays that permit an information-processing system to selectively omit signal information without risking the introduction of alias artifacts is a striking example of the utility of randomization in information processing. Whereas the introduction of noise is generally—and properly—viewed as undesirable, there are certain instances where the appropriate use of unpredictable processes enables the information processing system to discriminate between arbitrary signals and those that possess coherent structure.

Now let us use the uncertainty relation for measurement to formalize the duality between noise and aliasing. Signals that correspond to the expenditure of an infinite amount of energy over an infinite domain of variation of the independent variable—typically time for temporally varying signals, or a 2-dimensional surface for an image signal—are an idealization of what can be realized in practice, but they constitute the appropriate generalization of the typical case of signals that involve a relatively small fluctuation in their *power* extended through a relatively long interval of time or area on a surface in space. For signals of this type, which have finite power but infinite energy consumption, the traditional theory of harmonic analysis and the FOURIER transform are not applicable. Norbert WIENER generalized harmonic analysis and the FOURIER transform to apply to this situation. We will summarize his ideas, restricting the formal presentation to 1-dimensional signals for notational convenience, but our primary application will be to 2-dimensional image signals.

Let $t \mapsto f(t)$ be a real valued function defined on the field **R** of real numbers. We shall call f a "signal." The *autocorrelation* of f is the function

$$C(t) = \lim_{T \to \infty} \frac{1}{2T} \int_{-T}^{T} f(u + t)f(u) \, du. \tag{5.26}$$

WIENER proved that $C(0) \geq C(t) \geq 0$ for all real numbers t. It is easy to verify that the autocorrelation function will be zero if the signal f has finite energy, i.e., if

$$\int_{-\infty}^{\infty} f(u)^2 \, du < \infty,$$

in which case traditional FOURIER analysis can be applied. So let us assume that the autocorrelation is not the zero function but that it, rather than the signal, is square-integrable, that is:

$$\int_{-\infty}^{\infty} C(t)^2 \, dt < \infty.$$

Then the FOURIER transform of the autocorrelation function C exists[4] (although the FOURIER transform of the signal f itself does not):

$$P(\omega) = \int_{-\infty}^{\infty} e^{-2\pi i \omega t} C(t) \, dt. \qquad (5.27)$$

P is called the *spectral power density* corresponding to the signal f, and the variable ω is the *frequency* corresponding to the signal variable t.

Different signals f_1 and f_2 may correspond to the same autocorrelation function, and hence possess the same spectral power density: the properties of the signal are only partially represented by the properties of the auto-correlation or spectral power density.

A signal is said to be *white noise* in the frequency interval $\omega_0 \leq \omega \leq \omega_1$ if $P(\omega) = $ constant for ω in that interval. In particular, if $P(\omega)$ is constant on the entire real line, then the autocorrelation function $C(t)$ must be proportional to the impulse distribution $\delta(t)$, which implies that

$$\lim_{T \to \infty} \frac{1}{2T} \int_{-T}^{T} f(u + t)f(u) \, du = 0$$

for $t \neq 0$: $f(t)$ is uncorrelated with itself for all shifts of the signal variable $u \mapsto u + t$ except for $t = 0$; in the latter case, $f(t)$ is of course perfectly correlated with itself. To say that a function or sequence is entirely uncorrelated with itself is just another way of saying that it is random, or "noisy." Thus the degree of noisiness of a signal can be measured by the deviation of its spectral power density from a constant function over the frequency range of interest, with "whiter" noise corresponding to smaller deviations.

Suppose that the spectral power density function is band-limited, i.e., concentrated on a finite interval (which we may take to be centered at $\omega = 0$). The corresponding autocorrelation function will have unbounded support, which means that correlations of arbitrarily long range will exist in the signal.

Now suppose that f denotes a *sampled* signal. We know that the sampled signal will exhibit aliases if the sampling rate is less than the NYQUIST rate *and if the samples are regularly spaced*. The aliases will correspond to long range correlations that do not reflect periodicities in the original unsampled signal. Hence, if we consider an ensemble of signals undersampled by a common sampling that is not necessarily regular, we will find that the broader the support of the autocorrelation function, the greater the aliasing; and the broader the support of the spectral power density, the noiser the sampled signal.

This informal duality between noise and aliasing can be quantified by

---

[4] The formula requires that $C(t)$ be integrable; the situation is more complicated otherwise.

applying the definition of the half-width of a function and its FOURIER transform introduced in eqs(4.35, 4.36) to the present circumstances. Define

$$(\Delta t)^2 = \int_{-\infty}^{\infty} t^2 C(t)^2 \, dt \Big/ \int_{-\infty}^{\infty} C(t)^2 \, dt$$

and

$$(\Delta \omega)^2 = \int_{-\infty}^{\infty} \omega^2 P(\omega)^2 \, d\omega \Big/ \int_{-\infty}^{\infty} P(\omega)^2 \, d\omega.$$

Then the inequality of (4.39), expressing the uncertainty relation for measurement, reads

$$\Delta t \, \Delta \omega \geq \frac{1}{4\pi}. \tag{5.28}$$

If we adopt Δt and $\Delta \omega$ as measures of the size of the support of the autocorrelation and spectral power density functions, respectively, then (5.28) is the quantitative expression of the duality between noise and aliasing, for it shows that either the support of the autocorrelation, and hence aliasing, or the support of the spectral power density, and hence noise, can be made as small as desired but not both. Aliases can be traded for noise or noise for aliases, but it is inherent in the measurement process that one of the two must be relatively great if the other is relatively small.

CHAPTER 6

Pattern Structure and Learning

the texture of sensory fields appears to be an adaptation favoring the tendency to categorize, which means to impose invariances on a world that sends us changing stimulation.

—Richard J. HERRNSTEIN

6.1 Purpose of This Chapter

In the largest sense this chapter is concerned with the problem of categorization.

Section 6.2 investigates the problem of segmenting visual images in order to determine the shapes of objects based upon variations in the *texture* of the image. The theory of *textons* developed by Bela JULESZ in 1981 provides the foundation upon which we erect a more general mathematical model for analyzing and synthesizing textures.

In section 6.3 we turn to the study of edge detection. Although the previous section argues the insufficiency of edge detection paradigms for identifying the boundaries of objects and regions, it is evident that there are important circumstances where edge detection rather than texture variation plays the crucial role in segmenting an image. The traditional editorial cartoon, a line drawing which provides a sharply defined intensity gradient against a background field of uniform texture, provides the simplest example.

Texture discrimination and edge detection as mechanisms for segmenting images are both normally thought of as procedures that belong to the domain of "low level" image processing, close to the signal processing techniques that acquire and modify an image signal for its complex journey through to the upper, cortical, reaches of the brain or to the more complex pattern classification and recognition modules of a machine vision system. Yet these early processes perform critical functions of selective omission and re-organization of the signal, often utilizing hierarchical data structures and always subject to the schizophrenic competition imposed by the uncertainty principle for measurement, that play an important direct role in the process of categorization. Thus we prefer to conceive these processes as categorical functions rather than as mere transformations of the data without loss of information. The final section considers processes that are more generally construed as "high level" in that their inputs are the outputs of "low level" processes as well as certain other information, such as stored knowledge about various domains and prior experience. Section 6.4 briefly examines the twin problems of memory and learning in the context of mathematical models that have their origins in the statistical thermodynamics of BOLTZMANN and find their most felicitous expression in terms of measures and costs of information. These so-called *connectionist* models address the problem of how complexity can emerge from simplicity as a collective property of aggregates. Most modern theories of intelligent behavior are based upon ideas that ultimately depend upon mechanisms for transforming the aggregated response of simple components, such as idealized neurons, into complex patterns of behavior. They consequently have the potential to reveal profound relationships between the emergence of complexity and order in physical systems and the development of the corresponding features in biological information processing systems.

6.2 Texture and Textons

> preattentive texture perception is an early warning system which triggers the attentive perceptual system. . . . it is my contention that it is the local nonlinear features of this system that are the building blocks of form perception.
>
> —Bela JULESZ

The word *texture* refers to the minute molding of a surface and ultimately is derived from the same root as *textile*, having to do with the superficial irregularities generated by the process of weaving. It is difficult to give a definition which more precisely characterizes the range of surface structure the term is intended to describe. Indeed, provision of a suitable definition or characterization of texture is still one of the principal research problems connected with this topic although considerable progress has been made in recent years.

Various materials are distinguished by their texture, which may be described as *rough* or *smooth* or by other adjectives, but the wealth of discriminable textures is so vast that vocabulary is insufficient for an adequate description just as vocabulary fails to suffice for the accurate description of perceived colors.

Since texture refers to the *molding* of a surface, which may be regular or irregular, one might believe that a detailed description of the configuration of the surface would be equivalent to a description of its texture but this would not be correct. Although texture can be derived from the detailed microscopic properties of a surface just as temperature can be derived from knowledge of the complete microscopic motions of the molecules of a gas, texture is not itself a microscopic property. Rather, it is, as we shall see, a *statistical* property of certain characteristic features of the surface. The characteristic features are called *textons*. They are the primitive elements—the atoms—from which perceived textures are constructed.

If it be asked why texture exists, that is, why a vision system detects the statistical distribution that constitutes texture rather than simply directly sensing the conformation—the molding—of the surface, the answer is easy to find. Just as knowledge of the statistical parameters of temperature and pressure suffice to describe the important macroscopic properties of a gas, so does the statistical description provided by texture suffice to describe the macroscopic properties of a surface that are important for pattern recognition and classification by the human visual processing system. Moreover, the amount of information required to describe the detailed characteristics of the molding of a surface is vastly greater than the amount of information necessary to define a texture. The introduction of texture as a component of an information-processing system makes it possible to greatly reduce the information load on the system. Thus texture can be considered primarily as a means for selectively omitting information.

Although texture is often thought of in the context of a visual stimulus, the tactile modality is also sensitive to textural stimuli. When a surface is spoken of as rough or smooth, the terms used refer to the sense of touch first of all and only secondarily to the sense of sight. But the sensation of tactile texture involves motion of the sensitive portion of the skin relative to the textured surface, so the neural response confounds motion stimuli with textural stimuli. Recent studies of the tactile sense in the macaque monkey demonstrate that neural response to the spatial features of the surface results both from the movement of the sensitive region of skin sensor and from the pattern of the surface. It follows that texture can only be unambiguously represented in *populations* of neurons, but not by individual neurons, which is what would be expected if it were primarily a derived statistical property of the surface and if the effects of motion of the sensors were eliminated from the stimuli at the cortical processing level.

Similar circumstances apply to the visual perception of texture, where relative motion of the surface and the eye is eliminated from the stimulus by neural processing.

In human vision, differences in two textures presented side by side (or one embedded within the other) for intervals as short as 0.05 to 0.2 seconds can be detected. Such brief "preattentive" exposures prevent scanning eye motions and shifts of attention. It follows that the discrimination of texture is an information processing function which is distributed throughout the non-foveal part of the retina.

Our present understanding of texture is largely due to the investigations initiated by JULESZ nearly two decades ago which have led to particularly interesting results in the past few years. His methods rely on ingenious algorithms for the construction of synthetic textures having special properties which can be used to decisively distinguish between various theoretical hypotheses and alternatives. The synthetic textures are computed and displayed by a computer, which plays a crucial role in the work because without it the ability to design and execute experiments would be sharply curtailed. For this reason the serious study of texture had to await the development of the computer.

The theory of preattentive perception of texture has progressed through several stages. The tremendous general success of FOURIER analysis in communication engineering led to its application to human information processing. This point of view was reinforced by the accumulation of evidence for CAMPBELL and ROBSON's proposal that the vision system responds to spatial-frequency tuned filters selective to sinusoidal gratings rather than to localized features such as bars and edges. Thus, the earliest modern theories of texture adopted the position that global statistics derived from information provided by FOURIER analysis of an image was necessary and sufficient to distinguish textures. In particular, it was proposed that the human vision system could compute so-called *second order* statistics of an image but not statistics of higher order, and that images whose second order statistics coincide correspond to textures that are indistinguishable. *Second order*, or *dipole*, statistics describe the probability that a pair of randomly selected points in the image will have the same color. For binary images (which we may take to consist of black and white pixels), second order statistics determine the *autocorrelation function* of the image, and the FOURIER transform of the autocorrelation function is the so-called *power spectrum*, which does not distinguish phase (position) information that is crucial when images are viewed by focal attention. As a result of work by JULESZ it now appears that global FOURIER analysis does not play a role in texture perception. Instead, the perception and discrimination of texture seems to be based on the density (first order statistics) of local conspicuous features.

These local conspicuous features—the atoms of texture—are called *textons*. At the present time three types of texton parameters are known. They are:

1. Color,
2. Elongated blobs,
3. Terminator number.

Elongated blobs are quasi-rectangular configurations that are characterized by three parameters:

Figure 6.2.1   (a) Equivalent textons; (b) inequivalent textons.

- Orientation,
- Length,
- Aspect ratio.

The orientation parameter is applicable if the ratio of length to width (aspect ratio) of the elongated blob is significantly greater than 1; otherwise, it does not apply.

Terminators are the ends of the elongated blobs that have a large aspect ratio. The end of a line segment is an example. A sans-serif letter T has three terminators. The number of terminators is an invariant that distinguishes textons, all of whose other parameters coincide. For instance, the symbols shown in Figure 6.2.1(a) are elongated blobs which coincide as textons because their color, orientation, length, and aspect ratios are the same, but the symbols shown in Figure 6.2.1(b) are different textons.

It becomes increasingly difficult for the preattentive vision system to distinguish between textons which differ only in the number of terminators when the number of terminators is large and the difference small. Thus terminator number is significant primarily when it is small.

With the exception of the terminator, texton parameters are continuous variables: color varies in the 3-dimensional space of perceived colors; texton length and aspect ratio in the 1-dimensional space of positive real numbers; and orientation between 0 radians and π radians.

It is obvious that differences in color can be used to define edges or boundaries or regions in an image. This is certainly one of the most common methods employed by the vision system in routine perception; the boundary of a red apple seen against green grass or of a bright object against a dark background can most readily and directly be perceived by means of the difference in color. Textons can be used in general to define edges and boundaries.

The famous Dalmatian Dog (Illusion 5.7 in section 5.11) is a binary image in which only elongated blobs are available as stimuli (Figure 6.2.2).

From their statistical properties, the partly subjective contours of the dog can be discerned; this exquisite example suggests that texton densities are estimated over relatively small areas of the image, a conclusion which has been confirmed by experiment.

It appears to be accepted that the human vision system cannot calculate third order image statistics; consequently, neither they nor higher order statistics can play any role in texture perception.

Figure 6.2.2   Dalmation Dog. (From Thurston and Carraher, *Optical Illusions and the Visual Arts*, p. 18. Copyright 1966 Van Nostrand Reinhold. Reproduced with permission.)

Figure 6.2.3 displays an example of an indistinguishable pair of textures whose second order statistics are identical. This and other similar examples might lead one to the erroneous belief that in general textures having the same second order statistics are indistinguishable.

Let us verify that the textures in this figure do have the same second order statistics. They are composed of symbols

∖∣ and ∖∣

distributed with random orientation in a lattice of pixels. It is evident that the positions in which a pair of points can fall on one of these symbols is in one-to-one correspondence with the positions in which the pair of points can all on the other. Moreover, these corresponding point pairs which do not lie in congruent positions are related by a translation, which does not change the orientation of the line determined by the point pair (see illustration).

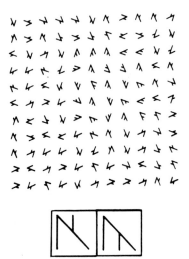

Figure 6.2.3   Indistinguishable textures with identical second order statistics. (From Julesz and Schumer, *Biological Cybernetics* Vol. 41, p. 132. Copyright 1981 Springer-Verlag. Reproduced with permission.)

Figure 6.2.4   Indistinguishable texture pairs formed from mirror image symbols. (From Julesz, Gilbert, and Victor, "Visual Discrimination of Textures with Identical Third-Order Statistics." *Biological Cybernetics* Vol. 31, p. 137. Copyright 1978 Springer-Verlag. Reproduced with permission.)

$\searrow\!\!\!\nparallel$ and $\nwarrow\!\!\!\nparallel$.

Hence the second order statistics of the two patterns must agree.

Observe that the five characteristic parameters of both symbols coincide; according to the theory of textons, the two textures should be indistinguishable.

The two textures in Figure 6.2.4 also have the same second order statistics and are indistinguishable. The symbols R and Я are mirror images which are distributed on the image with random orientations. Although one of them is very familiar to readers of English and the other is very unfamiliar (these characterizations would be reversed for most native readers of Russian), the relative familiarity of one of the symbols does not help the pre-attentive observer to distinguish the textures. It is easy to verify that the second order statistics of the textures agree, for if they are imagined as placed on a transparent background, then viewing the image from the rear transforms one type of symbol into the other. Distances are preserved, but orientations are changed. Since second order statistics are averaged over all possible orienta-

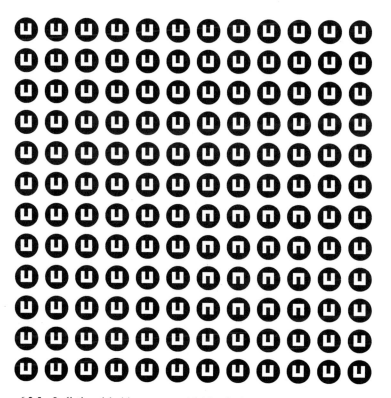

Figure 6.2.5   Indistinguishable textures with identical second order statistics. (Adapted from Julesz, Frisch, Gilbert, and Shepp, "Inability of Humans to Discriminate Between Visual Textures That Agree in Second-Order Statistics—Revisited." Copyright *Perception 2*, 1973.)

tions, the result will be independent of the symbol from which the pattern was constructed.

Since the symbols are mirror images, they obviously have the same length, aspect ratio, and terminator number (2). They also have the same color. Although the orientations of the mirror images are distinct, this parameter is eliminated by the construction procedure which distributes the symbols on the image with random orientations. Hence the two textures consist of textons which belong to the same equivalence class.

Figure 6.2.5 exhibits another example of indistinguishable textures whose second order statistics coincide. In this case the ∪ symbol which is used to form one texture is converted into the ∩ symbol which forms the other by a rotation by π radians. It follows immediately that the second order statistics of the two patterns are the same. The indistinguishability shows that the location of the arc which connects the two parallel vertical lines is not captured by preattentive vision. In this case the symbols belong to the same texton equivalence class because ∩ and ∪ have the same color, orientation, length,

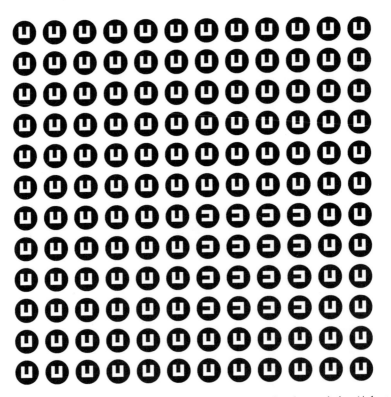

Figure 6.2.6 Distinguishable textures with different second order statistics. (Adapted from Julesz, Frisch, Gilbert, and Shepp, "Inability of Humans to Discriminate Between Visual Textures That Agree in Second-Order Statistics—Revisited." Copyright *Perception 2*, 1973.)

aspect ratio, and number of terminators (2). From this example, we learn that *orientation* refers to equivalence classes of parallel lines; that is, to undirected lines. In this use of the term (which differs from the usage common in mathematics), a given line has but one orientation although it can have two, opposite, directions.

By rotating the symbol ∪ through an angle θ, it is possible to construct a family of textures whose members vary continuously with the parameter θ. If the original texture corresponds to $\theta = 0$, then the texture corresponding to $\theta = \pi$ will be indistinguishable from it, but for all other values of θ the second order statistics of the two textures will be different and the textures will be distinguishable. Figure 6.2.6 displays the effect of rotation by $\pi/2$ radians; the two textures are easily distinguished. Note that the orientations being different, these patterns are composed of symbols which belong to different equivalence classes of textons and should consequently be perceived as distinct.

The second order statistics of the family of textures varies continuously with θ. For small θ, the difference in second order statistics between the

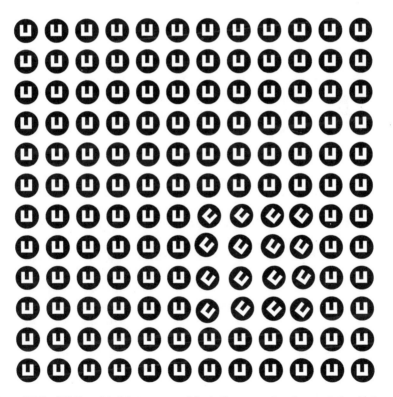

Figure 6.2.7 Distinguishable textures with similar second order statistics. (Adapted from Julesz, Frisch, Gilbert, and Shepp. "Inability of Humans to Discriminate Between Visual Textures That Agree in Second-Order Statistics—Revisited." Copyright *Perception 2*, 1973.)

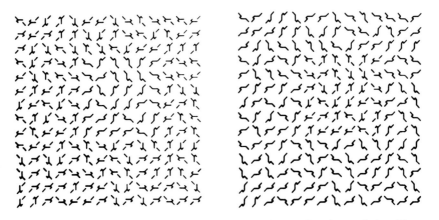

Figure 6.2.8   Distinguishable textures with identical second order statistics. (From Julesz and Schumer, *Annual Review of Psychology*, Volume 32. © 1981 by Annual Reviews, Inc. Reproduced with permission.)

transformed and the original patterns will be small. If texture discrimination were primarily due to differences in second order statistics, then one could expect the ability to distinguish textures for which the difference in θ decreases to diminish. But the textures shown in Figure 6.2.7, for which the difference is $\pi/9$, are easily distinguished. According to the theory of textons, discrimination should occur as long as the orientation of the textons which constitute the two patterns are perceptibly different, as indeed they are in the example.

It is not difficult to construct distinguishable textures that have identical second order statistics. Each of the images in Figure 6.2.8 are made up from randomly oriented copies of the symbols $\nearrow$ and χ and each displays a pair of distinct textures. But in each case the textures have identical second order statistics. Considered as textons, the symbols have the same color, orientation, length and aspect ratio parameters, but the number of terminators differs: it is 2 for one of the symbols, 5 for the other.

On the other hand, textures that have different second order statistics may not be distinguishable, as the interesting pattern in Figure 6.2.9 demonstrates.

The evidence suggests that the perception of texture differences does not have much to do with the statistics of patterns. This conclusion is reinforced by the existence of distinguishable textures that have the identical third order statistics (hence identical second order statistics and density of black pixels). These textures appear to differ in "granularity." Thus granularity is either a fourth order property, or a property of the density variation of local features. The latter possibility seem smore likely because it is difficult to imagine a mechanism which would enable the preattentive vision system to calculate fourth order statistics "on the fly." Figure 6.2.10 exhibits a typical example of a distinguishable pair of textures with identical third order statistics. The two textures clearly differ in the aspect ratio of the elongated blobs and,

Figure 6.2.9 Indistinguishable textures with different second order statistics. (Adapted from Julesz, Frisch, Gilbert, and Shepp. "Inability of Humans to Discriminate Between Visual Textures That Agree in Second-Order Statistics—Revisited." Copyright *Perception 2*, 1973.)

in certain subregions, in the length of the blobs. These textures appear to be very complicated, but in fact they can be generated using a simple recursion formula and can be displayed on a home computer. Following JULESZ, GILBERT, and VICTOR, we will describe the algorithm and then sketch a proof that the texture pairs it generates have identical third order statistics.

Let the plan be partitioned into a lattice of pixels which we assume are rectangular and let the pixels be associated with a pair of integer Cartesian coordinates (x, y). Consider binary images only, and write

$$c(x, y) = 1 \quad \text{if pixel } (x, y) \text{ is colored black}$$
$$0 \quad \text{if pixel } (x, y) \text{ is colored white.}$$

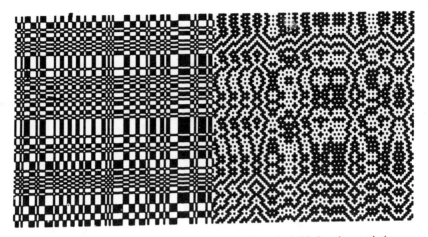

Figure 6.2.10   Distinguishable textures with identical third order statistics.

We will consider two types of colorings of the plane, called *even* and *odd*. Both colorings will express the color of pixel (x, y) in terms of the colors of pixels on the x-axis, i.e., those with coordinates $(x, 0)$, and of pixels on the y-axis, i.e., those with coordinates $(0, y)$, by means of a recursion formula. If the colorings of x- and y-axes are prescribed, a coloring of the entire plane of pixels will be determined.

An *even* coloring is defined by the recursion formula:

$$E: c(x, y) = c(x, 0) + c(0, y) + c(0, 0) \quad (\text{mod } 2) \tag{6.1}$$

whereas an *odd* coloring is defined by the recursion:

$$O: c(x, y) = c(x, 0) + c(0, y) + c(0, 0) + xy \quad (\text{mod } 2). \tag{6.2}$$

In these formulae the notation (mod 2) indicates that the right side of the equality has the value 1 if the arithmetical expression which precedes the symbol (mod 2) is odd, and the value 0 if that arithmetical expression is even.

If we consider four adjacent pixels forming a square, with coordinates (x, y), $(x + 1, y)$, $(x, y + 1)$, and $(x + 1, y + 1)$, then for the even coloring their colors satisfy the relation:

$$E: c(x, y) + c(x + 1, y) + c(x, y + 1) + c(x + 1, y + 1)$$
$$= 2c(x, 0) + c(x + 1, 0) + c(0, y) + c(0, y + 1) + 4c(0, 0) \quad (\text{mod } 2). \tag{6.3}$$

The arithmetical expression on the right side is a multiple of 2, hence always even. The coloring is called "even" for this reason. Every 4-pixel square colored according to the odd coloring formula has an odd number of black pixels because in that case:

$$O: c(x, y) + c(x + 1, y) + c(x, y + 1) + c(x + 1, y + 1)$$

$$= 2c(x, 0) + c(x + 1, 0) + c(0, y) + c(0, y + 1) + 4c(0, 0) \qquad (6.4)$$

$$= xy + (x + 1)y + x(y + 1) + (x + 1)(y + 1) \qquad \text{(mod 2)};$$

the arithmetical expression on the right side is the sum of an even number and the expression involving the x's and y's. The latter simplifies to $4xy + 2y + 2x + 1$, which is an odd number. Hence the sum of the colors of a 4-pixel square is odd for the odd coloring, i.e., every 4-pixel square has an odd number of black pixels.

The colors of the pixels on the two axes can be chosen arbitrarily. Once they are selected, either the odd or the even recursion formula can be used to propagate these initial values recursively throughout the plane. If the odd coloring formula is used to continue the initial values into the fourth quadrant and the even coloring formula is used to continue the initial values into the third quadrant, textures such as those illustrated in Figure 6.2.10 will result if the initial conditions are the result of a random assignment of colors on the two coordinate axes.

If can be proved that the even and odd colorings have the same third order statistics. The idea behind the proof is to select three arbitrary pixels in one of the colorings. Since each pixel can be either black or white, there are eight possible permutations of colorings. By counting the ways in which the eight colorings can arise (for either the even or the odd case) one finds that the given coloring of the triple can be obtained from exactly one-eighth of the number of the corresponding equally likely initial conditions. Hence, the eight possible colorings are equally likely regardless of whether the even or the odd recursion is used. Consequently, the third order statistics of the even and odd colorings are identical.

A striking example of the lack of significance of positional information about texton parameters is provided by textures constructed from the pair of symbols illustrated in Figure 6.2.1(a). In this case, one of the symbols is connected, whereas the other consists of two components, one of which is a closed curve. These symbols differ only in the position of two short line segments. From the texton standpoint, they belong to the same equivalence class because the color, orientation, length, aspect ratio, and terminator numbers are equal. Can textures built up from these very different elements be distinguished? Figure 6.2.11 leaves no doubt about the answer to this question.

The preceding theory of textons can be extended to include the case of textures formed by randomly intermixing representative symbols drawn from various classes of textons. Indeed, the seeds of this generalization are already contained in the foregoing discussion, for the preattentive indistinguishability of textures formed from distinct representatives of one texton equivalence class is a special case of the general situation.

We will motivate the theoretical discussion by studying several examples.

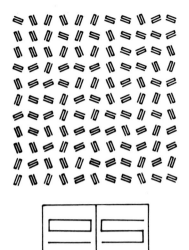

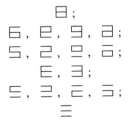

Figure 6.2.11   Indistinguishable textures demonstrating the irrelevance of terminator position in textons. (From Julesz and Schumer, *Biological Cybernetics*, Vol. 41, p. 132. Copyright 1981, Springer-Verlag. Reproduced with permission.)

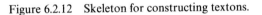

Figure 6.2.12   Skeleton for constructing textons.

Consider three parallel line segments of equal length as illustrated in Figure 6.2.12. Two are opposite sides of a square, and the third passes through the center of the square and is limited by the sides of the square perpendicular to it. Various symbols and texton classes of symbols can be constructed from this symbol by joining zero or more adjacent pairs of segment endpoints by straight line segments. The 16 symbols which can be constructed by this process fall into six texton classes that are characterized by their number τ of terminators. The symbols are displayed in Figure 6.2.13, grouped according to the number of terminators.

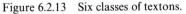

Figure 6.2.13   Six classes of textons.

All 16 patterns have the same length, aspect ratio, and orientation. Further-more, all patterns which have a given number of terminators also have the same color, where *color* refers to the perceived gray level, which is measured here by the amount of black in the pixel containing the symbol or, equiva-lently, by the total length of the line segments that form it. Observe that the $\tau = 6$ pattern has least black whereas the $\tau = 0$ pattern has most. Among the remaining textons classes, the symbols for which $\tau = 2$ and those for which $\tau = 3$ have the same length and therefore the same color. Hence, the *only* difference between representative symbols for textons with $\tau = 2$ and those with $\tau = 3$ is the number of terminators. Therefore, any mixture of symbols whose number of terminators is equal to 2 should be perceived as a single texture and any mixture of symbols whose average number of terminators is equal to 3 should be perceived as a single texture, but regions of these two types should be perceived as different textures.

If symbols for which $\tau = 6$ are randomly mixed with symbols for which $\tau = 2$, so that the texton classes are equinumerous, then the *average* number of terminators will be $\tau_{av} = 4$ and the *average* color will be the same as the color of a pattern with $\tau = 4$. Hence one might expect that a random combina-tion of, e.g., symbols $=$ and $\sqsubseteq$ in equal proportions would be perceptually equivalent to the texture consisting of, e.g., the symbol $\sqsubseteq$. It is easy to test this hypothesis experimentally; the result confirms our expectation if the symbols are small enough to permit the local averaging processes of pre-attentive vision but large enough to permit regions consisting of different textons to be perceived as distinct.

This conclusion suggests the generalization that any random mixture of symbols in some proportion will be perceived as the same texture as another such mixture, if and only if, the average number of terminators per symbol and the average color per symbol of the first mixture are respectively equal to the corresponding quantities for the second mixture. This is equivalent to the assertion that the perceived parameters of color and terminator number are linear combinations of the corresponding parameters of the constituent sumbols (ultimately, with variations of symbol density taken into account).

Consider mixtures of pairs of symbols drawn from the inventory given in Figure 6.2.13. Let a texton class be denoted by its terminator number τ. The possible mixtures of equinumerous pairs of textons can be arranged in a matrix. Only the entries on and above the diagonal are necessary since mixtures of texton representative do not depend on the order of the textons in the matrix. The average values τ_{av} of terminators per symbol and γ_{av} of color per symbol are indicated in each matrix cell in the form τ_{av}/γ_{av}. Here the color is expressed in multiples of one-half the side of the square which is the convex hull of the symbol. Table 6.2.1 displays the matrix.

Inspection of the matrix shows that the following mixtures of pairs of symbols have the same average number of terminators per symbol, and the same average color per symbol, where (n) denotes the texton whose terminator number is n, "$\oplus$" denotes the mixture of equinumerous representatives of the

Table 6.2.1 τ_{av}/γ_{av} for Equinumerous Mixture of Symbol Pairs

τ \ τ	0	1	2	3	4	6
0	0/10	0.5/9.5	1/9	1.5/9	2/8.5	3/8
1		1/9	1.5/8.5	2/8.5	2.5/8	3.5/7.5
2			2/8	2.5/8	3/7.5	4/7
3				3/8	3.5/7.5	4.5/7
4					4/7	5/6.5
6						6/6

texton classes, and "$=$" denotes indistinguishability of the textures on either side of the equality symbol.

According to our hypothesis, mixtures of symbols for which the average parameter values per symbol coincide should be indistinguishable as textures.

Evidence for the validity of these perceptual equivalences is provided by Figures 6.2.14 to 6.2.19.

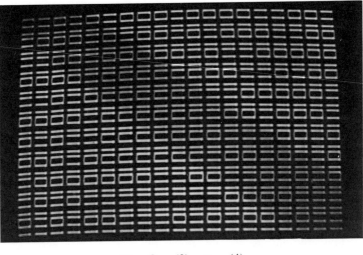

(6) $\oplus$ (2) $=$ (4)

Figure 6.2.14

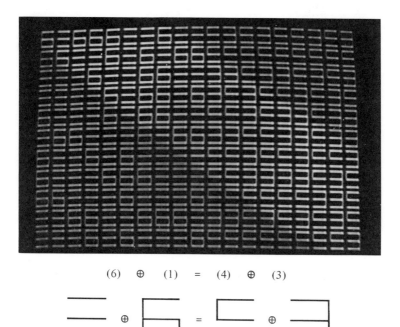

Figure 6.2.15

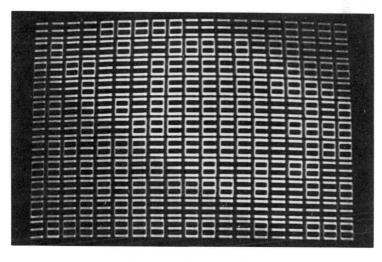

Figure 6.2.16

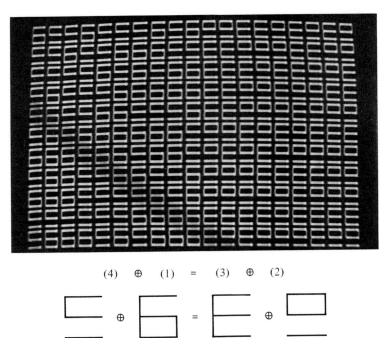

(4)   ⊕   (1)   =   (3)   ⊕   (2)

Figure 6.2.17

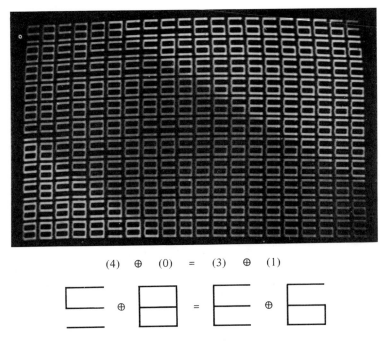

(4)   ⊕   (0)   =   (3)   ⊕   (1)

Figure 6.2.18

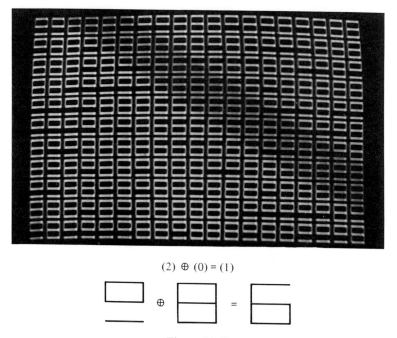

Figure 6.2.19

Mixtures of triples of symbols can be considered in a similar way. The equinumerous mixtures below should be indistinguishable:

$$ \boxminus \oplus \boxminus \oplus \boxminus = \boxminus \oplus \boxminus \oplus \boxminus $$

These mixtures are displayed in Figure 6.2.20. They have the common average terminator number per symbol and the common average color per symbol 8/3 and 8, respectively.

Let us formalize these results and extend them to nonequinumerous mixtures of symbols.

Suppose that **S** denotes a set of symbols (or patterns) from which various textures can be formed. We will think of the process of formation of a texture in the following way: symbols drawn from **S** will be positioned within congruent pixels that cover the visual plane completely and without overlap. A symbol will be placed on a pixel with a given orientation which may be variable or fixed depending on the placement algorithm under consideration.

Let a fixed symbol x be drawn from the inventory **S** and let copies of x be placed on the pixels of the visual plane, covering it entirely. This distribution defines a texture which we will denote $T(x)$. Let another symbol y be drawn from **S** and denote the corresponding texture by $T(y)$. $T(x)$ may be indistinguishable from $T(y)$. If this is true, we will write

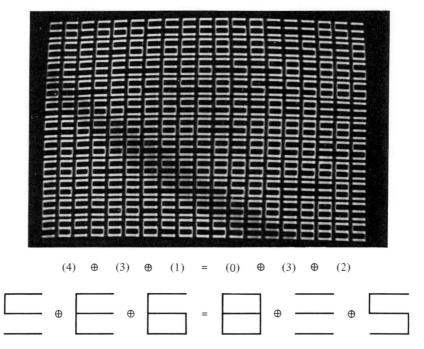

$$(4) \quad \oplus \quad (3) \quad \oplus \quad (1) \quad = \quad (0) \quad \oplus \quad (3) \quad \oplus \quad (2)$$

Figure 6.2.20   Mixtures of triples of textons.

$$T(x) = T(y),$$

where the equality sign is used to denote indistinguishability. Evidently $T(x) = T(x)$ for any symbol x, and $T(x) = T(y)$ implies $T(y) = T(x)$ for any symbols x and y: the indistinguishability of two textures does not depend on which side of the symbol "$=$" a texture appears.

We will assume that if x, y, and z are any three symbols in **S** and if $T(x) = T(y)$ and $T(y) = T(z)$, then $T(x) = T(z)$. This is a plausible assumption but it is not a logical consequence of the definition of texture. Indeed, just as in other areas of perception, two stimuli which are each not noticeably different from a third may nevertheless be noticeably different from each other. But apart from this universal problem of psychophysics, experiments verify that indistinguishability of textures is transitive in the same sense that color matching experiments define a transitive equivalence relation. In particular, this will be true for the classes of textures with which we will be concerned in what follows.

Let us accept the assumption of transitivity: that two textures indistinguishable from a third are indistinguishable from each other. Then the relation "$=$" is an *equivalence relation* which partitions the set **S** of symbols into equivalence classes, that is, into subsets whose elements produce pairwise equivalent textures. It follows that textures constructed from symbols drawn from different equivalence classes are not equivalent, and hence are distinguishable, from each other. An equivalence class of textures is a *texton*.

If X and Y are any two textons and if a and b are non-negative numbers such that $a + b = 1$, then define the *mixture* of X and Y with respective weights a and b to be that texton which is produced by placing representative symbols of X in pixels with probability a and representative symbols of Y in pixels with probability b so that the symbols of X and Y cover the visual plane. This new texton will be denoted Z:

$$Z = aX + bY.$$

The concept of mixing textures can be extended to any finite number of textures. If $X_1, \ldots, X_n$ are textons and $p_1, \ldots, p_n$ are nonnegative numbers satisfying $p_1 + p_2 + \cdots + p_n = 1$, then

$$Z = p_1 X_1 + \cdots + p_n X_n \tag{6.5}$$

denotes the texton obtained by distributing the symbols from X_k on pixels with probability p_k, $k = 1, \ldots, n$.

Extend the notion of multiplication of a texton by a nonnegative number between 0 and 1 to multiplication by an arbitrary real number by defining $aX + bY$ for arbitrary a and b which satisfy $a + b = 1$ as follows: If $a > 1$ (which we may assume without loss of generality by relabeling X and Y if necessary),

$$Z = aX + bY \tag{6.6}$$

is that (ideal) texton such that:

$$\frac{1}{a} Z + \frac{a - 1}{a} Y = X. \tag{6.7}$$

Notice that $b = 1 - a$ and that:

$$\frac{1}{a} + \frac{a - 1}{a} = 1; \tag{6.8}$$

since, by hypothesis, $a > 1$, $(1/a)$ and $((a - 1)/(a))$ are probabilities so the expression (6.7) is a *bona fide* mixture of textures.

We see that mixtures of textures $X_1, \ldots, X_n$ can be correspondingly defined for weights $p_1, \ldots, p_n$ subject only to the condition $p_1 + \cdots + p_n = 1$. If the textons $X_1, \ldots, X_n$ are thought of as points in some multidimensional Euclidean space, then the mixture $Z = p_1 X_1 + \cdots + p_n X_n$ runs through the points of the hyperplane spanned by the points corresponding to $X_1, \ldots, X_n$ as the coefficients $p_1, \ldots, p_n$ are varied.

Fix a texton X_n in this hyperplane and consider the differences $X - X_n$ as X varies through the textures on this hyperplane. If $Z = p_1 X_1 + \cdots + p_{n-1} X_{n-1} + p_n X_n$ with $p_1 + \cdots + p_n = 1$, then

$$Z = (p_1 X_1 + \cdots + p_{n-1} X_{n-1}) + (1 - p_1 - \cdots - p_{n-1}) X_n \tag{6.9}$$

so

$$Z - X_n = p_1(X_1 - X_n) + p_2(X_2 - X_n) + \cdots + p_{n-1}(X_{n-1} - X_n). \quad (6.10)$$

Since the numbers $p_1, \ldots, p_{n-1}$ can be chosen *arbitrarily* (the constraint is borne by p_n in the expression $p_1 + \cdots + p_n = 1$), we see that arbitrary linear combinations of the differences $X - X_n$ can be formed as X runs through the equivalence classes of textures. This shows that the set of differences $X - X_n$ is a *real vector space*.

Every vector space has a *dimension* and a *basis*. If we assume that the vector space of texture differences is finite dimensional, say of dimension n, then there is a basis of the vector space of texture differences that consist of n classes of textures, and every texture class in the vector space can be expressed as a mixture of the n basis vectors.

These ideas can be applied to the inventory **S** of symbols displayed in Figure 6.2.13. The 16 symbols fall into six texton classes characterized by the terminator number. Select one of these symbols and denote it by X_6. The set of all possible mixtures of differences $X - X_6$ forms a vector space. We will prove that the dimension of this vector space is 2. It will suffice to show that each difference $X_i - X_6$ can be expressed as mixture of two fixed differences.

We will denote a texture class by writing a representative's symbol. Consider the differences $\equiv - \boxminus$ and $\equiv - \boxminus$. An arbitrary linear combination of these elements of the vector space has the form:

$$Z - \boxminus = a(\equiv - \boxminus) + b(\equiv - \boxminus).$$

This relation is equivalent to

$$Z = a\boxminus + b\equiv + c\boxminus \qquad (6.11)$$

where $c = 1 - a - b$. Let us try to express the six texton classes as a linear combination of the type (6.11). Table 6.2.2 lists the coefficients a, b, c.

From this it follows that an arbitrary mixture of textures produced from the 16 symbols (six texton classes) shown in Figure 6.2.13 can be expressed as a linear combination according to (6.11) subject only to the condition:

$$a + b + c = 1.$$

Table 6.2.2 Representation of Textons

Symbol	(τ, γ)	$a\boxminus + b\equiv + c\boxminus$		
$\boxminus$	(0, 10)	1	0	0
$\boxminus$	(1, 9)	$\frac{1}{2}$	0	$\frac{1}{2}$
$\boxminus$	(2, 8)	0	0	1
$\boxminus$	(3, 8)	$\frac{1}{2}$	$\frac{1}{2}$	0
$\boxminus$	(4, 7)	0	$\frac{1}{2}$	$\frac{1}{2}$
$\equiv$	(6, 6)	0	1	0

Furthermore, the characteristic parameters τ and γ of the texture Z in (6.11) are given by the formulae

$$\begin{cases} \tau = & 6b + 2c; \\ \gamma = & 10a + 6b + 8c; \\ 1 = & a + b + c. \end{cases} \qquad (6.12)$$

Conversely, if a texton Z with parameters τ and γ is given, then Z can be expressed as a linear combination of the form (6.11) where the coefficients are given by the formulae:

$$\begin{cases} a = & \frac{1}{2}\tau + \gamma - 9; \\ b = & \frac{1}{2}\tau + \frac{1}{2}\gamma - 5; \\ c = & -\frac{1}{2}\tau - \frac{3}{2}\gamma + 15. \end{cases} \qquad (6.13)$$

In particular, we see that the terminator number need not be an integer: it can be an arbitrary real number; and the same is true of the color.

The accessible textures—the ones that can actually be constructed as a mixture of symbols using coefficients that are non-negative and hence can be interpreted as probabilities—occupy the interior and boundary of the triangle spanned by the basis texture classes which have representatives $\boxminus$, $\equiv$, and $\boxminus$ (Fig. 6.2.21). The ideal textures correspond to points in the plane determined by these three textons that lie outside the triangle. Any three noncollinear textons are a basis for the space of texture classes. In particular, since the medians of a triangle intersect at a point, we obtain the identity illustrated in Figure 6.2.20. But the basis $\boxminus$, $\equiv$, and $\boxminus$ is distinguished because the accessible textures are precisely the mixtures with nonnegative coefficients relative to these basis symbols.

Let us pursue the concept of accessible textures in greater detail. If $X_1, \ldots, X_n$ is a basis for a family of textures, then every texture belonging to the family

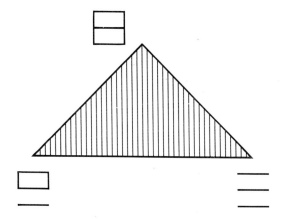

Figure 6.2.21   Accessible textures.

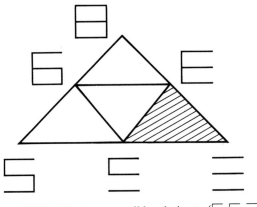

Figure 6.2.22   Textons accessible relative to $\{\boxminus,\boxminus,\equiv\}$.

can be expressed in the form $X = p_1 X_1 + \cdots + p_n X_n$ where $p_1 + \cdots + p_n = 1$. If each p_k is nonnegative, then $\{p_k: 1 \le k \le n\}$ is a probability distribution and the corresponding texture can be realized by mixing a representative of the texton X_k with probability p_k, $1 \le k \le n$. Such textures are said to be *accessible* but to be completely precise they should be called *accessible relative to* $\{X_1,\dots,X_n\}$.

If a texton X is not accessible relative to one basis it nevertheless may be accessible relative to some other basis. For example, the accessible textons relative to $\{\boxminus,\boxminus,\equiv\}$ are the textons that lie within and on the boundary of the triangle spanned by these three textons, shown hatched in Figure 6.2.22.

Textons that lie in or on the boundary of the triangle spanned by $\{\boxminus, \boxdot,\equiv\}$ are accessible relative to those three textons but are not accessible relative to $\{\boxminus,\boxminus,\equiv\}$ if they lie outside the triangle spanned by the latter three textons. This raises the question of determining the largest set of accessible textons.

Write the general texton for the family spanned by the textons listed in Figure 6.2.13 as

$$X = a\boxminus + b\equiv + c\boxdot$$

with $a + b + c = 1$. The textons accessible relative to $\{\boxminus,\equiv,\boxdot\}$ are shown hatched in Figure 6.2.23 drawn relative to Cartesian a- and b-axes.

Consider the texton X_0 for which $\tau(x) = \gamma(x) = 0$. From eq(6.13) we have

$$a = \tfrac{1}{2}\tau + \quad \gamma - \quad 9 = -9;$$
$$b = \tfrac{1}{2}\tau + \tfrac{1}{2}\gamma - \quad 5 = -5; \qquad\qquad (6.14)$$
$$c = \quad \tau + \tfrac{3}{2}\gamma + 15 = \quad 15.$$

That is:

$$X_0 = -9\boxminus - 5\equiv + 15\boxdot. \qquad\qquad (6.15)$$

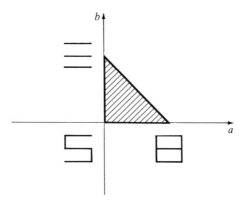

Figure 6.2.23   Relative accessible textons.

This is equivalent to

$$\text{⑤} = \frac{1}{15}X_0 + \frac{9}{15}\text{⊟} + \frac{5}{15}\text{=}. \qquad (6.16)$$

Hence X_0 must be the blank texton: no color, no terminator, since $\tau(X_0) = \gamma(X_0) = 0$ and the eq(6.16) asserts that X_0 occurs in the decomposition of ⑤ with probability 1/15. Hereafter denote X_0 by $\varnothing$, the symbol for the empty set.

The texton $\varnothing$ is the intersection of the lines defined by $\tau(X) = 0$ and $\gamma(X) = 0$. Since terminator number and color of an accessible texton must be nonnegative, the largest set of accessible textons is the intersection of the half-planes defined by $\tau(X) \geq 0$ and $\gamma(X) \geq 0$.

Since $\tau(\varnothing) = \tau(\text{⊟}) = 0$, the ray from $\varnothing$ through ⊟ defines one boundary of this region. The other boundary is the ray from $\varnothing$ continuing into the first quadrant which has the equation $0 = \gamma(X) = 10a + 6b + 8(1 - a - b) = 2a - 2b + 8$, i.e., $b = a + 4$.

The resulting maximal region of accessible textons is shown in Figure 6.2.24.

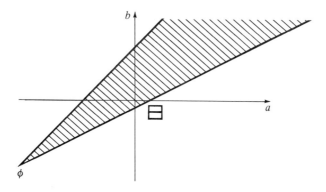

Figure 6.2.24   Maximal region of accessible textons.

Figure 6.2.25   Map of texton space: the (γ, τ)-plane.

In terms of the basis $\{\boxminus, \equiv, \varnothing\}$, an arbitrary texton (not necessarily accessible) can be written in the form

$$X = \frac{\tau}{6}\equiv + \left(\frac{\gamma - \tau}{10}\right)\boxminus + \left(\frac{30 - 2\tau - 3\gamma}{30}\right)\varnothing \qquad (6.17)$$

where $\tau = \tau(X)$ and $\gamma = \gamma(X)$.

Figure 6.2.25 displays a portion of the (γ, τ)-plane in the space of textons for the parameter range $0 \le \tau \le 8$, $6 \le \gamma \le 12$, where the color γ is expressed in multiples of a conventional unit.

6.3 Edge Detection, Uncertainty, and Hierarchy

Edges are the boundaries that separate regions in the visual plane. They may be defined by a curve of one color, say black, viewed against a uniformly colored, say white, background. This idealization of the general situation is one which is frequently encountered through the medium of printed documents and script writing. Cartoons sometimes provide particularly pure examples. More generally, edges generally correspond to portions of an image where the light intensity changes abruptly, although not all abrupt changes of intensity are interpreted as edges by the visual system nor do all perceived edges correspond to intensity changes, abrupt or otherwise. Sometimes edges result from changes in texture fields, including changes in color. However, color may be relatively less significant as a discriminator of the boundary of a visual object than other types of texture differences. Indeed, experiments have shown that boundaries and edges of objects are more readily distinguished in gray scale images than in images in which color varies but intensity is held constant.

The importance of edge detection in machine and biological vision is that it supports the identification of features and objects in the image, and thereby is assumed to play a fundamental role in laying the proper foundation for higher-level object recognition and identification. Yet there are striking examples of object recognition for which edge detection does not play a significant role. The Dalmatian Dog (Figure 6.2.2) is surely recognized without the benefits of edge detection; the prominent straight lines produced by the "venetian blinds" (Figure 6.3.1) do not aid perception of the racing horses; the six children (in Figure 6.3.2) are immediately evident although the lattice of edges carries information that describes the printing process rather than the imaged objects. Other objects, such as trees in leaf or the agitated surface of the sea or a shaggy dog, have boundaries so contrived and convoluted that the burden of computation and storage of the edges surely must outweigh the utility of the information derived from them, insofar as categorization and recognition are concerned. Indeed, the details of the position and direction of edges can change dramatically throughout such a scene without resulting in the slightest change in perceived features or in the categories of knowledge that result from them. Just as many different distributions of velocity of the molecules of air in a room are compatible with a given perception of temperature, so it is, most often, with the distribution of edges in a complex natural scene: some functional of the edge information that is stable with respect to certain classes of changes in that data often seems to be all that matters for analyzing and understanding the image.

Nevertheless, the situation is quite different for most objects from the manufactured world, whose boundaries are dictated by the symmetry groups that underly mass production technology and whose edges generally assume the appearance of long segments of straight lines or conic section that result from the perspective projection of circles. In a line drawing or cartoon all of the information is contained in the edges, and it is well known as a result of the work of HUBEL and WIESEL that the vision system of mammals contains tangent line detectors, i.e., means for identifying the point of tangency and direction of the tangent line to a curve which separates a region of high intensity from a region of low intensity. However, these differential edge

26   Advertisement (detail). Voigtlander Camera. Design: Albert E. Markarian. Kalmar Advertising. Courtesy the designer.

Figure 6.3.1 "Venetian blinds." (From Thurston and Carraher, *Optical Illusions and the Visual Arts.* Copyright 1966, Van Nostrand Reinhold. Reproduced with permission.)

Figure 6.3.2 Six children. (From Lanners, *Illusionen,* p. 32. Copyright 1973 Verlag C.J. Bucher. Reproduced with permission.)

detectors are not restricted to edge detection but probably also participate in the analysis of orientation and other parameters of image texture.

Thus it is of some importance to understand the principles that underly edge detection processes and the information that they provide for pattern recognition and categorization and, ultimately, for the extraction of those features that are used to recognize and remember objects. At the same time, we will want to bear in mind that the role of edge detection in image analysis is still far from clear: it is neither necessary for feature extraction and object recognition, nor is it necessarily the most efficient means to employ for these purposes even in those circumstances where it is applicable, yet means for computing edges are present in biological vision systems and edges have a natural interpretation in the decomposition of images into disjoint objects.

Edge detection has been employed in machine vision systems for decades but with variable effectiveness. One class of difficulties results from the unavoidable noisiness of image data in normal environments. Since the detection of edges is essentially a differential process, that is, it involves comparison of the intensity at nearby points of the image in order to estimate rates of change, it is very sensitive to errors of observation. The goal of localizing an edge in the image with high spatial resolution and the goal of capturing small edge features, i.e., obtaining sharply defined features, are in conflict, due to the uncertainty principle for measurement, but both goals are also sensitive to the presence of noise.

Mathematical processes that involve differentiation are unreliable unless the data is accurate, but integration processes are smoothing operators that spread the inaccuracies due to noise or to inadequacies of the measurement process over the collective ensemble of data. Thus deviations from the norm that are randomly distributed with opposing sign are given the opportunity to cancel each other, thereby improving the accuracy of the result. But smoothing operators necessarily rely on distributed data which, by the uncertainty principle, implies in the case of image analysis that the high spatial frequency information (corresponding to information-bearing features in the image, not to noise) will be lost. Nevertheless, interpolation methods can sometimes be employed to "sharpen" the boundaries of objects that have been isolated by integration processes that accumulate information. Since the computation of texture parameters involves the statistical aggregation of data collected from a neighborhood of an image point, that is, a smoothing operation, it should be expected that texture perception will be less sensitive to noise than edge detection based upon differentiation. Thus we may anticipate that texture variation might be used by biological systems, and perhaps should be used by machine vision systems, for the determination of boundaries separating objects in an image, and that the cognitive structures that support object recognition may, by employing this artifice, enjoy a loose dependence on precise details of edge location and orientation without a corresponding loss of effectiveness.

Recent work in edge detection falls into two main categories. The first extends the ideas of local determination of gradients and associated tangent directions in the intensity field to include smoothing operators to cope with the effects of image noise. In order to compensate for the decreased sensitivity to high spatial frequencies, multiple scales of resolution may be introduced with the hope that the coherent structural properties of meaningful image features will be highly correlated at many or even at all scales, thereby providing a basis for selectively omitting information and for interpolating from the lower resolution and less noisy scales to the higher resolution but noise-prone ones. These ideas appear to have originated with David MARR and his coworkers, especially Tomaso POGGIO and Ellen HILDRETH. The second category of ideas involves operators of integration in a more central way and thereby offers potential for diminishing the effects of noisy observations. These methods are based upon the use of a variational principle that takes into account the effect of the intensity at each point of the image on every other point. The work of POGGIO and his coworkers typifies this approach to the problem.

We have already remarked that the uncertainty principle for measurement implies that unlimited accuracy in spatial localization of an image signal and simultaneous resolution of details of the image that correspond to high spatial frequencies are incompatible: the product of the uncertainties is bounded from below. It follows that the best that can be accomplished in simultaneous localization and resolution of detail will result from selecting "filters" for measurement that minimize the product of the uncertainties. It is known that the GAUSSIAN function has this property. In 1 dimension it has the form

$$G_\sigma(x) = G(x) = \frac{1}{\sigma} e^{-x^2/2\sigma^2}, \tag{6.18}$$

where σ is the standard deviation. In the notation of (4.35), $\sigma = \Delta x$. The GAUSSIAN is not the only function that minimizes the uncertainty product: it is easy to see that

$$e^{iax} G(x)$$

has the same property for all constants a. These minimizing functions were introduced into the study of optics by Dennis GABOR, and their 2-dimensional analogues have been used by DAUGMAN with good effect in image analysis. Returning to the functions described by (6.18), we see that large values of σ correspond to poor localization but good smoothing, and find that the FOURIER transform of G, which by eq(9) of Table 4.4.1 is equal to

$$e^{-(1/2)(2\pi\sigma)^2 y^2},$$

is, in this case, well localized; i.e., most of the "energy" in the original function is found at low frequencies. If σ is small, the situation is reversed, with energy distributed throughout higher frequencies and the original function well localized. Thus the *scale* of localization can be adjusted by changing the value of the parameter σ, and this affords the opportunity to aggregate information from various scales and hence with various degrees of resolution by utilizing Gaussian filters of several sizes to analyse an image. It is suspected that the human vision system employs about four different scales of analysis with corresponding σ parameters increasing by a factor of 2 from one scale to the next, thereby providing a range of localization and resolution data as inputs to higher processes that each vary by as much as a factor of 16 in uncertainty.

Suppose then that a 1-dimensional intensity function $I(x)$ is observed by means of a filter $G(x)$. This means that if the filter is centered at a point a in the domain of the intensity function, then the convolution $\int I(x)G(x - a)\,dx = (I * G)(a)$ represents the value observed at a. Maxima in the rate of change of the intensity function correspond (in the absence of noise) to putative edges in the (1-dimensional) image, and for these points the second derivative $I''(x) = d^2I/dx^2$ is zero. Thus we are led to seek the zeros of the smoothed sampled version of I'', that is, the zeros of $I'' * G$. Integrating twice by parts and recollecting that G rapidly decreases to 0 as $x \to \infty$, we obtain $I'' * G = I * G''$, with

$$G''(x) = \frac{d^2G}{dx^2} = \frac{1}{\sigma^3}\left(\frac{x^2}{\sigma^2} - 1\right)e^{-x^2/2\sigma^2}. \tag{6.19}$$

The graph of G'' is displayed in Figure 6.3.3; it is evidently similar to the "on-center" simple receptive field response of receptors in the visual cortex illustrated in Figure 5.11.9, and this provides the connection between mathematical models for machine vision and neurophysiological models of the human vision system. In the analysis of biological vision systems the response of center-surround cells of the visual cortex has often been modeled as a difference of GAUSSIAN functions but the observations are not sufficiently precise to discriminate between the latter and G''.

Thusfar we have restricted our attention to 1-dimensional image signals (which may of course be thought of as the restriction to a line of the intensity function that describes an image). The generalization of the procedure for edge detection based on estimation of the zero crossings of I smoothed by G'' to 2-dimensional images is not entirely straightforward. The maximum gradient of the 2-dimensional image intensity function $I(x, y)$ may be oriented in any direction, which increases the computational load considerably, although we should recall that the ultimate goal is to determine the location in the image of the points where the smoothed gradient of the intensity achieves its maxima rather than to determine the values of the maxima themselves. This suggests that a procedure that is independent of orientation might provide the desired information in a more direct way.

The simplest differential operator that is independent of orientation is the operator of LAPLACE,

Edge   Thin bar   Wide bar

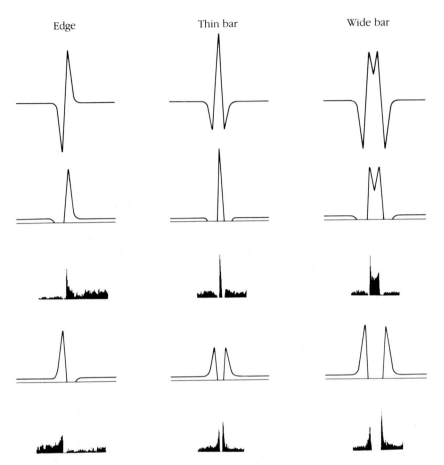

Figure 6.3.3   Comparison of the predicted responses of on- and off-center Y cells to electrophysiological recordings. The first row shows the response of $\partial/\partial t(\nabla^2 G*I)$ for an isolated edge, a thin bar (bar width = $0.5w_{1-D}$, where w_{1-D} is the width projected onto one dimension of the central excitatory region of the receptive field), and a wide bar (bar width = $2.5w_{1-D}$). The predicted traces are calculated by superimposing the positive (in the second row) or the negative (in the fourth row) parts of $\partial/\partial t(\nabla^2 G*I)$ on a small resting or background discharge. The positive and negative parts correspond either to the same stimulus moving in opposite directions, or stimuli of opposite contrast—for example, a dark edge versus a light edge—moving in the same direction. The observed responses (third and fifth rows) closely agree with the predicted ones, even in cases where both are elaborate (such as for the wide bar). (From Marr and Ullman, "Directional Selectivity and It's Use in Early Visual Processing." *Proceedings of the Royal Society of London B211*. Reproduced with permission from Ullman.)

$$\Delta I = \frac{\partial^2 I}{\partial x^2} + \frac{\partial^2 I}{\partial y^2}.$$

This operator will replace the second derivative for 2-dimensional images.

Regarding the generalization of the GAUSSian G, observe that the product $G_{\sigma_1}(x)G_{\sigma_2}(y)$ will be independent of orientation if and only if $\sigma_1 = \sigma_2$ and in this case

$$G(x, y) = G_\sigma(x, y) = G_\sigma(x)G_\sigma(y) = \frac{1}{\sigma^2} e^{-(x^2+y^2)/2\sigma^2}. \tag{6.20}$$

The development for the 2-dimensional case proceeds along the usual lines: repeated integration by parts yields $\Delta I * G = I * \Delta G$, with

$$\Delta G(x, y) = \frac{1}{\sigma^2}\left(\frac{r^2}{\sigma^2} - 2\right)e^{-r^2/2\sigma^2}, \tag{6.21}$$

where $r^2 = x^2 + y^2$. ΔG is the function of two variables that corresponds to the 2-dimensional center-surround response function.

Figure 6.3.4 exhibits edges detected by using the LAPLACian of the GAUSSian

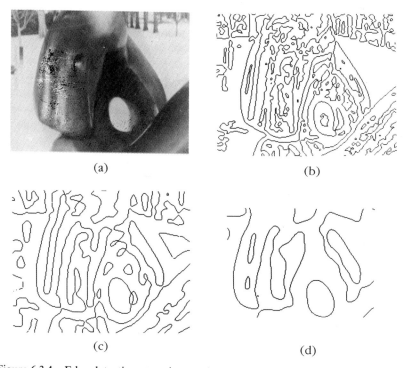

(a)

(b)

(c)

(d)

Figure 6.3.4   Edge detection at various scales. (From Marr and Hildreth, "Theory of Edge Detection." *Proceedings of the Royal Society of London B207.* Reproduced with permission from Hildreth.)

as a filter at various scales. It is evident that no one scale provides unambiguous information about the location of edges. As the scale is increased, the number of putative edges decreases. It has been suggested by WITKIN that the collective scale information can be systematically used to eliminate spurious edges by correlating edge information in the domain of the scale variable.

Suppose that $I(x)$ denotes the intensity function for a 1-dimensional image which is sampled at the discrete set of points $\{x_k: 1 \leq k \leq n\}$ and let $I_s(x)$ denote a smoothed version of $I(x)$ that is computed so as to minimize an expression of the form

$$E = \sum_{k=1}^{n} (I(x_k) - I_s(x_k))^2 + \lambda \int |I_s''(x)|^2 \, dx \qquad (6.22)$$

where λ is a parameter that governs the relative weight of each of the summands. The first term measures the deviation of the sampled values from the corresponding values of the smoothed intensity function, whereas the second term is large when the graph of the smoothed intensity deviates from straightness. The first summand is zero if the graph of the smoothed function passes through the data points. This can always be achieved in the absence of constraints, for a polynomial of degree $n - 1$ can always be made to assume n given values. However, the polynomial will have $n - 1$ turning points and consequently will become increasingly convoluted, and will not be stable, i.e., its degree and the values of its coefficients will be subject to large changes, as the number of sample points is increased. The second term will be zero if I_s'' is the zero function, i.e., if a linear function is used to fit the sample data. In general, the second summand will increase as the number of turning points of the graph of I_s increases. Thus, the two summands compete with each other and reflect the trade-off between smoothness of the intensity function and its agreement with the observations.

It can be shown that the solution to this variational problem is the convolution of the image intensity function with a cubic spline that is similar to the GAUSSIAN. Figure 6.3.5 compares the cubic spline and the GAUSSIAN filters and their first two derivatives for the 1-dimensional case. The relatively close agreement of these functions provides a partial bridge between use of a variational principle for detection of edges and use of local properties expressed in terms of differential operators.

Thus it appears that neurophysiological evidence, the use of optimal filters in the sense of the uncertainty principle for measurement, and conclusions drawn from the variational principle (6.21) are all in general agreement at the computational level, although at the theoretical level there are fundamental questions that remain to be resolved.

The employment of multiple scales of resolution to tease out edge information and other properties that feed the higher level pattern recognition processes carries with it an implicit hierarchical structure and thus epitomizes another general principle of information science. Suppose that a gray level scene is described by the intensity function $I(x, y)$, where x and y are variables

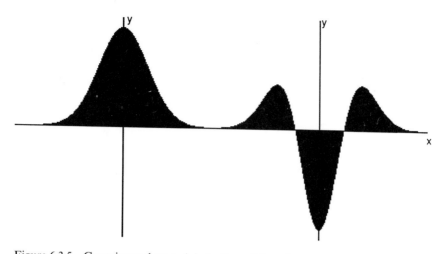

Figure 6.3.5   GAUSSIAN and second derivative of GAUSSIAN (maximum absolute value normalized to 1).

that range through the image plane, and that $I(x, y)$ is sampled by a filter of some fixed scale. This is roughly equivalent to a representation of the image by a corresponding discrete, or locally constant, function $I(i, j)$, where (i, j) denotes the $(i, j)^{th}$ *pixel* in the discrete array and the common size of the pixel is essentially the same as the scale of the filter. Now suppose that the image is filtered at a sequence of scales corresponding to pixel widths $l_1, \ldots, l_n$ arranged so that the ratio $l_{i+1}/l_i = k$ is an integer that is independent of the scale; for instance, $n = 4$ and $k = 2$ might characterize the scales of resolution for the human vision system. Working our way from the finest to the coarsest level of resolution, we can identify a $k \times k$ block of pixels at level i with a single pixel at level $i + 1$. Pixels at the ith level are k^i times as wide as pixels at the coarsest scale. Thus the multiscale data corresponds to n levels of a tree, or hierarchy, in which each node has fan-out k^2. The tree can be completed by sampling at increasingly coarse scales so that $n = \log_k N$, where the highest resolution version of the image consists of $N \times N$ pixels.

   This hierarchical data structure is frequently employed in computer graphics, where so-called *quadtrees* and *octrees* are used, respectively, to provide economical representations for 2-dimensional images and 3-dimensional objects for application to, e.g., computer aided design systems.

6.4 Pattern Structure and Learning

How do biological organisms impose order on the information that their sensory systems provide? Are the methods that different organisms use, and the categories that they induce, compatible instances of universal principles that apply to patterned structures in general rather than to particular biological machines? If so, are those principles an aspect of the underlying harmony of the general laws that govern the ebb and flow of the material content of the universe, or is there something special about pattern—about information—that is fundamentally separate from the laws of physics? Are the harmonies of physical law perhaps a reflection of some deeper laws of pattern? The reader will not find answers to any of these questions in this last section, but surely the ability of organisms to learn what they do not already know and the striking compatibility across species of the categories that they create, suggest that there may be principles of pattern classification and categorization that are very general and perhaps even universal. And, if that be true, then the difficulty in drawing a sharp line between the quick and the dead—between animate and inanimate matter—leads inexorably to the possibility that categorization and the creation of pattern may be a general and powerful feature of nature whose laws play a silent role in establishing and maintaining at every scale the order that we find in the universe around and within us.

Certainly there are surprises in the domain of animals. HERRNSTEIN and his colleagues have shown that pigeons have a remarkable ability to create abstract categories that cannot possibly correspond to genetically prewired patterns designed to confer upon them a specific competitive advantage. Indeed, the powers of categorization of pigeons as applied to visual information are in many respects comparable or superior to human performance. Thus it has been shown that when trained to discriminate printed letters of the roman alphabet, pigeons trained on one font are able to generalize to other fonts. In one experiment, the performance of a pigeon was closer to the mean performance of five human subjects than was the performance of any of the humans. Pigeons are able to generalize from a single exemplar of each of six species of leaf to the category of "oak leaf" and classify arbitrary samples with high accuracy. Since oak leaves differ greatly in number of lobes and relative dimensions, it seems clear that the pigeons have generalized to some abstract concept of oak leaf; perhaps we may say that they have *learned the concept* oak leaf. Pigeons do have something to do with trees so it is of course possible that the long course of evolution has built oak leaf recognition into the hardware of the pigeon's brain. But what are we to make of pigeons who have learned to recognize the presence of tropical fish in photographs of the fish in their natural habitat? Their common ancestor flourished more than 55 million years ago and the fish have changed shape since then, so surely it is not hard-wired for pigeons to classify tropical fish. In general, the pigeons seem to do well, achieving performance levels comparable to our own, for image categorization tasks that do not require what we may think of as a "world view"; they do less well otherwise. Thus they have far greater difficulty than we in recognizing that a fish whose body intersects the picture frame is a fish. It seems that we know what a picture frame is and can make the necessary compensation. Finally, let us note that the categories that the pigeon establishes seem in every respect to be compatible with those that a person would establish. When encouraged to sort photographs that exhibit bodies of water —in a glass or in a lake—it makes no difference whether the work has been performed by a professor or by a pigeon.

The pigeon's brain is a fraction of the size of the human brain yet in categorizing visual data it is capable of a level of performance that does not differ substantially in quality or in kind from human performance. This suggests that

- The algorithms that underlie the categorization of visual data are simple.
- The algorithms that underlie the categorization of visual data are universal.

It is tempting to infer that the word *visual* should be omitted from both assertions.

Recent research into the problem of learning and categorization has concentrated on models for systems that, like the brain, appear to consist of large numbers of relatively elementary components that can exchange information through a communication network that binds them together. This is a description of a "parallel" or concurrent computer as well as of a simple model of a brain, so one should expect that research in this direction may have implications for the design of thinking machines as well as for the analysis of the mental activities of organisms. Since the elementary components are assumed to be identical as well as simple, it follows that complex behavior can only arise as a consequence of collective properties of the system mediated by the topology of the communication network and the ease with which its links can transmit information. But the details of the network topology cannot play a central role either, for as neurons die, the topology of the brain may change dramatically without noticeably affecting its performance. No doubt the statistical properties of the topology of interconnections play the central role, with sufficient redundancy built in to ensure a mean time before failure that is comparable to the expected life span of the body that houses the brain. These models, often called "connectionist" to stress their collective nature and dependence on the communication network for the emergence of complexity, are currently under active investigation by HOPFIELD, HINTON, and others. Although it is too early to assess the role that such theories will play, it already is evident that connectionist models provide a rich structure that is closely related to the themes that underlie statistical thermodynamics. Indeed, let us take note of the irregularities inherent in the patterns of interconnections of biological information processing systems and of their statistical significance. Although the obvious gross anatomical similarities among structures serving similar functions in one organism, and the same functions in different organisms, usually capture our attention, the details of the "wiring diagram" fluctuate substantially. The manufacturing process evidently lacks the vigorous quality control necessary to ensure that pairs of vertices in the graph of the communication network, i.e., neurons, are connected if and only if the "blueprint" calls for it. Nature has had to make a virtue of necessity by building in means for processing information that feature the random element in interconnection topology. Thus the mechanisms of brain operation must to a considerable extent be independent of certain local interconnection details: we observe complex behavior emerging from the unpredictable combination of grossly similar simple parts. The collective properties of interacting neurons constitutes a kind of "statistical neurodynamics" to which the classical statistical thermodynamics of BOLTZMANN, which is nothing but a theory of information in disguise, is broadly applicable. Thus it appears that these models necessarily have an equally close relationship to the measures and structures of information developed in the earlier chapters of this book.

Let us pursue this train of ideas, following HOPFIELD's work on memory and the work of HINTON and SEJNOWSKI on learning.

Consider the network that corresponds to a collection of simple elements, which we will call "neurons" although they might be elementary processors in a concurrent computer. Each element will be assumed to exist in one of two possible states. The state of the ith neuron will be denoted S_i, where $S_i = 0$ if the neuron is "off" and $S_i = 1$ if it is "on" (which will be interpreted as firing at its maximum rate). The state of the neural network will be represented by the vector $S = (S_i)$. The interactions between pairs of neurons will be described by the strength of the excitation or inhibition between vertices linked by an edge: denote the strength or weight of the connection between the ith and j^{th} neurons by $W_{i,j}$; $W_{i,j} = 0$ if the neurons are not connected. The weights determine the computational capabilities of the network. Thus they must be supposed to change as the network learns, but to be relatively more fixed than the set of states of the system. A network with a fixed set of weights can be thought of as a kind of special purpose computer; by modifying the weights, the computer and its program (which are in fact one) are correspondingly modified.

If t_i denotes the threshold for neuron i, then we may suppose that i changes state according to the rule

$$S_i \to 1 \quad \text{if} \quad \sum_j W_{i,j} S_j > t_i$$

$$S_i \to 0 \quad \text{if} \quad \sum_j W_{i,j} S_j \le t_i.$$

Changes of state may occur randomly and asynchronously but at a fixed average rate.

In order to bring the full power of the machinery of statistical thermodynamics to bear, we must introduce a generalized cost or "energy" function, and also the information or "entropy" associated with the system. We will proceed along the lines of the discussion in section 4.3.

The energy E of a configuration of the network, that is, corresponding to some fixed set of states of the neurons and weights of the connections, is a quadratic form in the elements of the vector S:

$$E = -S'WS,$$

where $S = (S_i)$ and $W = (W_{i,j})$, $0 \le i, j \le n$ with $W_{i,0} = W_{0,i} = -t_i/2$.

Such a system can "remember" configurations and it can also associate one of its memories with an arbitrary new configuration to which it is exposed. Thus it can serve as a *content addressable memory*. For instance, let $\{S^m: m = 1, \ldots, k\}$ be configurations to be stored as memories. Then selection of the weights according to the formula

$$W_{i,j} = \sum_{m=1}^{k} (2S_i^m - 1)(2S_j^m - 1), \qquad i \neq j$$

defines an energy function E for which the states S^m are local minima. HOPFIELD finds that about $0.15n$ configurations can be remembered before recall becomes severely compromised.

The values of the components of S can be taken to represent the acceptance or rejection of some hypothesis: $S_i = 1$ may represent acceptance of the hypothesis and $S_i = 0$ its rejection. Suppose that W is a symmetric matrix. Then the hypotheses represented by i and j are mutually supportive if $W_{i,j} > 0$ and mutually antisupportive (i.e., acceptance of one hypothesis supports rejection of the other) if $W_{i,j} < 0$. If $W_{i,j} = 0$, then the hypotheses are independent: hypotheses communicate only to support or reject one another. With this interpretation, the energy of a configuration represents a measure of the aggregated support for the collection of hypotheses.

If we think of the multidimensional surface defined by the energy as a function of the state vector elements, then the "memories" S^m correspond to local minima—to valleys in the surface—which can be located by standard gradient descent methods. If an input state lies close to some local minimum, then gradient descent will locate that minimum and thereby identify the input state with it. In this sense the network can be said to be capable of storing new memories (by changing the energy function to include them) and of identifying a new impression with a stored memory although the two may not be identical. The model may therefore be capable of accounting for the puzzling ability of the brain to organize percepts that are dissimilar in their details into classes that share some important gross similarity.

A more complex and subtle situation arises when network models are applied to evaluate hypotheses, for in this case what is desired is not a local minimum, which would represent acceptance or rejection of the hypothesis relative to nearby states of the configuration, but rather a global determination of whether the hypothesis should be accepted that takes into account the totality of support and rejection, direct and indirect, that is represented by the combination of topology and weights of the connections. It is well known that local methods of analysis, such as gradient descent, are in general inadequate for finding a global minimum. It is at this point that the measure of information enters the analysis in the context of a statistical thermodynamical exploration of all states. The time evolution of thermodynamical systems carries them forward toward a stable state that is characterized by minimum information. The system will fluctuate about this minimum due to random variations in the values of the state variables. Let us index the possible states of the system by Greek subscripts. Denote the probability of occurence of state S_α by p_α; then the measure of information associated with the collection of possible states is

$$I = -\sum_\alpha p_\alpha \log_2 p_\alpha. \tag{6.23}$$

In the equilibrium configuration of minimum information the probabilities are related to the cost (i.e., the energy) function by the BOLTZMANN distribution formula

$$\frac{p_\alpha}{p_\beta} = e^{-(E_\alpha - E_\beta)/T}, \tag{6.24}$$

where E_α is the energy corresponding to the state S_α and T denotes a parameter which is determined by the structure of the system and plays a role analogous to the temperature in thermodynamics. When T is large, the relative probability of finding the system in a state of relatively high energy, that is, cost, increases; if T is close to zero, then the system spends most of its time in the lowest available energy state. The computational method for determining a state that corresponds to a global minimum of the information function constitutes the numerical version of the physical process of annealing: just as a physical system at high temperature can be made to jump from one "valley" to another on the energy surface and will, as the temperature is decreased, settle as it "crystallizes" into a configuration whose energy is, with very high probability, the global minimum, so too can the neural network model be made to "crystallize" into the state that corresponds to the global acceptance of a hypothesis as the consequence of a large number of small random interactions throughout the network.

Whether such a model can be realized by biological neural networks remains to be seen. It also remains to be seen whether it provides a practical means for enabling machines to learn. Nevertheless, it does provide a stimulating approach to the development of an information-based theory of pattern recognition and learning that can in principle provide an account of the large number of weakly interacting simple functional units that are konwn to occur in the only systems that are presently able to learn and categorize.

Biographical Sketches

Ludwig BOLTZMANN was born in Vienna on 20 February 1844 and died, by his own hand, near Trieste on 5 September 1906. Although the reasons for his suicide are unknown, it is thought to have been attributable in part to his despair over what he viewed as the rejection of his life's work by the scientific community, which at that time had turned away from atomistic theories although but a few years later the existence of atoms was definitely established by experimental studies of BROWNian motion.

BOLTZMANN was educated in Austria; he was awarded the doctorate in physics from the University of Vienna under Josef STEFAN in 1867 and later was professor at Graz, Vienna, Munich, and Leipzig.

Already in 1866 he sought a direct connection between the second law of thermodynamics and the mechanical principle of least action. By 1868 he had become acquainted with the statistical approach introduced by MAXWELL and introduced the notion that energy was distributed with equal probability over all elementary states of a system. By assuming that energy comes in small but finite quanta (and thereby anticipating part of PLANCK's great discovery more than a quarter of a century later), BOLTZMANN was able to reduce the problem of determining the distribution of states of a system to a combinatorial calculation.

In 1877, he showed that thermodynamic entropy is related to the statistical distribution of molecular configurations, with increasing entropy corresponding to increasing randomness. The relationship was made precise by BOLTZMANN's celebrated formula $S = k \log W$, where S denotes the entropy, W is the number of possible molecular configurations (modern term: *microstates*) having a given energy which correspond to a given macroscopic configuration, and k is a universal constant of proportionality known today as *BOLTZMANN's constant* which converts from the numerical measure log W to the physical units of entropy. The formula $S = k \log W$ appears as an epitaph on BOLTZ-MANN's tombstone.

Sadi CARNOT (1796–1832), a French physicist, was the inventor of thermodynamics. His father was the minister of war for the first French Republic and for Emperor NAPOLEON I, as well as a mathematician whose work has been quoted in our own time. Sadi's younger brother Hippolyte was minister of education in the Second Republic and a sociologist, and his nephew became president of the third French Republic in 1887 and was assassinated by an Italian anarchist.

CARNOT published his pioneering work on thermodynamics in 1824 at the age of 28, and in the years that followed he refined his early ideas and penetrated to the statistical point of view which lies at the root of a deeper understanding of the phenomenological properties of thermodynamics. His thoughts were recorded in a notebook which passed into Hippolyte's personal library until about 1878, when it was given to the French Academy of Sciences. It was not until then that it was recognized how far CARNOT had proceeded in determining the mechanical equivalent of heat and sketching the ideas of a kinetic theory of heat. But full publication of his work did not come until 1927.

The major contributions of CARNOT include the definition of entropy in thermodynamical processes and the enunciation of the *second law of thermodynamics*, which in essence states that the entropy of a closed system cannot decrease and in general increases as the system evolves with time. These ideas provided a scientific basis for speculations about the nature of time, in particular, why time appears to be directed, by enabling the future direction to be identified with the increase in entropy of the universe as a whole. As the reader will suspect, this set the stage for innumerable controversies, publications, and silliness as well as a few profound thoughts. Since then, time's arrow has often been identified with the increase of entropy; in terms of the modern interpretation of entropy as a measure of information or organization, the unidirectionality of time can be identified with a continuing decrease of organization in the universe as a whole. Still, many thorny questions raised by this interpretation remain unresolved.

CARNOT died in 1836 in a cholera epidemic in Paris, depriving the world of one of its most original minds.

Gustav Theodore FECHNER (1801–87) was a precocious child who had learned Latin by the age of 5. He took his M.D. at the age of 21 but never studied medicine. His first writings were metaphysical pieces of a fantastic and sometimes brilliant nature that attacked materialism, which was popular in Germany at that time.

FECHNER's first scientific work was in physics, concerned with electricity. Later he was the victim of a protracted and serious neurotic illness which began with partial blindness brought on by viewing the sun in the course of experiments on colors and afterimages. After living for 3 years in the dark, and despairing of ever seeing again, he unwound his bandages in his garden and found his vision regained and abnormally powerful. This highly emotional experience led him to philosophical speculations about the unity of mind and matter, which he thought of as two aspects of the same thing. He set himself the problem of finding how to pass from measures of the one to measures of the other. On the morning of October 22, 1850, while he was still in bed, he conceived the "solution." It was to make the relative increase of physical stimulation the measure of the corresponding psychological response. That day is commemorated by some psychophysicists as "FECHNER Day." His classic work, published in 1858, incorporates a decade of thought and experiment and founded psychophysics as a scientific discipline. It contains the famous "law" of FECHNER, that the measure of subjective response is proportional to the logarithm of the measure of physical stimulus.

Jean Baptiste Joseph FOURIER (1768–1830) was a French politician and administrator who somehow found the time to do brilliant work as a mathematician and mathematical physicist. His life illustrates the vicissitudes of fortune and the permanence of mathematical genius.

By the age of 9 FOURIER was an orphan and was placed in the military school in his birthplace of Auxerre where he discovered his passion for mathematics. He hoped to join the artillery or military engineers but was refused and sent instead to a Benedictine school to prepare him to pursue his interests at its seminary in Paris. The French Revolution interfered with this plan, and he returned to Auxerre in 1789 to take up a teaching position in the school in which he had been a student. This simple life was not to last long. FOURIER was involved in local political affairs and was arrested in 1794 because he had defended victims of the Terror. His timing was good, however, for although a personal appeal to Robespierre availed naught, he was released shortly after Robespierre's execution. His enforced visit to Paris proved fortuitous, for the next year he was appointed assistant lecturer in the École Polytechnique. Soon thereafter he was arrested again, this time as an ostensible supporter of the very Robespierre who had refused to free him, but his new colleagues were able to obtain his release.

In the thick of activity and associated with the academic community upon which NAPOLEON relied, FOURIER was selected to join the Egyptian campaign. He became secretary of the Institut d'Egypte, conducted negotiations between NAPOLEON and the wife of the chief bey, and held other diplomatic posts.

FOURIER intended to resume his post at the École Polytechnique upon his return to France in 1801, but NAPOLEON had recognized FOURIER's administrative abilities and appointed him prefect of the department of Isère centered at Grenoble and extending to what was then the Italian border. In 1808 NAPOLEON elevated him to a barony.

After NAPOLEON's fall in 1814, the route of his return march on Paris in

1815 took him through Grenoble, where FOURIER was still prefect. Fulfilling his duties, FOURIER ordered the preparation of defenses, which he knew would be useless, and left the city by one gate as NAPOLEON entered by another. The two friends met at Bourgoin. NAPOLEON elevated FOURIER to count and appointed him prefect of the department of the Rhône.

In protest against the severity of the regime of the Hundred Days, FOURIER resigned his new title and prefecture and went to Paris to take up research as a full time activity. After a difficult period without a job, his political reputation and scientific standing began to rise. He was elected to the reconstituted Académie des Sciences but Louis XVII rejected the nomination because of FOURIER's connection with NAPOLEON; his nomination in 1817 was unopposed and in 1822 he was elected perpetual secretary of that body and, in 1827, to membership in the Académie Française.

FOURIER's mathematical achievements occur in the context of his interest in the diffusion of heat. In a long paper presented to the Academy in 1807 he already used what today are called FOURIER series to represent temperature distributions in solid bodies. This paper was rejected because of strong opposition by the famous mathematician LAGRANGE and was never published. In 1810 a prize problem on heat conduction was proposed and FOURIER submitted a revised version of his 1807 paper which included the FOURIER integral theorem for a function and its FOURIER transform. His paper won the competition. The mathematical parts were expanded into the book *Théorie analytique de la chaleur*, which is one of the great classics of mathematical physics.

Hermann Günther GRASSMANN (1809–77), the third of his parents' twelve children, was born in Stettin, Germany (now Szczecin, Poland), to a family of scholars. His father studied theology, mathematics, and physics and taught the latter two subjects at the Stettin Gymnasium after having been a minister for a short time. Hermann followed in his father's footsteps, first studying theology at Berlin in addition to classical languages and literature. After returning to Stettin in 1830 the younger GRASSMANN studied mathematics and physics independently. In 1831 he failed to win a teaching certificate; the examiners stated that he had to display greater knowledge to be qualified to teach the highest grades of secondary school. He passed the qualifying examination in 1839, including a written assignment on the tides, for which he gave a simplified treatment of LAPLACE's work based upon a new geometric calculus that GRASSMANN had developed. By 1840 he had decided to concentrate on mathematical research.

After writing several secondary school textbooks for the study of German and Latin he published the first edition of his geometric calculus, the *Ausdehnungslehre*, whose importance was not understood by his contemporaries. The famous mathematician MÖBIUS declined to review the book and it was totally disregarded by the experts. Today, it is recognized as the first presentation of the abstract theory of vector spaces over the field of real numbers, and of what we now know as the *Grassmann algebra* of a vector space. In 1862 GRASSMANN published a new version of the *Ausdehnungslehre*, which fared no better than the first. Disappointed, he turned away from mathematics and concentrated his intellectual activity on linguistic research and comparative philology, which were well received by scholars.

In 1852 GRASSMANN succeeded his father as professor at the Stettin Gymnasium and the next year he wrote his important paper on color, which corrected errors of HELMHOLTZ and mathematically formulated the fundamental laws of color mixture. In a letter to his friend MÖBIUS written the same year he described his work on color as an application of his general geometrical calculus, concluding with the observation,

> So bieten sich nach und nach immer mehr Beziehungspunkte zwischen der geometrischen Analyse und den Gesetzen der Natur dar, wie dies allerdings zu erwarten stand, wenn jene Analyse eine naturgemässe war.

Werner Karl HEISENBERG (1901–76) was the inventor of quantum mechanics and discovered the uncertainty principle. On the basis of the work of physicists, especially Max PLANCK and Neils BOHR, in the previous two decades, HEISENBERG took the decisive step that created a revolution in our understanding of physical reality. He studied physics at Munich, obtaining his Ph.D. in 1923. Two years later he invented matrix mechanics and was able to give the first precise mathematical description of the structure of the atom as an application of his general theory.

In 1927 HEISENBERG discovered the *uncertainty principle* (in fact, it was a consequence of his earlier theory), which states that the product of the uncertainty in a simultaneous measurement, e.g., of the position and momentum of a physical system, must be not smaller than a certain universal constant. This violates cause and effect and thereby called for a far-reaching reassessment of the traditional interpretation of the nature of the physical world and the extent to which scientific theories can describe it. Thus his work profoundly influenced philosophers as well as physical scientists. In the same year, at the age of 26, HEISENBERG was offered the chair of theoretical physics at Leipzig, where he remained until 1941. From 1942 until his death he was director of various Max PLANCK Institutes for Physics, first in Berlin until 1945; thereafter in Göttingen and Munich. HEISENBERG directed Germany's atomic research during World War II. It has been said that the lack of progress was due in part to his efforts to divert the program from the development of atomic weapons.

Hermann von HELMHOLTZ (1821–94) was one of the broadest gauge scientific personalities of the nineteenth century. His interests turned early to physics but his father, a teacher at the Potsdam Gymnasium, could not afford to send HELMHOLTZ to the university and persuaded him to study medicine for which the state provided financial aid. In 1837 HELMHOLTZ received a 5-year stipend for medical study in Berlin in return for which he committed

himself to serve as an army surgeon for 8 years. While studying medicine at
the Friedrich Wilhelm Institute he also took courses in chemistry and psychology at the University of Berlin, the latter under Johannes MÜLLER, and
privately read the works of the mathematicians and mathematical physicists
LAPLACE, BIOT, and Daniel BERNOULLI.

HELMHOLTZ received the M.D. degree in 1842 and was appointed surgeon
to the regiment at Potsdam. In 1847 he read his path-breaking memoir *Über
die Erhaltung der Kraft*, which set forth the mathematical principle of the
conservation of energy, to the recently formed Physical Society in Berlin. In
1848 he was released from his military obligation and appointed associate
professor of physiology at Königsberg, where he produced a succession of
important works, including the first measurement of the speed of transmission
of nerve impulses (which, prior to his work, was generally considered to be
too great to admit of measurement), papers on acoustics and optics, and, in
1851, the invention of the ophthalmoscope.

In 1855 HELMHOLTZ was appointed to the chair of anatomy and physiology
at Bonn, where he continued his work in sensory physiology, publishing the
first volume of his celebrated *Handbuch der Physiologischen Optik* in 1856. In
1858 he published a seminal article on the hydrodynamics of vortex motion.

HELMHOLTZ was becoming the most famous German scientist while his
colleagues were reporting to the ministry of education that his lectures in
anatomy were incompetent. It was therefore not surprising that he accepted
the offer of a chair and a new physiology institute at Heidelberg, which was
then a highly regarded scientific center.

At Heidelberg, in 1862, HELMHOLTZ published his important and wide-
ranging *Die Lehre von der Tonempfindungen als Physiologische Grundlage fur
die Theorie der Musik*. He wrote on the physics of organ pipes and violin
strings. By 1869, he published articles on the foundations of geometry based
upon questions arising from the study of visual perception of depth and
magnitude.

By 1866 his interests had turned away from sensory physiology and centered on physics, and in 1871 he accepted the prestigious chair of physics and
directorship of a new institute at Berlin. HELMHOLTZ turned his attention to
the theory of electromagnetism recently formulated by MAXWELL and inspired
the work of his student Heinrich HERTZ, who demonstrated the existence of
the electromagnetic radiation predicted by MAXWELL's theory, which was a
critical step in the practical development of radio and other electromagnetic
telecommunications.

In 1891 and 1892 HELMHOLTZ turned his attention to color perception. He
was perhaps the first to realize that RIEMANN's theory of differential geometry
on higher dimensional spaces could be used to model psychophysical phenomena. HELMHOLTZ considered a 3-dimensional space with a RIEMANNian but
non-EUCLIDean metric whose points were identified with perceived lights (i.e.,
triples consisting of a line, a saturation, and an intensity). The RIEMANNian
distance between two points of this color space was identified with the quanti-

tative measure of their subjective difference. For lights which differed only in intensity HELMHOLTZ retrieved FECHNER's logarithmic psychophysical function as the distance measured from his model.

James Clerk MAXWELL (1831–79) died at the age of 48 of abdominal cancer. His first paper, published in the *Proceedings of the Royal Society of Edinburgh* at the age of 14, presented a construction of Cartesian ovals by a method that generalized the well known pencil and string construction of ellipses. In his later writings he continued to display his flair for combining the abstract with the concrete, the experimental with the theoretical. He made contributions to the theory of Saturn's rings, geometrical optics, photoelasticity, and the theory of servomechanisms, as well as to thermodynamics, statistical mechanics, color vision, and the theory of electromagnetism, which was the crowning achievement of nineteenth century physics. He created quantitative colorimetry and thereby immortalized the eye of his wife, who was the subject of many of his experiments. MAXWELL revitalized YOUNG's three-color theory of color perception and made fundamental contributions to the understanding of color blindness. A giant among theoretical physicists, MAXWELL was the first professor of experimental physics at Cambridge University and founded the Cavendish Laboratory.

Leo SZILARD (1898–1964) was born in Budapest and died in La Jolla, California. He has been called "one of the most profoundly original minds of this century." SZILARD contributed to statistical mechanics, nuclear physics and engineering, molecular biology, and genetics. He began his studies in electrical engineering but World War I intervened. In 1920 he went to Berlin to complete his education and was attracted by theoretical physics, which was in a particularly exciting stage of development. SZILARD received his doctorate in 1922 under Max VON LAUE at the University of Berlin for a dissertation concerned with the second law of thermodynamics. His interest in thermodynamics continued and led to his famous 1929 paper in which he "established the connection between entropy and information and foreshadowed modern cybernetic theory."

SZILARD had an inventive genius and patented a series of devices related to nuclear particle accelerators as well as an electromagnetic pump for liquid refrigerants coinvented with Albert EINSTEIN, which revealed the principles for the pumps which are now used to circulate liquid metal coolants in nuclear reactors.

SZILARD left Germany for England when HITLER rose to power in 1933. There he conceived the idea that a nuclear chain reaction might be possible and realized that nuclear fission would release large amounts of energy. In 1938 he immigrated to the United States and, with Enrico FERMI, organized the work that led to the first successful nuclear chain reaction at the University of Chicago in 1942.

It was SZILARD who arranged for the letter from EINSTEIN to President

Roosevelt that culminated in the establishment of the Manhattan Project. During the last months of World War II he attempted to convince President Truman to approve a nonlethal demonstration of the atomic bomb; his efforts were not successful.

After World War II Szilard turned his energy to research in biology. His last paper, titled "On Memory and Recall," was published posthumously.

Alan Mathison Turing (1912–54) was born, and died, in England. After taking a degree from Cambridge University he spent 1936–38 with the mathematical logician Alonzo Church at Princeton University and, while there, published an important contribution on computability in which "he analyzed the processes that can be carried out in computing a number" and arrived at a concept of a theoretical "universal" computing machine (known today as the *Turing machine*). After returning to Cambridge, his research was interrupted by World War II. From 1939 to 1948, he was employed in the communication department of the Foreign Office, where he worked on cryptologic problems. After the war, he worked on the design of an "automatic computing engine" (ACE) and later was assistant director of the Manchester automatic digital machine (MADAM).

In 1950, Turing turned his attention to the question of whether a machine can think making a valuable contribution by proposing a test (the "Turing test") for distinguishing intelligent from nonintelligent responses.

Turing died of what has been characterized as "accidental poisoning" in 1954.

John von Neumann (1903–57), born Johann in Budapest, Hungary, died in Washington, D.C., after a remarkably varied and active life which included intellectual contributions to mathematics, mathematical physics, economics, and computer science in addition to contributions to various scientific projects during World War II and public service as a member of the Atomic Energy Commission beginning in 1954. He died of cancer 3 years later.

von Neumann was Privatdozent at Berlin from 1927 to 1929 and at Hamburg in 1929–30. After 3 years at Princeton University he was appointed to the newly opened Institute for Advanced Study in Princeton in 1933 as its youngest permanent member.

His most famous work in mathematical physics was his 1932 axiomatization of quantum mechanics which included an extensive and original analysis of the problem of measurement.

von Neumann become interested in computers in connection with the need for numerical solutions to complex problems of hydrodynamics and nuclear power reactor design. The computing equipment then available was inadequate. This led him to analyze the problems of machine computation from a general and fundamental point of view, producing numerous important contributions, including the idea of the stored program computer and its realization at the Institute for Advanced Study in 1952.

In his final years VON NEUMANN was concerned with the general theory of automata, particularly focusing on the problems of designing reliable machines from unreliable components, and the construction of self-reproducing machines, both of which had previously been solved by Nature.

Ernst Heinrich WEBER (1795–1878), one of the founders of psychophysics, was the eldest of three scientist brothers whose "greatest achievements lay in applying the modern exact methods of mathematical physics to the study of the functioning of various systems of higher animals and man." Born the third of thirteen children in Wittenberg, Germany, Ernst received an M.D. degree in 1815. In 1821 he was appointed to the chair of human anatomy at Leipzig, to which was added, in 1840, the chair of physiology.

In 1826 WEBER began a series of systematic investigations of the sensory functions. He provided sensory physiology with a fresh impulse toward quantitive methods. His most memorable contribution is undoubtedly his claim that two stimuli are just noticeably different if the ratio of the difference of the stimuli to the stimulus intensity is a constant, approximately equal to 1/30. This "law" applies to most sensory modalities only over a limited range of stimulus intensities, but it was the key piece of experimental evidence that FECHNER used to formulate his psychophysical function, which expresses the measure of subjective response to a given level of stimulus intensity.

Norbert WIENER (1894–1964) was a child prodigy who found his true intellectual calling only after several false starts. He was educated by his father until, at the age of 9, he entered high school. After 4 years at Tufts University he went to graduate school at Harvard University to study zoology but was dissatisfied and studied philosophy at Cornell University instead. In 1913 he received a Ph.D. from Harvard University with a dissertation whose subject was the boundary between philosophy and mathematics. A Harvard University traveling fellowship exposed him to some of the greatest European mathematicians, but, at this stage of his development, he was not infected. Upon returning to the United States he tried various jobs teaching philosophy, engineering, and mathematics and ultimately settled, in 1919, in the Department of Mathematics at the Massachusetts Institute of Technology, where he remained until his retirement.

The mathematical problems that arose in engineering proved to be the touchstone for WIENER's most important mathematical contributions, including generalized harmonic analysis and the theory of prediction, interpolation, and smoothing of time series. The latter grew out of military applications of radar during World War II. His interest in prediction and interpolation led him to a more comprehensive interest in communication theory and its implications for biological systems as well as machines. WIENER's far-ranging ideas and hopes for an integrated theory of communication and control in "the animal and machine" were set forth in his widely read and influential book, *Cybernetics*, which appeared in 1948.

References

Abramov, I., Gordon, J., Hendrickson, A., Hainline, L., Dobson, V., and La Bossiere, E.: "The retina of the newborn human infant," *Science 217* (1982), 265–267.

Ackley, David H., Hinton, Geoffrey E., and Sejnowski, Terence J.: "A learning algorithm for Boltzmann machines," *Cognitive Science 9* (1985), 147–169.

Albrecht, D.G., DeValois, R.L., and Thorell, L.G.: "Receptive fields and the optimal stimulus," *Science 216* (1982), 204–205.

Albrecht, D.G., DeValois, R.L., and Thorell, L.G.: "Visual cortical neurons: Are bars or gratings the optimal stimuli?" *Science 207* (1980), 88–90.

Anderson, Norman H.: "Algebraic rules in psychological measurement", *American Scientist 67* (1979), 555–563.

Atlan, Henri: "On a formal definition of organization," *J. Theor. Biol. 45* (1974), 295–304.

Attneave, Fred: "Some informational aspects of visual perception," *Psych. Rev. 61* (1954), 183–193.

Autrum, I.H., editor: *Comparative Physiology and Evolution of Vision in Invertebrates. Part B: Invertebrate Visual Centers and Behavior,* Springer-Verlag, New York, 1981.

Badler, N.I. and Smoliar, S.W.: "Digital representations of human movement," *Computing Surveys 11* (1979), 19–38.

Ball, Karlene and Sekuler, Robert: "A specific and enduring improvement in visual motion discrimination," *Science 218* (1982), 697–698.

Banks, Edwin R.: "Information processing and transmission in cellular automata," Technical Report *MAC TR-81,* Massachusetts Institute of Technology, Project MAC (January 1971).

Barrow, Harry G. and Tenenbaum, J. Martin: "Recovering intrinsic scene characteristics from images," Technical Note 157, SRI International, April 1978, 25 pp.

Beadle, G.: "The new biology and the nature of man," Dewey F. Fagerburg memorial lecture, University of Michigan, 1963: *Phoenix, 2 (1),* (1963), 3–4.

Bekenstein, J.D.: "Black holes and entropy." *Physical Review D 7,* 2333–2346 (1973).

Bell, Curtis C.: "An efferency copy which is modified by reafferent input," *Science 214* (1981), 450–452.

Bellugi, Ursula, and Klima, Edward S.: "Language: perspectives from another modality," *Brain and Mind*, Ciba Foundation Series 69 (new series), Excerpta Medica, October 1969.

Bellugi, Ursula and Studdert-Kennedy, M.: "Signed and spoken language: biological constraints on linguistic form," *Life Sciences Research Report 19*, Dahlem Konferenzen 1980, Verlag Chemie GMBH, 3–12.

Benioff, Paul: "Quantum mechanical hamiltonian models of Turing machines that dissipate no energy," *Int. J. Theoretical Physics 21* (1982), 117–202.

Bennett, C.H.: "Logical reversibility of computation," *IBM Journal of Research and Development 6* (1973), 525–532.

Bennett, Charles H.: "The thermodynamics of computation—a review," *Int. J. Theoretical Physics 21* (1982), 905–940.

Berliner, A. and Berliner, S.: "The distortion of straight and curved lines in geometric fields." *American Journal of Psychology 61*, 153–166 (1948).

Blaivas, A.S.: "A psychological aspect of expansion of a group theory vision model into color vision information processing," *Brain Theory Newsletter 3* (1978), 94–96.

Blaivas, A.S.: "Visual analysis: Theory of Lie group representations," *Mathematical Biosciences 28* (1975), 45–67.

Block, Ned (Ed.): *Imagery*, The MIT Press, Cambridge, 1981.

Boltzmann, L.: "Studien über das Gleichgewicht der Lebendigen Kraft zwischen bewegten materiellen Punkten," *Wien. Ber. 58*, 517: *Abhandlungen*, Volume 1, Barth, Leipzig, 1909, p. 49.

Bossomaier, T.R.J., Snyder, A.W., and Hughes, A.: "Irregularity and aliasing: solution?" *Vision Res. 25* (1985), 145–147.

Brady, Michael: "Computational approaches to image understanding", *Computing Surveys 14* (1982), 3–71.

Branscomb, Lewis M.: "Information: the ultimate frontier," *Science 201* (1979), 133–117.

Bremermann, H.J.: "Optimization through evolution and recombination," in M.C. Yovits, G.T. Jacobi, and G.D. Goldstein *editors, Self-Organizing Systems*, Spartan Books, Washington, D.C., 1962, 93–106.

Bremermann, H.J.: "Quantum noise and information," in *Proc. 5th Berkeley Symp. Mathematical Statistics and Probability*, vol. 4, University of California Press, Berkeley and Los Angeles, 1967, pp. 15–20.

Bremermann, H.J.: "Minimum energy requirements of information transfer and computing," *Inter. J. Theoretical Physics 21* (1982).

Brentano, F.: "Über ein optisches Paradoxen." *Journal of Psychology 3*, 249–258 (1982).

Brillouin, Leon: *Science and Information Theory*, Academic Press, New York, 1956, 2d edition 1962.

Brillouin, Leon: *La vie, la materie et la théorie de l'information*, Albin-Michel, Paris, 1959.

Brillouin, Leon: *Scientific Uncertainty and Information*, Academic Press, New York, 1964.

Broomell, George and Heath, J. Robert: "Classification categories and historical development of circuit switching topologies," *ACM Computing Surveys 15* (1983), 95–133.

Brown, Roger and Herrnstein, Richard J.: "Icons and images," reprinted an *Imagery*, Ned Block, *editor*. Cambridge, MA: The MIT Press, 1981.

Bruns, Volkmar: "Functional anatomy as an approach to frequency analysis in the mammalian cochlea," *Verh. Stsch. Zool. Ges.* (1979), 141–154.

Bruns, Volkman and Godbach, Margit: "Hair cells and tectorial membrane in the cochlea of the greater horseshoe bat," *Anatomy and Embryology 161* (1980), 51–63.

Bruns, Volkman and Schmieszek, Edeltraut: "Cochlear innervation in the greater horseshoe bat: demonstration of an acoustic fovea," *Hearing Research 3* (1980), 27–43.

Bullock, Theodore H.: "Recognition of complex acoustic signals," *Life Sciences Research Report 5*, Dahlem Konferenzen, (1977), 111–126.

Burgess, A.E., Wagner, R.F., Jennings, R.J., and Barlow, H.B.: "Efficiency of human visual signal discrimination", *Science 214* (1981), 93–94.

Burns, B.D. and Pritchard, R.: "Geometrical illusions and the response of neurones in the cat's visual cortex to angle patterns." *Journal of Physiology 213*, 599–616 (1971).

Caelli, T. and Julesz, B.: "On perceptual analyzers underlying visual texture discrimination: Part I," *Biol. Cybernetics 28* (1978), 167–175.

Caelli, T., and Julesz, B.: "On perceptual analyzers underlying visual texture discrimination: Part II," *Biol. Cybernetics 29* (1978), 201–214.

Caelli, T. and Julesz, B.: "Psychophysical evidence for global feature processing in visual texture discrimination," *J. Optical Soc. Amer. 69* (1979), 675–678.

Campbell, F.W. and Robson, J.G.: "Application of Fourier analysis to the visibility of gratings." *Journal of Physiology 197*, 551–566 (1968).

Carnap, Rudolph: *Two Essays on Entropy*, University of California Press, Berkeley, 1977.

Carnot, Sadi: *Reflexions surs las puissance du feu et sur les machines propres a developer cette puissance*, Bachelier, Paris, 1824.

Carpenter, Gail A. and Grossberg, Stephen: "Neural dynamics of category learning and recognition: Attention, memory consolidation, and amnesia", to appear in *Brain Structure, Learning, and Memory*, J. Davis, R. Newburgh, and E. Wegman (Eds.), AAAS Symposium Series, 1985.

Chang, J.J. and Carroll, J.D.: "Three are NOT enough: an INDSCAL analysis suggesting that color space has seven (± 1) dimensions," *Color Research and Application 5* (1980), 193–206.

Chernoff, Herman: "The use of faces to represent points in k-dimensional space graphically," *American Statistical Association 68* (1973), 361–368.

Chernoff, Herman and Haseeb Rizvi, M.: "Effect of classification error of random permutations of features in representing multivariate data by faces," *American Statistical Association 70* (1975), 548–554.

Cherry, E.C.: "A history of the theory of information," *Proc. IEEE 98* (1951), 383–393.

Christie, L.S. and Luce, R.D.: "Decision structured time relations in simple choice behavior," *Bull. Math. Biophysics 18* (1956), 89–112.

Church, Alonzo: "An unsolvable problem in elementary number theory". *American Journal of Mathematics 58* (1936), 345–363.

Cleveland, W.S., Diaconis, P., and McGill, R.: "Variables on scatterplots look more highly correlated when the scales are increased," *Science 216* (1982), 1138–1141.

Cohen, Yoav, Landy, Michael S., and Pavel, Misha: "Hierarchical coding of binary images," *IEEE Transactions on Pattern Analysis and Machine Intelligence PAMI-7* (1985), 284–298.

Colby, Kenneth Mark: "Mind models: An overview of current work," *Math. Biosciences 39* (1978), 159–185.

Cooper, D.B., Elliott, H., Cohen, F., Reiss, L., and Symosek, P.: "Stochastic boundary estimation and object recognition," *Computer Graphics and Image Processing 12* (1980), 326–356.

Cooper, D.B., Sung, F., and Schenker, P.S.: "Toward a theory of multiple-window algorithms for fast adaptive boundary findings in computer vision," Brown University Technical Report #ENG PRMI 80-3, April 1980, revised July 1980, 54 pp.

Coren, Stanley and Girgus, Joan Stern: *Seeing Its Deceiving: The Psychology of Visual Illusions*, Lawrence Earlbaum Associates, Hillsdale, N.J., 1978.

Cornsweet, Tom N.: "Changes in the appearance of stimuli of very high luminance," *Psych. Rev. 69* (1962), 257–273.

Cornsweet, T.N.: *Visual Perception*. New York: Academic Press, 1970.

Courtines, M.: "Les fluctuations dans les appareils de measures," *Congr. Intern. d'Electricite*, Volume 2, Paris (1932).

Cover, Thomas M. and King, Roger C.: "A convergent gambling estimate of the entropy of English," *IEEE Trans. Information Theory 24* (1978), 413–421.

Crane, H.D. and Clark, Michael R.: "Three-dimensional visual stimulus deflector," *Applied Optics 17* (1978), 706–714.

Crane, H.D. and Steele, C.M.: "Accurate three-dimensional eyetracker," *Applied Optics 17* (1978), 691–705.

Crane, Hewitt D. and Piantanida, Thomas P.: "On seeing reddish green and yellowish blue," *Science 221* (1983), 1078–1079.

Crick, F.H.C., Marr, D.C., and Poggio, T.: "An information-processing approach to understanding the visual cortex," in *The Organization of the Cerebral Cortex*, E.O. Schmitt, F.G. Worden, and G.S. Dennis, *editors.*, MIT Press, Cambridge, 1981, pp. 505–533.

Cross, George R. and Jain, Anil K.: "Markov random field texture models," *IEEE Trans. on Pattern Analysis and Machine Intelligence PAMI-5* (1983), 25–39.

Daugman, John G.: "Spatial visual channels in the Fourier plane," *Vision Res. 24* (1984), 891–910.

Daugman, John G.: "Uncertainty relation for resolution in space, spatial frequency, and orientation optimized by by two-dimensional visual cortical filters," *J. Optical Soc. Amer. A 2* (1985), 1160–1169.

Davies, P.C.W.: *The Physics of Time Asymmetry*, University of California Press, 1974.

Davies, P.C.W.: "Inflation and time asymmetry in the Universe." *Nature 301*, 398–400 (1983).

de Monasterio, F.M., Schein, S.J., and McCrane, E.P.: "Staining of blue-sensitive cones of the macaque retina by a fluorescent dye," *Science 213* (1981), 1278–1280.

Descartes, René: *Metéores*. Paris, 1638.

DeSolla Price, Derek: *Gears from the Greeks*, Science History Publications, New York, 1975.

Deutsch, David: "Is there a fundamental bound on the rate at which information can be processed?" *Physical Review Letters 48* (1982), 286–288.

Dolby, J.L.: "On the notions of ambiguity and information loss," *Information Processing and Management* (1977), 69–77.

Dolby, J.L. and Resnikoff, H.L.: "On the multiplicative structure of information storage and access systems," *Bull. Inst. of Management Sciences 1* (1971), 23–30.

Dyson, Freeman: *Disturbing the Universe*. New York: Harper & Row, 1979.

Eddington, Arthur Stanley: *The Nature of the Physical World*. Cambridge: Cambridge University Press, 1928.

Einthoven, W.: "Eine einfache physiologische Erklärung fur verschiedene geometrisch-optische Täuschungen." *Pflugers Archiv für Physiologie 71* (1898), 1–43.

Englestatter, R., Vater, M., and Neuweiler, G.: "Processing of noise by single units of the inferior colliculus of the bat *Rhinolophus ferrumequinum*," *Hearing Research 3* (1980), 285–300.

Euler, Leonhard: "Essai d'une explication physique des colours engendrees sur des surfaces extrement nunces." *Mém. de l'academie des sci. de Berlin 8*, 262–282 (1752) = *Opera omnia ser. (3), 5*, No. 209, 156–171.

Eyring, Henry and Urry, Dan W.: "Thermodynamics and chemical kinetics," in *Theoretical and Mathematical Biology*, T.H. Waterman and H.J. Morowitz, *editors*, Blaisdell Publ. Co., New York, 1965.

Fechner, G.T.: "Über ein wichtiges psychophysiches Grundgesetz und dessen Beziehung zur Schäzung der Sterngrössen," *Abh. k. Ges. Wissensch., Math.-Phys. K1.*

4 (1858).

Feldman, J.A. and Ballard, D.H.: "Connectionist models and their properties," *Cognitive Science 6* (1982), 205–254.

Feynman, Richard P.: "Simulating physics with computers," *Inter. J. Theoretical Physics 21* (1982), 467–488.

Fick, A.: *De errone quodam optic asymmetria bulbi effecto*. Marburg: Koch, 1851.

Fisher, R.A.: "On the mathematical foundations of theoretical statistics," *Phil. Trans. Roy. Soc. London*, sec A, *222* (1922).

Flanagan, J.L.: "Computers that talk and listen: man-machine communication by voice," *Proc. IEEE 64* (1976), 405–415.

Fourier, J.: "Mémoire sur la Théorie Analytique de la Chaleur." *Mémoires de l'Academie des Sciences, Tome VIII*, 581–622. Paris, 1829.

Fraser, J. "A new illusion of direction." *British Journal of Psychology 8* (1908), 49–54.

Frautschi, Steven: "Entropy in an expanding universe," *Science 217* (1982), 593–599.

Fredkin, Edward and Toffoli, Tommaso: "Conservative logic," *Inter. J. Theoretical Physics 21* (1982), 219–253.

Frome, F.S., Piantanida, T.P., and Kelly, D.H.: "Psychophysical evidence for more than two kinds of cone in dichromatic color blindness," *Science 215* (1982), 417–418.

Gabor, D.: "Theory of communication," *J. Inst. Elec. Engrs.*, *93* Part III, (1946), 429–457; *94*, III (1947), 369.

Gerrits, H.J.M.: "Apparent movements induced by stroboscopic illumination of stabilized images," *Exp. Brain Res. 34* (1979), 471–488.

Gibbs, W.: *Elementary Principles of Statistical Mechanics*, Yale University Press, New Haven, 1902.

Gödel, Kurt: "Über formal unentscheidbare Sätze der Principia Mathematica und verwandte Systeme I." *Monatshefte für Mathematik und Physik 38* (1931), 173–198.

Gödel, Kurt: On Undecidable Propositions of Formal Mathematical Systems. Mimeographed lectures notes. Princeton, N.J.: Institute for Advanced Study, 1934.

Goldberg, S.H., Frumkes, T.E., and Nygaard, R.W.: "Inhibitory influence of unstimulated rods in the human retina: evidence provided by examining conde flicker," *Science 221* (1983), 180–181.

Goldstine, Herman H.: *The computer from Pascal to von Neumann*, Princeton University Press, Princeton, 1972.

Gombrich, E.H.: *Art and Illusion*, The A.W. Mellon Lectures in the Fine Arts, 1956. Bollingen Series XXX05, Princeton University, Princeton, N.J., 1960. [I thank Professor Michael Godfrey for bringing this reference to my attention.]

Good, I.J.: "Distribution of word frequencies," *Nature 179* (1957), 595.

Grassmann, Hermann Günther: "Zur Theorie der Farbenmischung," *Ann. Physik und Chemie 89* (1853), 69–84.

Grassmann, Hermann Günther: "Bemerkungen zur Theorie der Farbenempfindungen," supplement to W. Preyer, *Elementen der reinen Empfindungslehre*, Hermann Dufft Verlag, Jena, 1877.

Green, D. and Swets, A.: *Signal Detection Theory and Psychophysics*, John Wiley and Sons, Inc, New York, 1966.

Gregory, R.L.: *Eye and Brain: The Psychology of Seeing*, World University Library, Weidenfeld and Nicholson, London, 1966.

Gregory, R.L.: *The Intelligent Eye*. London: Weidenfeld & Nicholson, 1970.

Gregory, R.L.: *Concepts and Mechanisms of Perception*, Charles Scribner's Sons, New York, 1974.

Grenander, U.: *Lectures in Pattern Theory Volume I: Pattern Synthesis*, Springer-Verlag, New York, 1976.

Grenander, U.: *Lectures in Pattern Theory Volume II: Pattern Analysis*, Springer-Verlag, New York, 1978.

Grenander, U.: *Lectures in Pattern Theory Volume III: Regular Structures*, Springer-Verlag, New York, 1981.

Grimson, W.E.L.: "A computer implementation of a theory of human stereo vision," Massachusetts Institute of Technology, Artificial Intelligence Memo 565, January 1980, 59 p.

Grimson, W.E.L.: "Differential geometry, surface patches and convergence methods," Massachusetts Institute of Technology, Artificial Intelligence Memo 510 (1979), 37 pp.

Grimson, W.E.L. and Hildreth, E.C.: "Comments on 'Digital step edges from zero crossings of second directional derivatives,'" *IEEE Trans. Pattern Analysis and Machine Intelligence PAMI-7* (1985), 121–126.

Grossberg, Stephen: *Studies of Mind and Brain*, D. Reidel, Dordrecht, 1982.

Grossberg, Stephen and Mingolla, Ennio: "Neural dynamics of perceptual grouping: textures, boundaries, and emergent segmentations," *Perception and Psychophysics 38* (1985), 141–171.

Haber, Ralph Norman (Editor): *Contemporary Theory and Research in Visual Perception*, Holt, Rinehart and Winston, Inc., New York, 1968.

Haber, Ralph Norman: "Visual perception," *Ann. Rev. Psychol.* 59 (1978), 31–59.

Haber, Ralph Norman and Wilkinson, Leland: "Perceptual components of computer displays," *IEEE Computer Graphics and Applications*, May 1982, 23–25.

Haken, Hermann: *Synergetics*, 3d edition, Springer-Verlag, Berlin, 1983.

Han, Te Sun: "Projective-geometrical and information-theoretical approach to the scaling problem of the auditory sensation," *Memoirs of Sagami Institute of Technology 12* (1978), 49–66.

Handler, Phillip, (Ed.): *Biology and the Future of Man*, Oxford University Press, 1970.

Haralick, Robert M.: "Statistical and structural approaches to texture," *Proc. IEEE 67* (1979), 787–804.

Harmuth, Henning F.: *From the Flat Earth to the Topology of Space-Time*, lecture notes, no date.

Harter, Stephen P.: "Optimum number of cards per file guide assuming binary searching," *J. Amer. Soc. Information Sci.*, (March–April 1971), 140.

Hartley, R.V.L.: "Transmission of information," *Bell System Tech. J.* 7 (1928), 535–563.

Hartline, H.K. and Ratliff, F.: "Inhibitory interaction of receptor units in the eye of *Limulus*," *J. Gen. Physiol. 40* (1957), 357–376.

Hartline, H.K., Ratliff, F., and Miller W.H.: in *Nervous Inhibition*, E. Florey, *ed.*, Pergamon, New York, 1961.

Hartline, H.K., Wagner, H.G., and Ratliff, F.: "Inhibition in the eye of Limulus." *Journal of General Physiology 39*, 651–673 (1956).

Hawking, S.W.: "Black holes and thermodynamics." *Physical Review D 13*, 191–197 (1976).

Hecht, S.: "The visual discrimination of intensity and the Weber-Fechner law," *J. Gen Physiol. 7* (1924), 235–267.

Hecht, S.: "A theory of visual intensity discrimination," *J. Gen. Physiol. 18* (1935), 767–789.

Heisenberg, W.: "Über den anschaulichen Inhalt der quantentheoretischen Kinematik und Mechanik." *Zeitschrift für Physik 43* (1927), 172–198.

Helmholtz, Hermann von: "Über die Theorie der zusammengesetzten Farben." *Annalen der Physik und Chemie 87* (1852), 45–66.

Helmholtz, Hermann von: *Handbuch der Physiologischen Optik*. Leipzig: Voss, Part I (1856), Part II (1860), Part III (1866). (Translated by G. Southall, 1925 and republished, New York: Dover, 1962.)

Helmholtz, Hermann von: "Versuch einer erweiterten Anwendung des Fechnerschen Gesetzes im Fabensystem," *Z. Psychol. Physiol. Sinnesorg. 2* (1891), 1–30.

Helmholtz, Hermann von: "Versuch, das psychologische Gesetz auf die Farbenuntershiede trichromatischer Augen anzuwenden," *Z. Psychol. Physiol. Sinnesorg. 3* (1892), 1–20.

Helmholtz, Hermann von: "Kurzeste linien im Farbensystem," *Z. Psychol. Physiol. Sinnesorg. 4* (1892), 108–122.

Herrnstein, R.J.: "Acquisition, generalization, and discrimination reversal of a natural concept," *Journal Experimental Psychology: Animal Behavior Processes 5* (1979), 166–129.

Herrnstein, R.J.: "Stimuli and the texture of experience." *Neuroscience and Biobehavioral Review 6*, 105–117 (1982).

Heymans, G.: "Quantitative Untersuchungen über das optische Paradoxen." *Zeitschrift für Psychologie 9*, 221–255 (1896).

Hildreth, Ellen C.: "The detection of intensity changes by computer and biological vision systems," *Computer Vision, Graphics, and Image Processing 22* (1983), 1–27.

Hildreth, Ellen C.: "Edge detection," Artificial Intelligence Laboratory Memo No. 858, M.I.T., September, 1985, 21 pp.

Hinton, G.E. and Sejnowski, T.J.: "Optimal perceptual inference," *Proc. IEEE Computer Soc. Conference on Computer Vision and Pattern Recognition*, Washington, D.C., June 1983, pp. 448–453.

Hirsch, Joy and Hyton, Ron: "Orientation dependence of visual hyperacuity contains a component with hexagonal symmetry," *J. Opt. Soc. Amer. A 1* (1984), 300–308.

Hirsch, J. and Hylton, R.: "Quality of the primate photoreceptor lattice and limits of spatial vision," *Vision Res. 24* (1984), 347–356.

Hooke, Robert: *Micrographia.* Reprint of edition of 1665. New York: Dover Pub., 1961.

Hopfield, J.J.: "Neural networks and physical systems with emergent collective computational abilities," *Proc. Nat. Acad. Sci. USA. 79* (1982), 2554–2558.

Hopfield, J.J.: "Collective processing and neural states," in *Modeling and Analysis in Biomedicine*, C. Nicolini, *editor*, World Scientific Publishers, New York, 1984.

Hopfield, J.J., Feinstein, D.I., and Palmer, R.G.: "'Unlearning' has a stabilizing effect in collective memories," preprint, n.d.; 10 pp.

Hopfield, J.J. and Tank, D.W.: "'Neural' computation of decisions in optimization problems," *Biol. Cybern. 52* (1985), 141.

Horton, C.W., Sr.: *Signal Processing of Underwater Acoustic Waves*, United States Government Printing Office, Washington, D.C., 1969.

Hubel, D.H. and Wiesel, T.N.: "Receptive fields of single neurons in the cat's striate cortex," *J. Physiol. 148* (1959), 574–591.

Hubel, D. and Wiesel, T.N.: *J. Physiol.* (London) *160* (1962), 106.

Hubel, D.H. and Wiesel, T.N.: "Receptive fields and functional architecture of monkey striated cortex", *J. Physiol. 195* (1968), 215–243.

Indow, Tarow: "Alleys in visual space," *J. Math. Psych. 19* (1979), 221–258.

Indow, Tarow: "An approach to geometry of visual space with no *a priori* mapping functions: multidimensional mapping according to Riemannian Metrics," Social Sciences Research Reports 25, School of Social Sciences, University of California, Irvine, (1979), 56 pp.

Ingarden, R.S.: "Information theory and thermodynamics of light. Part I. Foundations of information theory," *Fortschritte der Physik 12* (1964), 567–594; "Part II. Principles of information thermodynamics," *ibid. 13* (1965), 755–805.

Irwin, Eleanor: *Color Terms in Greek Poetry*, A.M. Hakkert, Ltd., Toronto, 1974.

Jacobson, H.: "Information and the human ear," *J. Acoustical Soc. America 23* (1951), 464–471.

Jain, Ramesh, and Aggarwal, J.K.: "Computer analysis of scenes with curved objects," *Proc. IEEE 67* (1979), 805–812.

Jastrow, J.: "A study of Zoellner's figures and other related illusions." *American Journal of Psychology 4*, 381–398 (1891).

Jastrow, J.: "On the judgement of angles and positions of lines." *American Journal of*

Psychology 5, 214–221 (1892).

Jaynes, E.T.: "Information theory and statistical mechanics," *Physical Rev. 106* (1957).

Jaynes, E.T.: "Information theory and statistical mechanics, II," *Physical Rev. 108* (1957), 171–190.

Jonides, J., Irwin, D.E., and Yantis, S.: "Integrating visual information from successive fixations," *Science 215* (1982), 192–194.

Julesz, B.: *Foundations of Cyclopean Perception.* Chicago: University of Chicago Press, 1971.

Julesz, Bela: "Experiments in the visual perception of texture," *Scientific American,* April 1975, 34–43.

Julesz, Bela: "Spatial nonlinearities in the instantaneous perception of textures with identical power spectra," *Phil. Trans. R. Soc. Lond. B. 290* (1980), 83–94.

Julesz, Bela: "Textons, the elements of texture perception, and their interactions," *Nature 290* (12 March 1981), 91–97.

Julesz, Bela: "A theory of preattentive texture discrimination based on first-order statistics of textons," *Biol. Cybern. 41* (1981), 131–138.

Julesz, Bela and Caelli, Terry: "On the limits of Fourier decompositions in visual texture perception," *Perception 8,* (1979), 69–73.

Julesz, B. Frisch, H.L., Gilbert, E.N. and Shepp, L.A.: "Inability of humans to discriminate between visual textures that agree in second-order statistics—revisited," *Perception 2* (1973), 391–405.

Julesz, B., Gilbert, E.N., and Victor, J.D.: "Visual discrimination of textures with identical third-order statistics," *Biol. Cybern. 31* (1978), 137–140.

Julesz, Bela and Schumer, Robert A.: "Early visual perception," *Ann. Rev. Psychol. 1981,* 32: 575–627.

Julesz, B. and Bergen, J. R.: "Textons, the fundamental elements in preattentive vision and the perception of texture," *Bell System Tech. J. 62* (1983), 1619–1645.

Kanal, Laveen: "Patterns in pattern recognition: 1968–1974," *IEEE Transactions on Information Theory 20* (1974), 697–722.

Kanal, Laveen: "Interactive pattern analysis and classification systems: a survey and commentary," *Proc. IEEE 60* (1972), 1200–1215.

Kanisza, G.: "Marzini quasi-perceptivi in campi con stimolazione omegenea." *Revista di Psicologia 49* (1955), 7–30.

Karatsuba, A.A.: *Doklady Akad. Nauk SSSR 145*, 293–294 (1962).

Keyes, Robert W.: "Physical limits in digital electronics," *Proc. IEEE 63* (1975), 740–767.

Kelly, D.H.: "Disappearance of stabilized chromatic gratings," *Science 214* (1981), 1257–1258.

Khinchin, A.I.: *Mathematical Foundations of Statistical Mechanics* (trans. from the Russian), Dover Publ., New York, 1949.

Klima, Edward S. and Bellugi, Ursula: *The Signs of Language,* Harvard University Press, Cambridge, 1979.

Knowlton, K.: "Progressive transmission of grey-scale and binary pictures by simple, efficient, and lossless encoding schemes," *Proc. IEEE 68* (1980), 885–886.

Knuth, Donald E.: *The Art of Computer Programming. Volume 2/Seminumerical Algorithms.* Reading, MA: Addison-Wesley, 1981.

Kolmogorov, A.: "Interpolation and Extrapolation von stationären zufäligen Folgen," *Bull. de l'acad. Sci. U.R.S.S. 30* (1941), 13–17.

Krantz, D.H., Luce, R.D., Suppes, P., and Tversky, A.: *Foundations of Measurement,* Academic Press, New York, 1971.

Kuffler, Stephen, W., Nicholls, John G., and Martin, A. Robert: *From Neuron to Brain,* second edition, Sinauer Associates, Inc., Sunderland Mass., 1984.

Landauer, Rolf W.: "Irreversibility and heat generation in the computing process," *IBM J. Res. Dev. 5* (1961), 183–191.

Landauer, Rolf W.: "Fundamental physical limitations of the computational process," *Ann. New York Acad. Sci. 426* (1985), 161–170.

Landauer, Rolf W. and Woo, J.W.F.: "Minimal energy dissipation and maximal error for the computational process," *J. Applied Physics 42* (1971), 2301–2308.

Lappin, Joseph S. and Fugua, Mark A.: "Accurate visual measurement of three-dimensional moving patterns," *Science 221* (1983), 480–482.

Laska, W.: Über einige optische Urtheiltäuschungen." *Archiv für Anatomie und Physiologie 14* (1890), 326–328.

Layzer, David: "A unified approach to cosmology," in *Relativity Theory and Astrophysics, Lectures in Applied Mathematics, Volume 8*, Ehlers, Jürgen, *editor*, American Mathematical Society, Providence, 1967.

Leiserson, Charles E.: *Area-Efficient VLSI Computation*, The MIT Press, Cambridge, Mass., 1983.

Lettvin, Jerome: "Experiments in perception," *Tech. Engineering News* (November 1964), 30–34.

Lettvin, Jerome: "Filling out the forms," *Science 214* (1981), 518–520.

Lettvin, Jerome: "On seeing sidelong," *The Sciences* (July/August 1976), 10–20.

Lettvin, J.Y., Maturana, H.R., McCulloch, W.S., and Pitts, W.H.: "What the frog's eye tells the frog's brain," *Proc. IRE 47* (1959), 1940–1959.

Lewin, Roger: "How did humans evolve big brains?" *Science 216* (1982), 840–841.

Leyton, Michael: "Perceptual organization as nested control," *Biol. Cybern. 51* (1984), 141–153.

Leyton, Michael: "A theory of information structure: I. General principles," *preprint*, 1985.

Leyton, Michael: "A theory of information structure: II. A theory of perceptual organization," *preprint*, 1985.

Linden, L.L. and Lettvin, J.T.: "Colors that come to mind," preprint, 7 October 1982, 36 pp.

Lipetz, Ben-Ami and Song, Czetong T.: "How many cards per file guide? Optimizing the two-level file," *J. Amer. Soc. Information Sci.* (March–April 1970), 140–141.

Lipps, T.: *Raumästhetik und geometrisch-optische Täuschungen*. Leipzig: Barth, 1897.

Luce, R. Duncan, *et al.*: "Behavioral and linguistic research bearing on information science," National Science Foundation, Division of Information Science and Technology, Washington, D.C., 1980.

Maffei, L. and Fiorentini, A.: "Electroretinographic responses to alternating gratings before and after section of the optic nerve," *Science 211* (1981), 953–955.

Mandelbrot, B.: "On the theory of word frequencies and on related Markovian models of discourse," *Structures of Language and its Mathematical Aspects, Proc. Symp. Applied Math. 12* (1961), 120–219.

Mandelbrot, B.: *The Fractal Geometry of Nature*, W.H. Freeman and Co., San Francisco, 1982.

Marr, D.: "Early processing of visual information," *Phil. Trans. Royal Soc. B 275* (1976), 483–524.

Marr, David: *Vision*, W.H. Freeman, San Francisco, 1982.

Marr, D.: "Artificial intelligence—a personal view," *Artificial Intelligence 9* (1977), 37–48.

Marr, D.: "Visual information processing: the structure and creation of visual representations," *Phil. Trans. R. Soc. Lond. B 290* (1980), 199–218.

Marr, D. and Hildreth, E.: "Theory of edge detection," *Proc. R. Soc. Lond. B 207* (1980), 187–217.

Marr, D. and Nishihara, H.K.: "Visual information processing: artificial intelligence and the sensorium of sight," *Technology Review 81* (1978), 23–49.

Marr, D., Poggio, T., and Hildreth, E.: "Smallest channel in early human vision," *Optical Soc. Amer. 70* (July 1980), 868–870.

Marr, D. and Poggio, T.: "A theory of human stereo vision," MIT Artificial Intelligence Laboratory Memo 451 (1977).

Marr, D. and Poggio, T.: "Some comments on a recent theory of stereopsis," MIT Artificial Intelligence Laboratory Memo 558, July 1980, 8 pp.

Marr, D., Poggio, T., and Ullman, S.: "Bandpass channels, zero-crossings, and early visual information processing, *J. Optical Soc. Amer. 69* (1979), 914–916.

Matsubara, Joanne A.: "Neural correlates of a nonjammable electrolocation system," *Science 211* (1981), 722–725.

Maxwell, James Clerk: "Theory of perception of colors." *Trans. Royal Scottish Soc. Arts 4* (1856), 394–400.

Maxwell, James Clerk: "Theory of compound colors, and the relations of the colors of the spectrum." *Proc. Royal Soc. London 10* (1860), 404–409.

Maxwell, James Clerk: "On color vision." *Proc. Royal Inst. Great Britain 6* (1872), 260–271.

Maxwell, James Clerk: *Theory of Heat.* London: Longmans, Green & Co., 1908.

Mazur, James E.: "Optimization: A result or a mechanism?" *Science 221* (1983), 976–977.

McCrane, F.M., de Monasterio, F.M., Schein, S.J., and Caruso, R.C.: "Non-fluorescent dye staining of primate blue cones," *Investigative Ophthalmology and Visual Science 24* (1983), 1449–1455.

McCulloch, Warren S.: "What is a number, that a man may know it, and a man, that he may know a number?" (the Ninth Alfred Korzybski Memorial Lecture), *General Semantics Bulletin, Nos. 26 and 27* (1961), 7–18, Lakeville, Conn.: Institute of General Semantics.)

Mead, Carver and Conway, Lynn: *Introduction to VLSI Systems,* Addison-Wesley Publ. Co., Reading, Mass., 1980.

Melzak, Z.A.: *Mathematical Ideas, Modeling and Applications,* John Wiley and Sons, New York, 1976.

Meredith, M. Alex and Stein, Barry E.: "Interactions among converging sensory imputs in the superior colliculus," *Science 221* (1983), 389–391.

Mesereau, Russell M.: "The processing of hexagonally sampled two-dimensional signals," *Proc. IEEE 67* (1979), 930–949.

Miller, George A.: "The magical number seven, plus or minus two: some limits on our capacity for processing information," *Psych. Rev. 63* (1956), 81–96.

Miller, George A. and Johnson-Laird, Philip N.: *Language and Perception,* Belknap Press, Harvard University, Cambridge, Mass., 1976.

Miller, James Grier: *Living Systems,* McGraw-Hill, New York, 1978.

Minsky, Marvin: "Steps toward artificial intelligence," *Proc. IRE 49* (1961), 3–30.

Mower, George D.; Christen, William G.; and Caplan, Caren J.: "Very brief visual experience eliminates plasticity in the cat visual cortex," *Science 221* (1983), 178–180.

Muller-Lyer, F.C.: "Optische Urteilstäuschungen." *Dubois-Reymonds Archive für Anatomie und Physiologie,* Supplement volume, 263–270 (1889).

Mumford, David and Shah, Jayant: "Boundary detection by minimizing functionals, I," *IEEE Proc. Computer Vision and Pattern Recognition,* 19–23 June 1985, San Francisco, CA., pp. 22–26.

Neuweiler, Gerhard: "Auditory processing of echoes: peripheral processing," *Animal Sonar Systems, editors,* Rene-Guy Busnel and James F. Fish, Plenum Publishing Corp., 1980.

Neuweiler, Gerhard: "How bats detect flying insects," *Physics Today,* August 1980, 34–40.

Neuweiler, Gerhard: "Recognition mechanisms in echolocation of bats," Dahlem Workshop on Recognition of Complex Acoustic Signals, Berlin 1976, *Life Sciences Research Report 5,* Theodore H. Bullock, ed., 111–126.

Neuweiler, G., Bruns, V. and Schuller, G.: "Ears adapted for the detection of motion, or how echolocating bats have exploited the capacities of the mammalian auditory system," *Acoustical Soc. Amer. 68* (1980), 741–753.

Neuweiler, G. and Vater, M.: "Response patterns to pure tones of cochlear nucleus units in the CF-FM bat, *Rhinolophus ferrumequinum*," *Comparative Physiology 115* (1977), 119–133.

Newall, A., J.C. Shaw and H.A. Simon: "A variety of intelligent learning in a General Problem Solver." In *Self-Organizing Systems*, M.C. Yovits and S. Cameron (*Eds.*). New York: Pergamon Press, 1960.

Newall, A. and H.A. Simon: "GPS, a program that simulates human thought." In *Computers and Thought*, E.A. Feigenbaum and J. Feldman (*Eds.*). New York: McGraw-Hill, 1963.

Newman, Eric A. and Hartline, Peter H.: "Integration of visual and infrared information in bimodal neurons of the rattlesnake optic tectum," *Science 213* (1981), 789–791.

Newman, Thomas G.: "A group theoretic approach to invariance in pattern recognition," *Pattern Recognition and Image Processing 1979, IEEE*, 407–412.

Newman, Thomas G. and Demus, David A.: "Lie theoretic methods in video tracking", preprint, n.d., 8 pp.

Newton, Isaac: "New theory about light and colors." *Phil. Trans. Royal Soc. London 80* (1671), 3075–3087.

Newton, Isaac: *Opticks*. London: Smith and Walford, 1704.

Nilsson, Nils J.: *Principles of Artificial Intelligence*, Tioga Publ. Co., Palo Alto, 1980.

Oppel, J.J.: "Über geometrisch-optische Täuschungen." *Jahresbericht des Frankfurter Vereins, 37–47* (1854–1855).

Osterberg, G.: "Topography of the layer of rods and cones in the human retina," *Acta Ophthal., Suppl. 6* (1935), 1–103.

Pasztor, V.M. and Bush, B.M.H.: "Impulse-coded and analog signaling in single mechanoreceptor neurons," *Science 215* (1982), 1635–1637.

Pattee, Howard H., (*Ed.*): *Hierarchy Theory*, George Braziller, New York, 1973.

Peierls, Rudolf: *Surprises in Theoretical Physics*, Princeton University Press, 1979.

Pentland, A.P.: "Fractal-based description of natural scenes," *IEEE Trans. Pattern Analysis and Machine Intelligence PAMI-6* (1984), 661–674.

Pepperberg, D.R., P.K. Brown, M. Lurie and J.E. Dowling: "Visual pigment and photoreceptor sensitivity in the isolated skate retina." *Journal of General Physiology 71* (1978), 369–396.

Picard, É.: *Biographie et Manuscrit de Sadi Carnot*, French Academy of Sciences, Paris, 1927.

Pierce, John R.: *An Introduction to Information Theory*, second, revised edition, Dover Publ. Inc., New York, 1980.

Pinker, S.: "Mental imagery and the visual world," Occasional Paper 4, Center for Cognitive Science, MIT, no date.

Pirenne, M.H.: "Vision and art," *Handbook of Perception, vol. V*, Edward D. Carterette and Morton P. Friedman, *editors*, Academic Press, New York, 1975, 433–490.

Pitts, Walter and McCulloch, Warren S.: "How we know universals: the perception of auditory and visual forms," *Bull. Math. Biophysics 9* (1947), 127–147.

Planck, M.: "Über eine Verbesserung der Wien'schen Spektralgleichung." *Verh. D. Phys. Ges. 2* (1900), 202–204.

Poggio, T., Voorhees, H., and Yuille, A.L.: "A regularized solution to edge detection," MIT Artificial Intelligence Laboratory Memo No. 773 (1984).

Pollen, Daniel A. and Ronner, Steven F.: "Phase relationships between adjacent simple cells in the visual cortex," *Science 212* (1981), 1409–1411.

Polyak, S.L.: *The Retina*, University of Chicago Press, Chicago, 1941.

Prigogine, I.: *Introduction to Thermodynamics of Irreversible Processes*, Charles C

Thomas, Springfield, Ill., 1955.

Prigogine, I.: *Non-Equilibrium Statistical Mechanics*, Interscience, New York, 1962.

Quastler, H. (Editor): *Information Theory in Psychology: Problems and Methods*, Free Press, Glencoe, Ill., 1955.

Rademacher, Hans: "On the roundest oval," *Trans. New York Acad. Sci., Ser. 2, 29* (1967), 868–874.

Rakic, Pasko: "Development of visual centers in the primate brain depends on binocular competition before birth," *Science 214* (1981), 928–931.

Ramachandran, V.S. and Cavanagh, P.: "Subjective contours capture stereopsis," *Nature 317* (1985), 527–530.

Ratliff, F.: *Mach Bands: Quantitative Studies in Neural Networks in the Retina*, Holden-Day, San Francisco, Calif., 1965.

Reichardt, Werner: "Nervous processing of sensory information," Chapter 14 of *Theoretical and Mathematical Biology*, Talbot H. Watterman and Harold J. Morowitz, *editors*, Blaisdell Publishing Co., New York, 1965.

Regan, D. and Beverley, K.I.: "How do we avoid confounding the direction we are looking and the direction we are moving?" *Science 215* (1982), 194–196.

Resnikoff, H.L.: "Differential geometry and color perception," *J. Math. Biology 1* (1974), 97–131.

Resnikoff, H.L.: "On the geometry of color perception," *Lectures in the Life Sciences 7* (1974), 217–232.

Resnikoff, H.L.: "On the psychophysical function," *J. Math. Biology 2* (1975), 265–276.

Resnikoff, H.L.: "Concurrent computation and models of biological information processing," in *Advances in Cognitive Science*, Westview Press, 1987.

Resnikoff, H.L.: "Linear superposition of textons," preprint, 1986.

Resnikoff, H.L.: "Cost-effectiveness of concurrent supercomputers," *J. Supercomputing 1* (1987), 231–262.

Resnikoff, H.L.: "The duality between noise and aliasing and human image understanding," *SPIE vol. 758: Image Understanding and the Man-Machine Interface* (1987), 31–38.

Richards, Whitman: "Why rods and cones?" *Biol. Cybernetics 33* (1979), 125–135.

Richards, Whitman: "Quantifying sensory channels: generalizing colorimetry to orientation and texture, touch, and tones," *Sensory Processes 3* (1979), 207–229.

Richter, J. and Ullman, S.: "A model for the spatio-temporal organization of X and Y-type ganglion cells in the primative retina," MIT Artificial Intelligence Laboratory Memo 573, April 1980, 58 pp.

Riggs, L.A., Ratliff, F., Cornsweet, J.C., and Cornsweet, T.N.: "The disappearance of steadily fixated visual test objects," *J. Optical Soc. America 43* (1953), 495.

Rock, Irvin: *The Logic of Perception*, MIT Press, Cambridge, Mass., 1983.

Rogers, Brian J. and Graham, Maureen E.: "Anisotropies in the perception of three-dimensional surfaces," *Science 221* (1983), 1409–1411.

Rohrlich, Fritz: "Facing quantum mechanical reality," *Science 221* (1983), 1251–1255.

Rosenblatt, F.: *Principles of Neurodynamics*, Spartan Books, Inc., Washington, D.C., 1962.

Rothstein, Jerome: "Organization and entropy," *J. Applied Physics 23* (1952), 1281.

Rothstein, Jerome: "Information and thermodynamics," *Physical Rev. 85* (1952), 135.

Rothstein, Jerome: "Information, measurement, and quantum mechanics," *Science 114* (1951), 171–175.

Rovner, J.S. and Barth, F.G.: "Vibratory communication through living plants by a tropical wandering spider," *Science 214* (1981), 464–466.

Rubin, John M. and Richards, W.A.: "Color vision and image intensities: when are changes material?" MIT Artificial Intelligence Laboratory Memo No. 631 (1981), 31 pp.

Sands, Stephen and Wright, Anthony: "Monkey and human pictorial memory scanning," *Science 216* (1982), 1333–1334.

Schrödinger, Erwin: "Grundlinien einer Theorie der Farbenmetrik in Tagessehen," *Ann. Physik 63* (1920), 397–426, 427–456, 481–520.

Schrödinger, Erwin: *What is Life?* Camridge University Press, Cambridge, 1944.

Schrödinger, Erwin: *Mind and Matter*, Cambridge University Press, 1958.

Schrödinger, Erwin: *Statistical Thermodynamics*, Cambridge University Press, Cambridge, 1964.

Schuller, Gerd: "Coding of small sinusoidal frequency and amplitude modulations in the inferior colliculus of 'CF-FM' bat, *Rhinolophus ferrumequinum*," *Exp. Brain Research 34* (1979), 117–132.

Schuller, Gerd: "Echo delay and overlap with emitted orientation sounds and Doppler-shift compensation in the bat, *Rhinolophus ferrumequinum*," *J. Comparative Physiol. 114* (1977), 103–114.

Schuller, Gerd: "Hearing the characteristics and Doppler shift compensation in South Indial CF-FM bats," *J. Comparative Physiol. 139* (1980), 349–356.

Schuller, Gerd: "Laryngeal mechanisms for the emission of CF-FM sounds in the Doppler-shift compensating bat, *Rhinolophus ferrumequinum*, *J. Comparative Physiol. 107* (1976), 253–262.

Schuller, Gerd: "The role of overlap of echo with outgoing echolocation sound in the bat *Rhinolophus ferrumequinum*," *Die Naturwissenschaften 61* (1974), 171–72.

Schuller, Gerd: "Storage of Doppler-shift information in the echolocation system of the "CF-FM" bat, *Rhinolophus ferrumequinum*," *J. Comparative Physiol. 105* (1976), 9–14.

Schuller, Gerd: "Vocalization influences auditory processing in collicular neurons of the CF-FM bat, *Rhinolophus ferrumequinum*," *J. Comparative Physiol. 132* (1979), 39–46.

Schuller, Gerd and Pollak, George: "Disproportionate frequency representation in the inferior colliculus of doppler-compensating greater horseshoe bats: Evidence for an acoustic fovea," *J. Comparative Physiol. 132* (1979), 47–54.

Segal, I.E.: "A note on the concept of entropy," *Mathematics and Mechanics 9* (1960), 623–629.

Shannon, C.E.: "A note on the concept of entropy," *Bell System Tech. J. 27* (1948), 379–423.

Shannon, Claude E.: "Communication in the presence of noise," *Proc. IRE 37* (1949), 10–21.

Shannon, Claude E.: "Prediction and entropy of printed English," *Bell System Tech. J. 30* (1951), 50–64.

Shannon, Claude E. and Weaver, Warren: *Mathematical Theory of Communication*, University of Illinois Press, Urbana, 1949.

Shapiro, H.S. and Silverman, R.A.: "Alias-free sampling of random noise." *J. Soc. Industrial and Applied Mathematics 8* (1960), 225–248.

Simon, Herbert A.: *The Sciences of the Artificial*, The MIT Press, Cambridge, Mass., 1969.

Simon, Herbert A.: "The organization of complex systems," in *Hierarchy Theory*, Howard H. Pattee, *ed.*, George Braziller, New York, 1973.

Smith, Alan Jay: "Cache memories," *Computing Surveys 14* (1982), 473–530.

Smith, David A.: "A descriptive model for perception of optical illusions," *J. Math. Psychology 17* (1978), 64–85.

Stevens, Kent A.: "Occlusion clues and subjective contours," MIT Artificial Intelligence Laboratory Memo No. 363, (1976), 19 pp.

Stevens, Kent A.: *Surface Perception From Local Analysis of Texture and Contour*, MIT Artificial Intelligence Laboratory Report AI-TR-512 (February 1980), 122 pp.

Stevens, Kent A.: "The visual interpretation of surface contours," *Artificial Intelligence 17* (1981), 47–73.

Stevens, S.S.: "On the theory of scales of measurement," *Science 103* (1946).

Stevens, S.S.: "To honor Fechner and repeal his law," *Science 133* (1961), 80–86.

Stokoe, W., D. Casterline, and C. Croneberg: *A Dictionary of American Sign Language on Linguistic Principles.* Washington, D.C.: Gallaudet College Press, 1965.

Suga, N., Neuweiler, G., and Moller, J.: "Peripheral auditory tuning for fine frequency analysis by the CF-FM bat, *Rhinolophus ferrumequinum,*" *Comparative Physiol. 106* (1976), 111–125.

Szentágothai, John and Arbib, Michael: "Conceptual models of neural organization," *Neurosciences Research Program Bulletin 12* (1977), 305–510, The MIT Press, Cambridge, Mass.

Szilard, Leo: "Über die Entropie verminderung in einern theormodynamischen System bei Eingriffen intelligenter Wesen," *Zeitschr. Physik 53* (1929), 840–865.

Thiery, A.: "Über geometrisch-optische Täuschungen." *Philosophische Studien 12,* 67–126 (1896).

Thurston, Jacqueline B. and Carraher, Ronald G.: *Optical Illusions and the Visual Arts,* Van Nostrand Reinhold Co., New York, 1966.

Toffoli, Tommaso: "Computation and construction universality of reversible cellular automata," *J. Comp. Syst. Sci. 15* (1977), 213–231.

Toffoli, Tommaso: "Physics and computation," *Inter. J. Theoretical Physics 23* (1982), 165–176.

Toffoli, Tommaso: "Reversible computing," Laboratory for Computer Science, MIT Report number MIT/LCS/TM/-151, 1980, 36 pp.

Toom, A.L.: *Doklady Akad. Nauk SSSR 150* (1963), 496–498.

Torre, V. and Poggio, T.: "On edge detection," *IEEE Trans. Pattern Analysis and Machine Intelligence PAMI-8* (1986), 147–163.

Traub, J.F., Wasilkowski, G.W., and Woźniakowski, H.: *Information, Uncertainty, Complexity,* Addison-Wesley, Reading, Mass., 1983.

Turing, A.M.: "On computable numbers, with an application to the Entscheidungs-problem." *Proceedings of the London Mathematical Society (ser. 2) 42,* 230–265 (1936–1937); correction, *ibid., 43,* 544–546 (1937).

Uesaka, Yoshinori: "On a complexity and segmentation of line drawings," *Proc. Fourth International Joint Conf. on Pattern Recognition,* 7–10 November 1978, Kyoto, pp. 411–413.

Ullman, Shimon: "Against direct perception," MIT Artificial Intelligence Laboratory Memo No. 574 (1980), 42 pp.

Ullman, Shimon: "Analysis of visual motion by biological and computer systems," *Computer,* August 1981, 57–69.

Ullman, S.: "Filling-in the gaps: the shape of subjective contours and a model for their generation," *Biol. Cybern. 25* (1976), 1–6.

Ullman, S.: "Interfacing the one-dimensional scanning of an image with the applications of two-dimensional operators," MIT, Artificial Intelligence Memo 591 (1980), 13 pp.

Valiant, L.G.: "A theory of the learnable," *Comm. ACM 27* (1984), 1134–1142.

Valiant, L.G.: "Deductive learning," *Trans. Royal Soc. London 312* (1984), 441–447.

Vitz, Paul, C. and Glimcher, Arnold B.: *Modern Art and Modern Science,* Praeger Pub., New York, 1984.

von der Heydt, R., Peterhans, E., and Baumgartner, G.: "Illusory contours and cortical neuron responses," *Science 224* (1984), 1260–1262.

von Neumann, John: *The Computer and the Brain,* Yale University Press, New Haven, 1958.

von Neumann, John: "The general and logical theory of automata," In L.A. Jeffries (Editor), *Cerebral Mechanisms in Behavior: The Hixon Symposium,* Wiley, New York, 1951, 1–31.

von Neumann, John: *Mathematical Foundations of Quantum Mechanics* (translation and revision of the original German edition), Princeton University Press, 1955.

von Neumann, John: *Theory of Self-Reproducing Automata*, edited and completed by Arthur W. Burks, University of Illinois Press, Urbana (1966).

Watson, J.D. and F.H.C. Crick: "A structure for deoxyribose nucleic acid." *Nature 171* (1953), 737–738.

Weber, E.H.: *De pulsu, resorptione, auditu et tactu: annotationes anatomicae et physiologicae*. Leipzig, 1834.

Weinhold, Frank: "Thermodynamics and geometry," *Physics Today*, March 1976, 23–30.

West, Gerhard and Brill, Michael H.: "Necessary and sufficient conditions for Von Kries chromatic adaptation to give color constancy," *J. Math. Biol. 15* (1982), 249–258.

Weyl, Hermann: *Symmetry*, Princeton University Press, 1952.

Weyl, Hermann: *The Theory of Groups and Quantum Mechanics* (translated from the second revised German edition), Dover Publ., New York, n.d.

Wheeler, John Archilbald: "The computer and the universe," *Inter. J. Theoretical Physics 21* (1982), 557–572.

Wheeler, John Archibald and Zurek, Wojciech Hubert, (Editors): *Quantum Theory and Measurement*, Princeton University Press, Princeton, 1983.

Whittaker, E.T.: "On the functions which are represented by the expansion of interpolating theory." *Proc. Royal Soc. Edinburgh 35* (1915), 181–195.

Wickens, Christopher, Kramer, Arthur, Vaness, Linda, and Douchin, Emmanuel: "Performance of concurrent tasks: a psychological analysis of the reciprocity of information processing resources," *Science 221* (1983), 1080–1082.

Wiener, Norbert: *Cybernetics: or Control and Communication in the Animal and the Machine*, second edition, The MIT Press, Cambridge, Mass., 1961.

Wiener, Norbert: *Extrapolation, Interpolation, and Smoothing of Stationary Time Series*, Technology Press of MIT Cambridge, Mass., John Wiley, New York, 1950.

Wiener, Norbert: "A new concept of communication engineering," *Electronics 22* (1949), 74–77.

Williams, David R.: "Aliasing in human foveal vision," *Vision Res. 25* (1985), 195–205.

Williams, David R., Collier, Robert J., and Thompson, Brain J.: "Spatial resolution of the short-wavelength mechanis," in *Colour Vision: Physiology and Psychophysics*, Academic Press, London, 1983.

Wilson, R. and Granlund, G.H.: "The uncertainty principle for image processing," *IEEE Trans. Pattern Analysis and Machine Intelligence PAMI-6* (1984), 758–767.

Winograd, Shmuel: *Arithmetic Complexity of Computations*, Society for Industrial and Applied Mathematics, Philadelphia, 1980.

Winston, P.H. and Brown, R.H.: *Artificial Intelligence: an MIT Perspective*, 2 vols., The MIT Press, Cambridge, Mass., 1979.

Witkin, Andrew P.: *Shape From Contour*, Doctoral Thesis, Massachusetts Institute of Technology. 22 February 1980, 99 pp.

Witkin, A.: "Scale-space filtering," *Proc. 8th Int. Joint Conf. Artificial Intelligence*, 1983, 1019–1022.

Woodward, P.M.: *Probability and Information Theory, with Applications to Radar*, McGraw-Hill, New York, 1953.

Woolf, Harry, (Ed.): *Quantification: A History of the Meaning of Measurement in the Natural and Social Sciences*, Bobbs-Merrill Co., Indianapolis, 1961.

Wundt, W.: "Die geometrisch-optische Täuschungen." *Akademie der Sachs. Wissenschaften, Leipzig Abhandlungen 24* (1898), 53–178.

Wyszecki, Günter and Stiles, W.S.: *Color Science*, John Wiley and Sons, New York, 1982.

Yau, Mann-May and Srihari, Sargur N.: "A hierarchical data structure for multidimensional digital images," *Comm. ACM F26* (1983), 504–515.

Yarbus, Alfred L: *Eye Movements and Vision,* Plenum Press, New York, 1967 (original Russian edition published in 1965).

Yellott Jr., John I.: "Spectral analysis of spatial sampling by photoreceptors: topological disorder prevents aliasing," *Vision Res. 22* (1982), 1205–1210.

Yellott, John I., Jr.: "Spectral consequences of photoreceptor sampling in the rhesus retina," *Science 221* (1983), 382–385.

Young, Thomas: "On the theory of light and colors." *Phil. Trans. Royal Soc. London 92* (1802), 20–71.

Zipf, G.K.: *Psycho-Biology of Language.* Boston: Houghton-Mifflin, 1935.

Zöllner, F.: "Über eine neue Art von Pseudoskopie und ihre Beziehung zu den von Plateau and Oppel bescriebenen Bewegungsphänomenon." *Annalen der Physik* (1860) 500–525.

Index

Entries in small upper case letters denote individuals. Italicized page numbers refer to figures or tables.

Accidents, 25
Acoustic fovea, 215–217
Acoustic speech waveform, 88, *89*
AESCHYLUS, 184
Aether, 187
ALBERTI, Leon Battista, 25
Algorithm
  NEWTON'S, 49
    for $\sqrt{2}$, *53*
Alias, 135, 246, 255
  in art, *248*
  duality between aliasing and noise, 262–263
  in motion pictures and television, 247 et seq.
  sampled image, *252*
  in vision, *254*
American Sign Language, 145
Americas, discovery of, 186
Angle
  cortical representation of, *230*
  perception of, *226*
Ant and chip, *95*
Antikythera, 14
Area 17 of the cerebral cortex, 179
ARISTOTLE, 184

Arithmetic
  efficient multiplication, 109
  positional notation, 108
Artificial intelligence, 140
ASCII symbols, 39
Atoms, *105*
  of texture (*see* Texton)
ATTNEAVE, Fred, 17, 166–168, 176, 183
  ATTNEAVE'S cat, *17*
  "Chesired" ATTNEAVE'S cat, *168*
Autocorrelation, 261
  and texture perception, 267

BABBAGE, Charles, 14
Babylonian
  invention, 108
  method for extracting square roots, 49
BARTH, 123
Basilar membrane response, *214*
Bat, echolating, 212 et seq.
BAUMGARTNER, 141, 179
BEADLE, George, 104
BEKENSTEIN, J.D., 83
BELLUGI, 145, 147
"Big bang", 83

Biological information processing sys-
    tems, 2
Bioware, 2
Bit, 118
BOLTZMANN, Ludwig, 5, 11, 37, 265,
    301–302
  Biographical sketch of, 305
  BOLTZMANN's constant, 5, 71, 74,
      104, 120
  BOLTZMANN distribution, 304
  BOLTZMANN's entropy formula, 4
BOSSERT, William, 123
BOSSOMAIER, 255
BRENTANO, 222, 226
Brightness
  and FECHNER's formula, 202
  of subjective contours, 179
BRILLOUIN, Leon, 18
BURNS, 229

CAMPBELL, 125, 267
CANTOR, Georg, 15
CARNOT, Sadi, 70–71
  biographical sketch of, 306
CASTEL, Bertrand, 186
Cat
  "Chesired" ATTNEAVE's, *168*
  visual cortex, 229
Categorization of visual data, 299–302
Cerebral cortex, area of, 17, 179
Channel capacity, 91
  American Sign Language, 148
  CHENEY's unpatentable machine illu-
      sion, *244*
  cochlear nucleus of a guinea pig, *93*
  codons, 102, *103*
  English, 147
  lateral geniculate of a cat, *93*
  neuron in optical cortex of a cat,
      92
  normal daylight vision, 91
  sequential vs. parallel decision making,
      94
  short term memory, 18
  telephone, 91
  television, 91
  typing performance, *94*
Chaos, emergence from order, 80
Christie, *92*

CHURCH, Alonzo, 15
CLEVELAND, W.S., 34, 36
Code, genetic, 16
Cognitive science
  and experiment, 3
  and information, 2
Color
  axioms, 196–197
  blindness, 184
  brightness
  circle, 188
  classical Greek perception of, 184–186
  color terminology, 184
  cone, 196, 207
  dependence on periodic time, 187
  frequency range, 190
  hue, 184
  metric, 200
  mixture, 185
  as mixture of black and white, 186
  perception of, 184
  photographic negative, 207
  and physical states, 186
  saturation, 188
  space, 192 et seq.
  texton parameter, 267
  theories of
    ARISTOTLE, 186
    DESCARTES, 187
    EULER, 189
    GOETHE, 186
    GRASSMANN, 196
    HELMHOLTZ, 184, 186, 189, 191–
        192, 196, 202–203
    HOOKE, 187–188
    MAXWELL, 189
    NEWTON, 184, 186, 188–189, 194–
        195, 203
    PLATO, 185
    YOUNG, 186
Communication engineering, 3
Completeness of mathematics, 15
Computation, and information, 71–72
Computer, 14
  bioware implementations, 16
  brain as a, 16
  digital, 14
  program, 16
  TURING machine, 15
  universal, 15

Concurrent processing (*see* Parallel, processing)

Conservation of information, 42, 66, 249

Consistency of mathematics, 15

Contours
concentration of information at endpoints, *180*
information content of, 166
subjective, 178

CONWAY, L., 74

Copilia, 208 et seq.

COREN, 179, 219, 229

Cossack Hetman, 186

Curvature, 17, 174–176, 180
discontinuous change of curvature, 183

CORNSWEET, Tom., 158, 194

CORTI, organ of, 125, 218

Cosines, law of, 137

Cost, of information, 119

CRANE, 158

CRICK, Francis, 18, 75

Cybernetics, 16

Dalmation Dog illusion, *245,* 268, *269,* 291

DARWIN, George, 184

DAUGMAN, John, 294

DAVIES, P.C.W., 81

Daylight, spectral composition of, *191*

DESCARTES, 184, 187

DIACONIS, P., 34, 36

DIDEROT's *Encyclopédie,* 14

Digital, speech spectrogram, *89*

DIRAC's impulse distribution, 25

Direction, information gain from measurement of a, 168

Directional feature detectors, *228, 229*

Divide and conquer, 98, 114

DNA (= deoxyribonucleic acid), 102, *103*

DOPPLER shift, 213, 217–218

Duality between
noise and aliasing, 262
space-time and momentum-energy, 68

DYSON, Freeman, 16

Echolocation, 212

EDDINGTON, A.S., 42, 66

Edge detection, 161, 264, 290

EINSTEIN, Albert, 82, 194

EINTHOVEN, 222

Electromagnetic radiation, 191

Electromagnetism, 65

Elementary particles, *105*

Energy, 96, 97
heat, 70
localization of, 294
and neurocomputer models, 302
thermal background, 104

English language, 39, 106

Entropy, 4, 12, 70–71, 96, 120, 302
and gravitation, 83

Equivalence relation, 284

ESCHER's impossible staircase, *243*

EULER, Leonhard, 189

Evolution, 96, 184

EXNER, 208

Eye, 141
human, cross-section, *150*
of *Limulus,* 176
PURKINJE surface, 158
Retina
area of, 150
"blind spot" = *lamina cribrosa,* 141, 150
location and shape, *154*
distribution of rods and cones, *151*
geometry of the retinal surface, *152*
neural microstructure of, *153*

FECHNER, Gustav Theodore, 9, 34, 192, 202
biographical sketch of, 306
FECHNER's law, 7, 34

Feedback, 16

FERMAT, 96

FICK's illusion, *220*

Finite energy signals (*see* Square integrable functions)

FISHER, Ronald, 13

FOURIER, Jean Baptiste Joseph
analysis, 40, 95, 125, 261
biographical sketch of, 307
"Fast FOURIER Transform", 130
series, 135
transform, 41, 67, 129, *130,* 246, 255–257, 263, 267, 294

FOURIER (*cont.*)
    of periodic sampling array, *256*
    of POISSON disk sampling array,
      259
    of POISSON sampling array, *258*
FRASER's twisted cord illusion, 222, *223*
FRAUTSCHI, S., 82, 84
Frequency, 40, 262
    spatial, 134
Frog, 18

GABOR, Denis, 294
GAUSS, K.F., 294–295
General relativity, 65, 83
Genetic code, 16, *102*
Geometric series, 46
GILBERT, 275
GIRGUS, 219, 229
GÖDEL, Kurt, 15
GOETHE, 184, 186
Graph, 112
GRASSMANN, Hermann Günther, 9, 142,
    196
    biographical sketch of, 308
Gravitation and entropy, 83
Greeks, classical, 184, 219
GREGORY, R.L., 140, 144, 179, 208,
    222
GREGORY's series, 44
Group, 57–64
    affine, 31
    circle, 56
    commutative, 58
    invariant, 20, 28, 31, 140
      measure, 56
    of linear transformations, 61
    matrix, 61
    of motions, 56
    rotations
      in the plane, 58
      in 3-dimensional space, 59
    theory, 16
    topological, 56

HAAR measure, 56
HAMILTON, William Rowan, 96
HARTLEY, R.V.L., 13
HARTLINE, 123

HARTLINE-RATLIFF equations, 176
HAWKING, S., 83
Heat, 70
HEISENBERG, Werner Karl
    biographical sketch of, 309
    and quantum mechanics, 11, 82
    uncertainty principle of, 15, 66, 139
HELMHOLTZ, Hermann von, 7–9, 66,
      125, 142, 155, 184, 186, 189,
      191–192, 196, 202–203, 218
    biographical sketch of, 309
HERRING's illusion, *236*
HERRNSTEIN, Richard J., 19, 264, 300
HEYMANNS, 222
Hierarchical
    information processing, 210
    network topology, 100
    structure, 98
      octree, 299
      of positional notation for numbers,
        107
      quadtree, 299
      tree, 112
Hierarchy, of material structures, 105
HILBERT, David, 15
HILDRETH, E., 294
HINTON, G., 302
HIRSCH, 255
HOMER, 184
Homogeneous space, 193
HOOKE, Robert, 184, 188
HOPFIELD, J., 302–303
HUBEL, D., 291
HUGHES, 255
HUMPHRY, George, 149
HUMPTY DUMPTY, 81
HUYGENS, Constantijn, 28
HYLTON, 255

Ignorance, 42
Illusions, 218 et seq.
    area, *235*
    BRENTANO's, 222, 226
    CHENEY's unpatentable machine, *244*
    classification of, 231
    continuity of the visual manifold,
      231
    curvature, *236*
    Dalmatian Dog, *245, 269,* 291

distortion, *238–239*
divided interval, *234*
ESCHER's impossible staircase, *243*
FICK's, *220*
FRASER's twisted cord, 222, *223, 240–241*
HERRING's, 236
impossible objects, *243–244*
moiré pattern (*see* Alias)
multivalued pattern continuation, *232*
MÜLLER-LYER's, *134*, 220, *221–222*, 226–227, 230, *234*
NECKER's cube, 156
parallel lines, *237*
PENROSE's impossible object, *243*
perspective, 242
POGGENDORF's, 220, *221*, 227, 230, *235*
of reality, 88
SCHRÖDER's staircase, 156
six children image, *292*
subjective
    contours, 178, *182*, 227, *228, 232–233*
    stereo image, *234*
"venetian blinds" image, *292*
visual, as an experimental tool, 18
WUNDT's, 236
ZÖLLNER's, 220, *221, 237–238*
Image
    granularity, 274
    high resolution, *251*
    information content of, 17
    intensity function, 295
    POISSON disk sampled, *260*
    segmentation, 264
    storage and manipulation, 18
    texture (*see* Texture)
    undersampled, *252*
Information, 1
    additivity of, 31
    approximation and, 25
    artifacts, 180
    average, 118
    and categorization, 299 et seq.
    and computation, 3, 44–54, 71–72
    content of
        calculation, 44 et seq.
        contours, 166
        images, 17

conservation of, 42–43, 66, 69, 249
cost of, 96, 97, 119
and curvature, 17, 174–176, 180
density of storage in DNA, 104
DIDEROT's *Encyclopédie,* 14
Duality between noise and aliasing in measurement, 262–263
and entropy, 12, 70–71, 83, 96, 120, 302
equivalence relations and, 2
gain, 26, 28, 170
    average information, 38–39
    formula for, 27
history of, 3
incremental, 29
independence of form of presentation, 1–2
mathematical processes and, 22
maximization, 140
measure of, 13, 16
    bit, 27
    and direction of time, 81–85
    generalized, 54
    money as, 22
    nat, 118
    numerical, 22 et seq.
measurement, 10 et seq. 75–81, 168, 263
    of length, *26*
    numerical, 25
minimization, 140
order and, 2
organization and, 2
probability interpretation of, 36 et seq.
propagation velocity in neurons, 7
psychophysical function and, 32
quantifiable identity, 1
quantity of, 96
relative nature of, 31
scatterplot presentation of, 34, *35*
science of, and abstraction, 1
selective omission of, 9 et seq., 140, 245 et seq., 265
and special relativity, 85
speed of propagation of, 66
and thermodynamics, 3 et seq.
transfer, 2
Information processing
    biological, 2, 16, 140, 265
    electronic, 16

Information processing (*cont.*)
   hierarchical organization, 140
   LOGAN's theorem, 18
   multiple scales of resolution, 294
   storage
      in proteins, 16
      videodisk, 106
Inhibition, lateral, 176
   Effect on neural firing frequency, *224*
Intelligence, artificial, 140
Interpolation, 87, 132
Invariant (*see* Group)
IRWIN, 185

JASTROW, 222
jnd (= just noticeable difference), 7, 32
JULESZ, Bela, 141, 231, 264, 265 et seq.

KANISZA, 179
KARATSUBA, 109
KLIMA, 147
Knowledge, 42
   of a number, 25, 44
   representation, 3, 18–19
KNUTH, Donald, 110
KOSSLYN, S., 18

LAGRANGE multiplier, 119
LAND, Edwin, 184
Language
   Akkadian, 146
   American Sign (ASL), 145
   Chinese, 146
   English, 106, 145
   Latin, 146
LAPLACE's
   differential operator, 165, 295, 297
   equation, 155
LASKA, 222
Learning, 19
   HINTON and SEJNOWSKI's work on,
      302
LEIBNIZ, 14
LETTVIN, J., 18, 157, 184
Library catalog system, 113
Life, nature of, 16, 74

Light
   and FOURIER analysis, 125
   speed of, 73
Limulus, 176
LIPPS, 222
LOGAN's theorem, 18
Logarithm function, and psychophysical
      response, 7
LUCE, R. Duncan, *92*

MARR, David, 18, 88, 294
Matrix
   invariant, 64
   model for color space, 207
   product, 61
MATURANA, 18, 157
MAUPERTUIS, 96
MAXWELL, James Clerk, 4–6, 82, 184, 189
   biographical sketch of, 311
   MAXWELL's demon, 5
McCULLOCH, Warren, 16–18, 20, 22
McGILL, R., 34, 36
MEAD, C., 74
Measure (*see* Information)
   Haar, 56
   and information, 54
   invariant, 56
Measurement (*see* Information)
   increasingly precise, *30*
   nested, *31*
   uncertainty principle for, 39
   unit of, 28
MELVILLE, Herman, 157
Memory
   capacity, 18
   content addressable, 303
   HOPFIELD's work, 302–303
MERSEREAU, 253
Metric, for color space, 200
MICHELSON, 66
MILLER, George, 17–18
MILLER, James G., 101
MINSKY, Marvin, 17
MOIRÉ pattern (*see* Alias)
Molecules, *105*
MORELY, 66
MÜLLER-LYER illusion, *134, 220, 221–
      222, 226–227, 230, 234*

Nat, 118
Network topology
    hierarchical, 100
    neural, 302
    star, 100
Neurocomputer models, 302
Neurons, velocity of information propa-
    gation, 7
NEUWEILER, 218
NEWELL, Allen, 17
NEWMAN, 123
NEWTON, Isaac, 49, 74, 90, 142, 184,
    188–189, 194–195, 203
Noise, 123, 246, 253
    duality between aliasing and, 262–263
    white, 262
Number, 22–23
    binary expansion of, 27
    decimal expansion of, 23
    rational, 23
    real, 23
NYQUIST frequency, 246, 249, 253, 260,
    262

Observation (*see* Information)
    limits to, 76–79
OPPEL, J.J., 219
Opsin, 191

Parallel
    lines, illusion of, 237
    processing, 94, 110, 210, 302
Parthenon, 219
PASCAL, Blaise, 14
Pattern
    categorization ability, 19, 300, 304
    continuation, 159
Patterned structures and information, 2
PENROSE's impossible object, 243
Perception
    color, 184, 192
    preattentive texture, 267 et seq.
    subjective contours, 178, 182
    theories of Greek color, 184
PETERHANS, 142, 179
Physics, breakdown of classical, 11
PICARD, É., 70

Pigeons and pattern categorization, 19,
    300
PINDAR, 184
PINKER, S., 18
PITTS, Walter, 16–18, 20, 157
PLANCK, Max, 11, 66, 68
PLANCK's constant, 68, 122, 123, 139,
    190
PLATO, 184–185
POGGENDORF's illusion, 220, 221, 227,
    230, 235
POGGIO, Tomaso, 18, 88, 294
POISSON disk sampling, 257 et seq.
Positional notation, 107
Principle
    of conservation of information, 42–43,
        249
    of extremization of information, 20
    of hierarchical organization of informa-
        tion, 21
    of selective omission of information,
        19, 245 et seq., 265
PRITCHARD, 229
Probability and information, 36 et seq.
Psychophysical function, 32
    of FECHNER, 34
    of STEVENS, 34

Quanta, 11, 66, 68
Quantum mechanics, 66
    harmonic oscillator, 122
    and information, 2
    theory of measurement in, 10 et seq.,
        75

RADEMACHER, Hans, 181, 183
RADEMACHER's ovals, 181, 183
Rainbow, 187
RATLIFF, 158
Relativity,
    general, 65, 83
    of information gain, 31
    special, 66, 85–86
Response
    of directional feature indicators, 229
    psychophysical, 6
    threshhold, 7

Retina (*see* Eye)
Retinal, isomeric forms of, *190*
Retinol, 191
Rhinolophus ferrumequinum, 208
RIEMANN's zeta function, 48, 121
RIGGS, 158
ROBSON, 125, 267
ROOD, 184
ROVNER, 123
RUSSELL, Bertrand, 15

Saturation, 188
S.B., 144
SCHICKARD, Wilhelm, 14
Scholasticism, 187
SCHRÖDER's staircase illusion, 242
SCHRÖDINGER, Erwin, 74, 82, 184
SCHWARZ' inequality, 137
SEJNOWSKI, T., 302
Selective omission of information, 9 et
        seq., 245 et seq., 265
    principle of, 19
Sequential search, 114
Series
    exponential function, 48
    geometric, 46
    GREGORY's, 46
    RIEMANN's zeta, 48, 121
Set theory, 15
SHANNON, Claude, 13, 16, 18, 39, 88,
        118
    SHANNON's sampling theorem, 125,
        132, 246, 253
SHAW, 17
Signal
    band limited, 246
    modulation, 95, 124
    processing, 88
        parallel, 210
        sequential, 210
    sampling
        continuous, 246
        discrete, 246
        hexagonal lattice, 253
        ideal, 255
        POISSON disk, 257
        regular, 262
        sound, 213 et seq.

SIMON, Herbert, 17, 21, 101
SNYDER, 255
Space
    absolute, 194
    homogeneous, 193
    of perceived colors, 192 et seq.
Space-time, 194
Spectral
    composition of daylight, *191*
    power density, 262
Speech
    spectrogram, *89*
    waveform, 88
Square integrable functions, 138
Statistical
    nature of texture, 266
        higher order statistics and percep-
            tion, 268
    thermodynamics, 265
        BOLTZMANN's constant, 104, 120
        BOLTZMANN distribution, 304
STEVENS, S.S., 7, 34
Stimulus, physical, 6
Striate cortex layer IVCβ, 18
Subjective contours, 178, *182, 227, 228*
Symmetry, 20
SZILARD, Leo, 5–6, 11, 70
    biographical sketch of, 311

Telephones
    number in United States, 99
Temperature, 4
Texton, 265
    definition of, 285
    map of texton space, *290*
    mixture, 279
    parameters, 267 et seq.
    representation of, *286*
Texture, 264 et seq.
Textures
    accessible, 287–290
    distinguishable, *272–274, 276*
    indistinguishable, *270–271, 275,
        278*
    mathematical theory of, 277 et seq.
    perceptual equivalence of, 279
    third order statistics and, 275–277
Thermal background energy, 104

Thermodynamics
  second law of, 82
  statistical, 265
  validity of, 4
THIERY, 222
Threshold, response, 7
Timaeus, 185
Time, 81–84
TOOM, 107
Tree
  information processing structure, 112
  library catalog system, 113
TURING, Alan, 16
  biographical sketch of, 312
  machine, 15
Twisted cord illusions, 240–241

ULLMAN, Shimon, 179
Uncertainty
  principle for measurement, 39
  principle of HEISENBERG, 67–69, 72, 139
  rectangle, *43*
  relation, 132, 136
Universals
  auditory and visual, 20

Vector space
  basis, 286
  and color perception, 90
  and matrices, 63
  and texture perception, 286
VICTOR, 275
Videodisk, 106
Vision
  chemical basis of, 190
  detection of discontinuities of curvature, 183
  directional feature detectors, *228, 229*
  edge detection, 291 et seq.
  eye as transducer for, 149
  hierarchical organization of, 142
  human, 8, *149, 150*
    channel capacity of, 8

hyperacuity, 18
  "off-center" and "on-center" receptors, 224, *225*
  machine, 293
  photopic, 142
  preattentive, 266
  purpose of, 184
  scotopic, 142
  stabilized, 157, 194
Visual
  cortex, 229
  data, categorization of, 299–302
  illusions (*see* Illusions)
  manifold, continuity of, 153
  texture, 264 et seq.
VON HEYDT, 142, 179
VON NEUMANN, John, 11, 15–16, 18, 66, 72
  biographical sketch of, 312

WATSON, J., 75
WEBER, Ernst, 6
  biographical sketch, 313
  fraction, 7
  WEBER's law, 230
WHEELER, John Archibald, 87
WHITTAKER, E. T., 133, 187
WIENER, Norbert, 13, 16, 261
  biographical sketch of, 313
WIESEL, 291
WILSON, Edward O., 123
WITKIN, A., 298
WUNDT, 222, *236*

YARBUS, 158
YELLOTT, Jr., John I., 245, 253
YOUNG, Thomas, 184

Zeta function of RIEMANN, 48, 121
ZIPF's law, 97, 121, *122*
ZÖLLNER's illusion, 220, *221, 237–238*